T0348896

Principles of Filtration

Principles of Filtration

Chi Tien
Syracuse University
Syracuse, NY

ELSEVIER

AMSTERDAM • BOSTON • HEIDELBERG • LONDON • NEW YORK • OXFORD
PARIS • SAN DIEGO • SAN FRANCISCO • SYDNEY • TOKYO

Elsevier

The Boulevard, Langford Lane, Kidlington, Oxford OX5 1GB, UK

Radarweg 29, PO Box 211, 1000 AE Amsterdam, The Netherlands

First edition 2012

Notice

No responsibility is assumed by the publisher for any injury and/or damage to persons or property as a matter of products liability, negligence or otherwise, or from any use or operation of any methods, products, instructions or ideas contained in the material herein. Because of rapid advances in the medical sciences, in particular, independent verification of diagnoses and drug dosages should be made

British Library Cataloguing in Publication Data

A catalogue record for this book is available from the British Library

Library of Congress Cataloging-in-Publication Data

A catalog record for this book is available from the Library of Congress

For information on all **Elsevier** publications
visit our web site at elsevierdirect.com

Printed and bound by CPI Group (UK) Ltd, Croydon, CR0 4YY

Transferred to Digital Print 2012

ISBN: 978-0-444-56366-8

Working together to grow
libraries in developing countries

www.elsevier.com | www.bookaid.org | www.sabre.org

ELSEVIER   BOOK AID International   Sabre Foundation

Dedication

To Julia C. Tien

Contents

Part I Cake Filtration

Preface

Filtration as a fluid–particle separation process is an engineering practice of long standing. Separation in filtration is effected by passing a fluid–particle mixture through a medium with particles retained by the medium and passage of clear fluid through it. To be more specific and descriptive of the process conducted, depending upon factors such as the operative particle retention mechanism, the type of medium used, the flow configuration and/or the system treated, different terminologies such as cake filtration, surface filtration, depth filtration, deep bed filtration, cross-flow filtration, aerosol filtration, water (liquid) filtration, granular filtration, fibrous filtration, fabric filtration, cartridge filtration, membrane filtration, etc. have been introduced over the years. As an unintended consequence, that these specific filtration processes are not totally different is often ignored, even among filtration experts.

The compartmentalization of filtration into different types of operation in terminology is reflected in published texts and monographs as well. In most unit operations textbooks for undergraduate teaching, discussion of filtration is invariably restricted to cake filtration of liquid suspensions. Similarly, textbooks on air pollution control engineering and water treatment limit their presentations mainly to fibrous or fabric aerosol filtration (in air pollution control) or deep bed filtration of aqueous suspensions (in water treatment). By considering these specific types of filtration as independent and unrelated subjects in teaching, the students are denied the opportunity of gaining a broader horizon and greater understanding of the subject matter and the advantage of applying results of one type of filtration to the solutions of others.

Technically speaking, one may well argue that separating and classifying filtration based on factors mentioned above is arbitrary and may not be realistic. For cake filtration, in which filter cakes formed play a major role in determining filtration performance, penetration of particles into filter medium cannot be ruled out in many cases. In fact, better understanding of the temporal evolution of filter medium resistance is a key factor to exact cake filtration analysis and accurate predictions of filtration performance, requires the knowledge of deep bed filtration (with the medium considered as a deep bed filter). Similarly, for proper design of deep bed filters of water treatment, in order to insure prolonged operation and to reduce the frequency of backwashing, knowledge of the variables affecting cake formation at deep bed inlet is necessary. Equally, if not more important, for membrane filtration carried out in the cross-flow mode, particle separation is often caused by a combination of particle deposition over membrane surfaces and particle retention within membrane pores; therefore, blurring the distinctiveness of one type of filtration from another. With these considerations, teaching cake filtration and deep bed filtration and treating them separately and independently in engineering texts and monographs have become an obsolete and ineffective practice.

The practice of analyzing filtration according to the medium used or the type of the system treated is equally questionable. In spite of the medium and/or system differences, problems arising from the flow of fluid suspensions (liquid or gas) through porous media (granular, fibrous, fabric) and the attendant particle deposition can be analyzed using the same set of equations and procedures. In other words, there exists a common core of knowledge and information, which form the basis for teaching filtration as a single subject.

I have designed this volume in order to provide a unified treatment of filtration operations in different guise. Major emphasis is placed on presenting basic and well-established principles for the description of the various types of filtration operations, some common procedures for data treatment and correlation, and a collection of filtration rate parameter correlations. A substantial number of illustrative examples are given throughout the text in order to demonstrate the principle/procedures discussed. The problems given at the end of each chapter provide further opportunities for readers to better understand these principles and procedures for their applications.

This book is written as a text/reference book for both engineering students and practitioners. Its purpose is to equip readers with certain basic knowledge and information of filtration at a level higher than those found in elementary texts and to enable them to undertake meaningful engineering work in filtration and/or begin their research in this area. As a textbook, it may be used for either undergraduate (senior) or graduate (first year) teaching for chemical, environmental, and mechanical engineering students. Different parts of the book can also be easily incorporated into courses such as fluid–solid separation, air pollution control, environmental engineering analysis, water treatment, membrane process and technology found in many engineering curricula. The level of presentation is consistent with what is taught at ABET accredited B.Sc. degree programs in chemical, environmental, and mechanical engineering. In spite of a fairly large number of mathematical equations present throughout the text, the book can be easily comprehended by students with elementary competence in calculus plus some knowledge of differential equations except perhaps those with extreme number-phobia.

In preparing the text, I have benefitted greatly from the discussions I have had with a large number of colleagues especially those with Professor B.V. Ramarao (SUNY-ESF). The preliminary draft of the work was read by a number of people. In particular, I would like to express my gratitude to Professor George G. Chase (University of Akron), Professor Rolf Gimbel (University of Duisburg-Essen), Professor W.P. Johnson (University of Utah), Professor Y.-W. Jung (Inha University), Professor Wallace W. Leung (Polytechnic University, Hong Kong), Professor Dominique Thomas (University of Nancy), Professor K.-L. Tung (Chun Yuan Christian University), and Professor Eugene Vorobiev (Universite de Technologie de Campiegne) for their insightful suggestions, critical comments, and moreover careful proofreading!!! Without their help and efforts, the book would not appear in its present form.

Finally, I would like to express my thanks to my editors, Dr. Kostas Marinakis and Dr. Anita Koch of Elsevier for their efforts and help which make prompt publication of this volume possible, Kathy Datthyn-Madigan for her efforts in assembling the manuscript, and last but not the least, my wife, Julia, for all the help and support she has given me during the past half century.

Chi Tien

1

Introduction

Notation

K_0 coefficient of Equation (1.1)
k exponent of Equation (1.1)
t time (s)
V cumulative filtrate volume per unit media surface area ($m^3\,m^{-2}$)

Filtration may be described as an operation in which solids (particles) present in a solid–fluid mixture are separated from the liquid by forcing the flow of the mixture through a supported mesh or cloth (Walker et al., 1937). The mixture is caused to flow by various forces: gravity, pressure, vacuum, or centrifugal force. The products of the separation consist of a fluid stream (filtrate) free or nearly free of particles, a solid phase with some entrained liquid and possibly a solid–fluid mixture with enhanced solid concentration (as in the case of crossflow cake filtration).

Giddings (1991) advanced the premise that separation of a mixture of several components is effected by the relative displacements of the various components present in the mixture. Earlier, King (1980) stated that the working of a separation process is accomplished by the application of a separating agent as shown in Fig. 1.1. The agent may be either energy or matter or both. Through the action of the agent, a feed is split into several streams of different compositions. Using King's description or Gidding's premise, filtration is a process employing energy (for the flow of the suspension to be treated) and matter (filter media) as separating agent, leading to a relative solid/fluid displacement from the flow of the suspension through the medium with particle retention at the surface of the medium, or particle deposition throughout the medium.

Filtration may be applied to both gas/solid and liquid/solid suspensions. In the following sections, a brief discussion on its use in certain industrial applications is presented as background information.

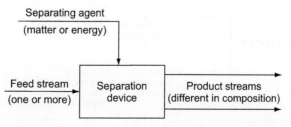

FIGURE 1.1 Representation of separation process.

Principles of Filtration, DOI: 10.1016/B978-0-444-56366-8.00001-3

1.1   Filtration as a Liquid–Solid Separation Technology

Liquid–solid separation technology, as the name implies, refers to a collection of processes for removing, separating, and recovering particles from liquid–solid mixtures. While the processes known as liquid–solid separation are too numerous to be cited individually, it is generally accepted that liquid–solid separation encompasses filtration, sedimentation, cycloning, thickening, flocculation, and expression. Tiller (1974) proposed a classification scheme based on liquid–solid separation functions (see Fig. 1.2), namely, liquid–solid separation may be viewed as a system consisting of one or more stages: pretreatment, solid concentration, solid separation, and post-treatment. According to this scheme, filtration is used in both the separation stage and post-treatment.

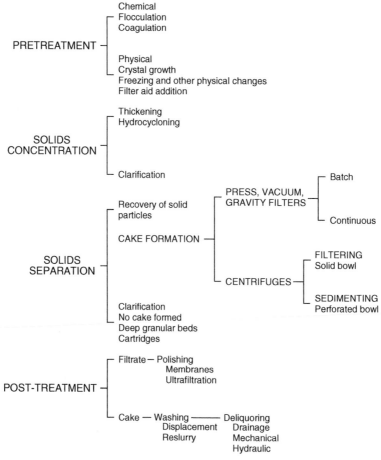

FIGURE 1.2 Stages of solid/liquid separation according to Tiller.

1.2   Classification of Filtration Processes

Over the years, for the purpose of describing the operation and/or specifying the type of medium used, a plethora of modifying terms have been added to filtration such as cartridge filtration, crossflow filtration, dead-end filtration, granular filtration, fabric filtration, centrifugal filtration, vacuum filtration, etc. While there are justifications for coining these terms, from the point of discussing the principles and analyses of filtration, one may classify filtration processes based on the mechanism of particle separation and the manner with which particle separation takes place. Generally speaking, separation of solids from liquid in filtration is effected through either the retention of particles at the surface of the filter medium or deposition of particles throughout the medium. Therefore, according to the manner of particle separation, filtration may be divided into two categories: cake filtration (or surface filtration) in which particles are retained at the media surface to form filter cakes and deep bed filtration (or depth filtration) in which particle removal is accomplished by particle deposition throughout the filter medium. With this classification scheme, granular filtration, cartridge filtration, and fibrous filtration are deep bed filtration, with their differences being the media used. On the other hand, dead-end filtration, crossflow filtration, fabric filtration, vacuum filtration, or any filtration process in which particle separation leads to the presence of growing filter cakes at media surface are cake (or surface) filtration. It also bears noticing that microfiltration, ultra-filtration, and nanofiltration are differentiated by the types of membranes used as filter media. However, they all belong to the cake filtration category.

Qualitatively speaking, occurrence of cake formation at medium surface vs. particle deposition throughout the medium is determined by the relative particle size to medium pore size. The empirical 1/3 law suggests cake formation if the particle size exceeds 1/3 of the pore size. While the value of 1/3 may not be exact, there is sufficient evidence indicating the occurrence of cake formation if particle size and pore size are of the same order of magnitude.

A schematic diagram illustrating the operating difference between deep bed filtration and cake filtration is shown in Fig. 1.3. For deep bed filtration, particle deposition takes place over the interior surface of the medium as the suspension flows through the medium. As a result, a filtrate free of particles (or with significantly reduced particle concentration) may be obtained. In cake filtration, the suspension to be tested may either flow through a medium or flow parallel to the medium with particle retention (and therefore cake formation) occurring at the upstream side of the medium. The mode of operation of the former may be described as "dead end" and that of the latter as "crossflow".

Conventional filter press and drum filters operate in the dead end mode. The so-called membrane filtration developed in more recent years operates mainly in the crossflow mode. For cake filtration in the crossflow mode, the feed stream to be treated split into three parts – filtrate, filter cake, and retentrate – with enhanced particle concentration if cake formation is significant.

For applications, cake filtration is used for treating suspensions with relatively high particle concentration. Deep bed filtration, on the other hand, is applied mainly for

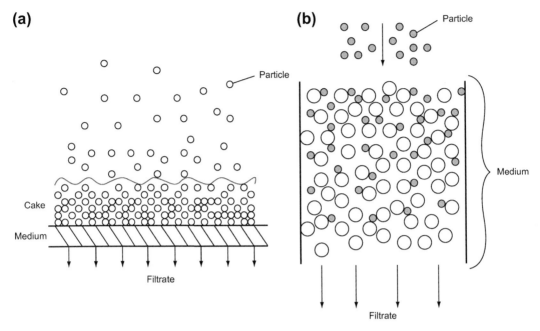

FIGURE 1.3 Cake filtration (a) vs. deep bed filtration (b).

clarifying suspensions of low particle concentration. For example, in water treatment with sand filters, the allowable particle concentration of the feed stream is limited to less than 100 parts per million (vol/vol) in order to avoid cake formation at the medium surface (or inlet of deep bed filters). This concentration difference, however, has become blurred with the advent of crossflow membrane filtration in recent years. This point will be discussed in some detail in a later section.

The difference in particle retention mechanisms between cake filtration and deep bed filtration does not imply that these two processes embody totally different and exclusive physical phenomena. Particles present in a suspension often cover a wide size range. Therefore, during the initial stage of cake filtration, finer particles may penetrate into filter medium leading to their deposition in the medium interior. At late stage, filter cake formed may function as a deep bed filter. Similarly, for a deep bed filter after prolonged period of operating, built-up of deposited particles at filter bed inlet may ultimately cause the formation of particle cake at the inlet of the filter, which, in turn, requires filter bed regeneration by backwashing. Proper and optimum design of either filtration system requires the knowledge and understanding of both cake and deep bed filtration.

1.3   Laws of Filtration

Hermans and Bredee (1935) postulated the so-called laws of filtration which classifies filtration operations according to different particle retention mechanisms. A later version of Hermans–Bredee's work was given by Hermia (1982). According to Hermia, four different particle retention mechanisms may be present in filtration: complete blocking (every retained particle acts to block a medium pore), intermediate blocking (there is a finite probability for a retained particle to block medium pores), standard blocking (particle retention taking place within medium resulting in a narrowing of medium pores), and cake filtration (particle retention results in cake formation and growth). It is clear that the "standard blocking" is the retention mechanism operative in deep bed filtration as discussed in 1.2 while complete blocking and intermediate blocking are present in the initial stage of cake filtration.

Using four assumed filtration rate expressions corresponding to each of the four retention mechanisms mentioned above, the dynamic behavior of filtration, according to the laws of filtration, may be expressed as

$$\frac{d^2t}{dV^2} = K \left(\frac{dt}{dV}\right)^k$$

$(1.3.1)^1$

where V is the cumulative filtrate volume per medium surface area collected m^3m^{-2} at time t under the condition of constant pressure. Both the coefficient K and exponent k are constant. In particular, the exponent k assumes the following values corresponding to the specific retention mechanism in operation:

Retention Mechanism	Value of k
Complete blocking	2
Standard blocking	3/2
Intermediate blocking	1
Cake filtration	0

According to the laws of filtration as expressed through Equation (1.3.1), there is a linear relationship between $\log(d^2t/dV^2)$ and $\log(dt/dV)$ with the slope of the line being the value of k. Therefore, based on constant pressure filtration data, a line or a series of line segments may be established by plotting the date present in the form of d^2t/dV^2 vs. dt/dV on the logarithmic coordinates. The slope of the line or those of the linear segments may assume any one of the k values given above. Therefore, the operative retention mechanism may be identified.[2]

The conceptual simplicity of the laws of filtration has attracted much attention from a number of investigators in recent years. The use of the laws has been found in a large number of recent publications for the purposes of identifying retention mechanisms and data interpretation. What is overlooked is a rather simple fact that this identification is not

---

[1] The instantaneous filtration rate is given by dV/dt. Therefore, dt/dV is the reciprocal of the instantaneous filtration rate $1/(dV/dt)$. On the other hand, d^2t/dV^2 can be shown to be $-(d^2V/dt^2)/(dV/dt)^3$ or the negative of the ratio of the time rate change of the filtration rate to the third power of the filtration rate.

[2] It should be noted that the laws of filtration allow the transition from one retention mechanism to another but do not allow the simultaneous presence of more than one medium.

based on direct observation, but upon agreement between filtration data and Equation (1.3.1) which, in turn, rests upon filtration rate expression proposed by Hermans and Bredee. Since these assumed rate expressions, in three cases, are incorrect or fail to agree with experimental data, the claim made regarding the value of the laws of filtration, therefore, cannot be justified. A more direct evidence of the lack of validity of the laws of filtration can also be seen from the simple fact that in many cases, the filtration data simply do not display the linearity between $\log(d^2 t/dV)$ and $\log(dt/dV)$ as required by Equation (1.3.1).

Wakeman and Tarleton (1999) pointed out that the validity of Equation (1) is limited only during the initial stage of filtration operation. Its use in design calculation is, therefore, limited. Based on a more general consideration, retention mechanism is largely dependent upon the relative particle to medium pore size. Considering the fact that both filter medium pores and suspended particles, generally speaking, are not uniform in size, but cover a range of values, it would be difficult to describe a given filtration process by a simple retention mechanism (Tien, 2006). Based on these considerations, the validity of and application of the laws of filtration are questionable if not fallacious.

Problem

1.1. Obtain expressions of V vs. t from Equation (1.3.1) for the four cases of filtration with $n = 0, 1, 3/2,$ and 2.

References

Giddings, R.W., 1991. Unified Separation Science. John Wiley & Sons, New York, pp. 11.

Hermans, P.H., Bredee, H.L., 1935. Zur Kennitnis der Filtrationgesetec. Rec. Trav. Chim. Des Pays-Bas 54, 680.

Hermia, J., 1982. Constant pressure blocking filtration laws – applications to power law non-newtonian fluids. Trans. Inst. Chem. Eng. 60, 183–187.

King, C.J., 1980. Separation Processes, second ed. McGraw Hill, Inc., New York, pp. 18.

Tien, C., 2006. Introduction to Cake Filtration: Analysis, Experiments, and Applications. Elsevier.

Tiller, F.M., April 29 1974. Chem. Eng., 117.

Wakeman, R.J., Tarleton, E.S., 1999. Filtration: Equipment Selection Modeling and Process Simulation. Elsevier Advanced Technologies.

Walker, W.H., Lewis, W.K., McAdams, W.H., Gilliland, E.R., 1937. Principles of Chemical Engineering, third ed. McGraw-Hill, Inc., New York, pp. 323.

PART

1

Cake Filtration

2

Cake Formation and Growth

Notation

A	defined by Equation (2.5.9a) (m^2 s^{-1}) or a measurable quantity (see Equation (2.7.5))
a_1, a_2	coefficients of Equation (2.2.16a)
b_1	coefficient of Equation (2.2.16b)
C_1	constant defined by Equation (2.6.10) (–)
$D(e)$	filtration diffusivity defined by Equation (xiv), Illustrative Example 2.3 (m^2 s^{-1})
d_p	particle diameter (m)
e	void ratio, defined as $(1 - \varepsilon_s)/\varepsilon_s$ (–)
F_p	force acting along the tangential direction (N)
F_q	force acting along the normal direction (N)
F_{q_1}, F_{q_2}	force defined by Equations (2.8.11) and (2.8.12), respectively (N)
f	function defined by Equation (2.7.4) (–)
f_1, f_2	hydrodynamic retardation factor (see Equations (2.8.9) and (2.8.11) (–)
F_{ij}	interaction force vector between phase j and phase I (N m^{-3})
f'	defined by Equation (2.2.10) (–)
h_{max}	maximum protrusion height (m)
k	cake permeability (m^2)
k^0	cake permeability at the zero-stress state (m^2)
L	cake thickness (m)
m_i	net mass transfer rate into phase 1 (kg m^{-3} s^{-1})
$\overline{m}$	wet to dry cake mass ratio (–)
n	total volume of particles per unit volume of suspending fluid (m^3 m^{-3}) or the exponent of Equation (2.2.15c)
p_A	parameter of the constitutive equations (Pa)
p_i	isotopic pressure of phase i (Pa)
p_ℓ	liquid pressure (Pa)
p_0	applied pressure (Pa)
p_s	compressive stress (Pa)
p_t	defined as $p_\ell + p_s$ (Pa)
p_{ℓ_m}	p_ℓ a cake/medium interface (Pa)
p_{s_m}	p_s at cake/medium interface (Pa)
Q	constant filtration rate (m^3 m^{-2})
q_ℓ	superficial liquid velocity (m s^{-1})
q_s	superficial particle velocity (m s^{-1})
q_{ℓ_m}	instantaneous filtration velocity (m s^{-1})
q_{ℓ_s}	liquid/particle relative velocity (m s^{-1})
R_m	medium resistance (m^{-2})
S_i	force vector due to stress tensor acting on phase i (N m^{-3})
s	particle mass fraction of feed suspension (–)
T_i	stress tensor acting on phase i (Pa)
t	time

Principles of Filtration, DOI: 10.1016/B978-0-444-56366-8.00002-5

t_c	time when deposition ceases (s)
t_m	quantity defined by Equation (2.5.9b) (s)
U_i	velocity vector of phase i (m s^{-1})
V	cumulative filtrate volume (m^3 m^{-2})
V_m	equivalent filtrate volume defined by Equation (2.5.7)
W_i	body force vector acting on phase i (N m^{-3})
w	cake mass per unit medium area (kg m^{-2})
x	distance measured away from medium (m)

Greek Letters

α	specific cake resistance (m kg^{-1})
α^0	specific cake resistance at the zero-stress state (m kg^{-1})
β	exponent of Equation (2.2.15a) (–) or fraction of particles being deposited
γ	adhesion probability
Δp_c	pressure drop across filter cake (Pa)
Δp_m	pressure drop across medium (Pa)
$\Delta \rho$	density different, $\rho_s - \rho$, (kg m^{-3})
δ	negative exponent of Equation (2.2.15b) (–)
δ	unit tensor
ε	cake (or suspension) porosity (–)
ε_i	volume fraction of phase i (–)
ε_s	volume fraction of particle, or solidosity (–)
ε_{s_0}	particle volume fraction of feed suspension (–)
ε_s^0	solidosity at the zero-stress state (–)
$\bar{\varepsilon}_s$	stress-averaged cake solidosity defined by Equation (2.5.1)
$\bar{\bar{\varepsilon}}_s$	average cake solidosity defined by Equation (2.3.6)
μ	fluid viscosity (Pa s)
ρ_i	density of phase i (kg m^{-3})
ρ	density of particle (kg m^{-3})
ρ_s	density of particle (kg m^{-3})
τ_w	shear stress at cake/suspension interface (N m^{-2})
τ_i	shear stress tensor of phase i
ϕ	objective function defined by Equation (2.7.5)

As mentioned in 1.2, cake filtration may be subdivided into two types of operations depending upon the flow configuration of the suspension to be treated. The traditional and still the most common operation is effected by forcing the suspension to be treated through a filter medium such that the flow of the suspension and that of the filtrate are in the same direction. Alternatively, the suspension may flow under pressure along the medium surface such that the direction of the suspension flow and that of the permeate (filtrate) are normal to each other. We focus our presentation on the first type of cake filtration (i.e. the dead end mode of operation) because of its long history of study, followed by a discussion on the second type of cake filtration (i.e. the crossflow mode).

2.1   Filtration Cycles

Actual operation of cake filtration equipment may be viewed as a sequence of processes including filtration, cake consolidation, deliquoring, cake washing, and cake discharge. The order of the sequence may vary from case to case and the number of the individual process involved is not fixed. Some typical situations given by Wakeman and Tarleton (1999) are shown in Fig. 2.1.

The physics of these difference processes is varied and complex. Filtration and consolidation may be described on a common basis, embracing liquid flow through

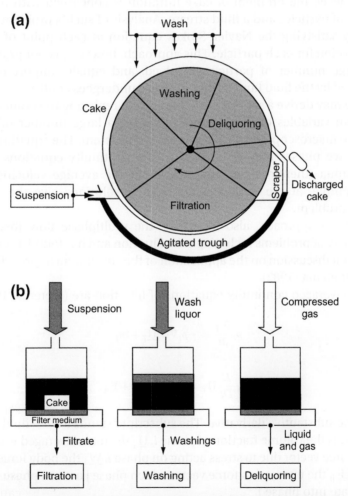

FIGURE 2.1 Two kinds of filtration cycle: (a) Rotary drum filter cycle; (b) Nutsche filter cycle. *[Reprinted from R.J. Wakeman and E.S. Tarleton, "Filtration: Equipment Selection, Modeling and Process Simulation", 1ˢᵗ Ed., Elsevier Advanced Technology, 1999, with permission of Elsevier].*

saturated media under growth and compression. However, deliquoring of saturated cake by air flow is governed by laws different from that of flow through saturated media. On the other hand, cake washing involves problems of solute mass transfer and displacement. Accordingly, these different phases of filtration will be discussed separately in the following chapter.

2.2   Analysis of Cake Filtration

A brief outline of cake filtration analysis based on the volume-averaged continuity equations is given below as background information for subsequent presentations. On a fundamental level, the problem of cake filtration is concerned with the motions of a large number of particles and a fluid stream. Analysis of such a problem can be made, in principle, by satisfying the Navier–Stokes equation at each point of the fluid and equations of motion for each particle. This approach, however, is not practical because of the very large number of particles involved, and equally important, the spatial domain occupied by the fluid (and particles) phase undergoes continuous change. As an alternative, one may derive a set of a few equations of replacing the point variables with their local mean variables over a region containing a large number of particles but smaller than the macroscopic scale of the intended problem. The equations obtained in the case with two phases may be viewed as the continuity equations of two inter-penetrary continua (therefore one can talk about the average velocities of the two phases, for example, at the same position). Their solutions give the results of the analysis of the problem.

This continuum approach, also known as the multiphase flow theory, has been applied to a variety of problems including cake filtration and has found a certain degree of success. A general discussion on the application of the continuum approach can be found in the work of Rietema (1982).

The volume-averaged continuity equations of filtration are (Rietema, 1982)

$$\rho_i \frac{\partial \varepsilon_i}{\partial t} = -\rho_i \nabla \cdot \varepsilon_i \underline{U}_i + m_i \tag{2.2.1}$$

$$\rho_i \varepsilon_i \frac{D}{Dt} \underline{U}_i = -\underline{S}_i + \underline{W}_i + \underline{F}_{ji} \tag{2.2.2}$$

where D/D_t is the substantive derivative. The subscript i stands for $\ell =$ fluid phase and $s =$ particle phase. ε_i is the volume fraction of phase i, $\underline{U}_i$ the mass-averaged velocity vector of phase i, $\underline{S}_i$ the force vector due to stress acting on phase i, $\underline{W}_i$ the body force vector acting on phase i, and $\underline{F}_{ji}$ the interaction force vector which phase j acts on phase i. m_i is the net mass transfer rate into phase i.

For one-dimensional cake filtration considered here, let $u_\ell = q_\ell/\varepsilon$ and $u_s = q_s/\varepsilon_s$. ε and ε_s are the volume fractions of the void and particle phase, respectively (or porosity and solidosity), with $\varepsilon + \varepsilon_s = 1$ and q_ℓ and q_s are the superficial liquid and particle velocities. The above equation becomes

$$\frac{\partial \varepsilon}{\partial t} = -\frac{\partial q_\ell}{\partial x} \tag{2.2.3a}$$

for $0 < x < L(t)$

$$\frac{\partial \varepsilon_s}{\partial t} = -\frac{\partial q_s}{\partial x} \tag{2.2.3b}$$

where $L(t)$ is the cake thickness.

From Equations (2.2.3a) and (2.2.3b), one has

$$\frac{\partial}{\partial x}(q_\ell + q_s) = 0$$

or

$$q_\ell + q_s = \text{constant} = q_{\ell m} \tag{2.2.4}$$

where $q_{\ell m}$ is the instantaneous filtration velocity.

A schematic diagram illustrating the physical situation is shown in Fig. 2.2.

Similarly, if the inertial effect and the presence of the body force can be ignored, by adding the equations corresponding to the two phases, namely, Equations (2.2.3a) and (2.2.3b) and noting $F_{ji} = -F_{ij}$, one has

$$\underline{S}_\ell + \underline{S}_s = 0 \tag{2.2.5}$$

where $S_i(i = \ell, s)$ is the force acting on phase i resulting from the stress tensor $\underline{\underline{T}}_i$ acting on the same phase. $\underline{\underline{T}}_i$ can be written as

$$\underline{\underline{T}}_i = \phi_i \underline{\underline{\delta}} + \underline{\underline{\tau}}_i \tag{2.2.6}$$

where p_i is the isotropic pressure of phase i and δ the unit tensor. τ_i is the shear stress tensor of phase i.

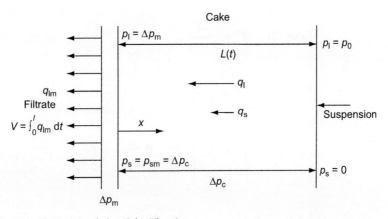

FIGURE 2.2 Schematic Diagram Depicting Cake Filtration.

Different expressions relating S_i and T_i have been proposed such as

$$S_i = \nabla T_i, \quad \nabla \varepsilon_i T_i \quad \text{or} \quad \varepsilon_i \nabla T_i \quad i = \ell \text{ or s} \tag{2.2.7}$$

In addition, for the dispersed phase (i.e. particle phase), the following relationship has also been suggested:

$$S_s = \varepsilon_s(\nabla \cdot T_\ell + \nabla \cdot T_s) \quad \text{or} \quad \varepsilon_s \nabla \cdot T_\ell + \nabla \cdot T_s \tag{2.2.8}$$

Thus by using different relationships of S_i vs. T_i, different results may be obtained from Equation (2.2.5). For the one-dimensional cake filtration case, the isotropic pressure terms (i.e. p_ℓ and p_s) are the dominant ones of the stress tensors. Some of the simplest possible relationships between p and p_s are (Tien et al., 2001)

$$\text{Type (1)} \qquad \mathrm{d}p_\ell + \mathrm{d}p_s = 0 \tag{2.2.9a}$$

$$\text{Type (2)} \qquad (1 - \varepsilon_s)\mathrm{d}p_\ell + \mathrm{d}p_s = 0 \tag{2.2.9b}$$

$$\text{Type (3)} \qquad (1 - \varepsilon_s)\mathrm{d}p_\ell + \varepsilon_s\mathrm{d}p_s = 0 \tag{2.2.9c}$$

$$\text{Type (4)} \qquad \mathrm{d}[(1 - \varepsilon_s)p_\ell] + \mathrm{d}[\varepsilon_s p_s] = 0 \tag{2.2.9d}$$

For a filter cake with $p_\ell = p_0$ at $x = L$, $p_\ell = \Delta p_m$ at $x = 0$, where Δp_m is the pressure drop across the medium.[1]

A general representation of the results given above may be written as

$$\frac{\partial p_\ell}{\partial p_s} = f' \tag{2.2.10}$$

and

$$\text{Case (1)} \qquad f' = -1 \tag{2.2.11a}$$

$$\text{Case (2)} \qquad f' = \frac{-1}{1 - \varepsilon_s} \tag{2.2.11b}$$

$$\text{Case (3)} \qquad f' = \frac{-\varepsilon_s}{1 - \varepsilon_s} \tag{2.2.11c}$$

$$\text{Case (4)} \qquad f' = \frac{(1 - \varepsilon_s^0)p_0 - p_s}{(1 - \varepsilon_s)^2}\frac{\mathrm{d}\varepsilon_s}{\mathrm{d}p_s} - \frac{\varepsilon_s}{1 - \varepsilon_s} \tag{2.2.11d}$$

[1]p_ℓ at the downstream side of the medium is assumed to be zero.

Equations Describing Cake Growth

Equations (2.2.3a), (2.2.3b), (2.2.4), and (2.2.10) are the starting points for analyzing cake filtration. To complete the description of cake filtration, two closing relationships will be introduced: (I) expression of liquid/particle relative velocity and (II) cake constitutive relationships.

(I) Expression of liquid/particle relative velocity. For liquid flow through a medium undergoing compression, the generalized Darcy's law (Shirato et al., 1969) expresses the liquid/particle relative velocity, $q_{\ell s}$, as

$$\frac{q_{\ell s}}{\varepsilon} = \frac{q_\ell}{\varepsilon} - \frac{q_s}{\varepsilon_s} = -\frac{1}{\varepsilon}\frac{k}{\mu}\frac{\partial p_\ell}{\partial x} \tag{2.2.12}$$

which reduces to the classical Darcy's law with $q_s = 0$. From Equation (2.2.4) and with $q_s = 0$ at $x = 0$, one has

$$q_\ell + q_s = q_{\ell_m} = \left[-\frac{k}{\mu}\frac{\partial p_\ell}{\partial x}\right]_{x=0} = \frac{-p_{\ell m}}{\mu R_m} \tag{2.2.13a}$$

where $p_{\ell m}$ is the filtrate pressure at $x = 0$ and R_m is the medium resistance. The negative sign is to account for the fact that filtrate flows in the negative x-direction. Equation (2.2.12) may be rewritten as

$$\frac{q_\ell}{\varepsilon} - \frac{q_{\ell_m} - q_\ell}{\varepsilon_s} = \frac{-1}{\varepsilon}\frac{k}{\mu}\frac{\partial p_\ell}{\partial x}$$

or

$$\frac{q_\ell}{\varepsilon\varepsilon_s} = \frac{q_{\ell_m}}{\varepsilon_s} - \frac{1}{\varepsilon}\frac{k}{\mu}\frac{\partial p_\ell}{\partial x}$$

Combining the above expression with Equation (2.2.13a) yields

$$q_\ell = -\varepsilon_s\frac{k}{\mu}\frac{\partial p_\ell}{\partial x} + (1 - \varepsilon_s)\left[-\frac{k}{\mu}\frac{\partial p_\ell}{\partial x}\right]_{x=0} \tag{2.2.13b}$$

(II) Cake constitutive relationships. In cake filtration analysis, the properties which characterize cake structure, namely, cake solidosity, ε_s, permeability, k, and specific cake resistance α $[= (k\varepsilon_s p_s)^{-1}]$, are assumed to be functions of the cake compressive stress, p_s, only. Specifically, the constitutive relationships may be expressed as

$$\varepsilon_s = \varepsilon_s(p_s) \tag{2.2.14a}$$

$$k = k(p_s) \tag{2.2.14b}$$

$$\alpha = \alpha(p_s) \tag{2.2.14c}$$

In particular, the following power-law expressions may be used to represent the constitutive relationships of ε_s vs. p_s; k vs. p_s; and α vs. p_s as

$$\varepsilon_s = \varepsilon_s^0 \left(1 + \frac{p_s}{p_A}\right)^{\beta} \tag{2.2.15a}$$

$$k = k^0 \left(1 + \frac{p_s}{p_A}\right)^{-\delta} \tag{2.2.15b}$$

$$\alpha = \frac{1}{\varepsilon_s \rho_s k} = \frac{1}{\varepsilon_s^0 k^0 \rho_s} \left(1 + \frac{p_s}{p_A}\right)^{\delta - \beta} = \alpha^0 \left(1 + \frac{p_s}{p_A}\right)^{n} \tag{2.2.15c}$$

For the above three equations, two are independent. Among the parameters, ε_s^0, k^0 and α^0 denote respectively the values of ε_s, k, and α at the zero-stress state (i.e. $p_s = 0$). p_A is the normalizing parameter of p_s and the exponents β, δ, n ($= \delta - \beta$) signify the compression effect due to p_s. Generally speaking, these parameters which can be obtained by fitting experimental data to the relevant equations are fitting parameters. Therefore, unless the data used for fitting cover appropriate ranges of p_s, the parameters ε_s^0, k^0 and α^0 may not have the physical significance as mentioned above.

In addition to the power-law expressions given above, other expressions may also be used to represent the constitutive relationships. In particular, soil scientists have proposed the use of the void ratio, e, defined as $e = \varepsilon/(1 - \varepsilon) = (1 - \varepsilon_s)/\varepsilon_s$ instead of ε_s, for characterizing cake structure. The cake compression behavior between e and the compressive stress, p_s, may be expressed as (Smiles, 1970)

$$e = a_1 \log(p_s) + a_2 \tag{2.2.16a}$$

As an estimate, the compression effect on cake permeability can be obtained from the Kozeny–Carman equation[2] which predicts the cake permeability being proportional to $(1 - \varepsilon_s)^3/\varepsilon_s^2 = e^3/(1 + e)$ or

$$k = b_1 \frac{e^3}{1 + e} \tag{2.2.16b}[3]$$

Equation (2.2.16b) can be used to estimate the compression effect on filtrate flow without the need of conducting direct experimental measurement. On the other hand, the accuracy of Equation (2.2.16b) depends upon the accuracy of Equation (2.2.16a) in relating e with p_s as well as the accuracy of the Kozeny–Carman equation in relating k with e. Whether or not the advantage gained from not carrying out permeability measurements in establishing the relationship between k vs. p_s is counterbalanced by the

---

[2]See 5.5 for more information of the Kozeny–Carman equation and the general flow rate-pressure drop relationships for flow through porous media.

[3]There is apparently a typographical error in the relationship between k and e given by Smiles (1970).

compromised accuracy, of course, depends upon the particular system being considered. However, if data of k vs. p_s are available, the data certainly should be used in establishing constitutive relationships of k vs. p_s. In such cases, the use of Equation (2.2.15b) perhaps gives a more simplified form of data representation.

Considering the case of using the power-law expression for the constitutive relationships, by combining Equations (2.2.3a), (2.2.10), (2.2.13), (2.2.15a), and (2.2.15b), one has

$$\frac{\partial \varepsilon_s}{\partial t} = \frac{\partial}{\partial x}\left[(-f')\frac{k^0}{\mu}\varepsilon_s\left(\frac{\varepsilon_s}{\varepsilon_s^0}\right)^{-\delta/\beta}\frac{\partial p_s}{\partial x}\right] - q_{\ell_m}\frac{\partial \varepsilon_s}{\partial x} \tag{2.2.17a}$$

where q_{ℓ_m} is liquid velocity at the cake/medium interface or the instantaneous filtration velocity. By definition [from Equation (2.2.12) at $x = 0$ and with $q_s = 0$)], q_{ℓ_m} is

$$q_{\ell_m} = -\left[\frac{k}{\mu}\frac{\partial p_\ell}{\partial x}\right]_{x=0} = \frac{-[p_\ell]_{x=0}}{\mu \cdot R_m} = \frac{-p_{\ell_m}}{\mu R_m} \tag{2.2.17b}$$

Equation (2.2.17a) is the basic equation of cake filtration. Its derivation is shown in Illustrative Example 2.1. The solution of Equations (2.2.17a) and (2.2.17b), corresponding to appropriate moving boundary conditions, initial and boundary conditions, give a complete description of cake filtration. These conditions are given below.

(I) Moving Boundary Condition

The moving boundary conditions describe cake growth and give cake thickness as a function of time. This condition may be obtained by mass balance considerations at cake surface (i.e. cake/suspension interface). Referring to Fig. 2.2, $x = L$ is the location of the cake/suspension interface (or L, the cake $q_\ell|_{L^-}$ thickness) such that $x < L^-$ is in the cake phase and $x > L^+$ in the suspension phase. If $q_\ell|_{L^+}$ and $q|_{L^-}$ denote the fluid velocities at $x = L^+$ and $x = L^-$, respectively, and $\varepsilon_s|_{L^+}$ and $\varepsilon_s|_{L^-}$ the corresponding solidosities, over a period of δt, the cake thickness increases by δL. Based on mass balance of fluid, one has

$$\frac{dL}{dt} = \frac{-q_\ell|_{L^+} - (-q_\ell|_{L^-})}{\varepsilon_s|_{L^+} - (\varepsilon_s|_{L^-})} \tag{2.2.18}$$

where $\varepsilon_s|_{L^-}$ is the cake solidosity at the cake surface where the compressive stress is zero. $\varepsilon_s|_{L^-}$ therefore can be taken to be ε_s^0, the solidosity at the zero-stress state. $\varepsilon_s|_{L^+}$ is equal to the particle volume fraction of the suspension, ε_{s_0}. Equation (2.2.18) therefore becomes

$$\frac{dL}{dt} = \frac{-q_\ell|_{L^+} - (-q_\ell|_{L^-})}{\varepsilon_{s_0} - \varepsilon_s^0} \tag{2.2.19}$$

From Equation (2.2.4), one has

$$q_\ell|_{L^+} + q_s|_{L^+} = q_\ell|_{L^-} + q_s|_{L^-} = q_{\ell m} = \left(-\frac{k}{\mu}\frac{\partial p_\ell}{\partial x}\right)_{x=0} \tag{2.2.20}$$

At $x = L^-$, by Darcy's law [Equation (2.2.12)]

$$\left. q_\ell \right|_{L^-} - \frac{1 - \varepsilon_s^0}{\varepsilon_s^0} \left. q_s \right|_{L^-} = \left[-\frac{k}{\mu} \frac{\partial p_\ell}{\partial x} \right]_{x=L^-} \tag{2.2.21a}$$

Also from Equation (2.2.20)

$$\left. q_\ell \right|_{L^-} = -\left. q_s \right|_{L^-} - \left(\frac{k}{\mu} \frac{\partial p_\ell}{\partial x} \right)_0 \tag{2.2.21b}$$

Solving for $\left. q_\ell \right|_{L^-}$ from the above two equations, one has

$$\left. q_\ell \right|_{L^-} = -\varepsilon_s^0 \left(\frac{k}{\mu} \frac{\partial p_\ell}{\partial x} \right)_{L^-} - \left(1 - \varepsilon_s^0 \right) \left(\frac{k}{\mu} \frac{\partial p_\ell}{\partial x} \right)_{x=0} \tag{2.2.21c}$$

At $x = L^+$, the suspended particles move at the same velocity as the suspending liquid, or

$$\frac{\left. q_s \right|_{L^+}}{\varepsilon_{s_0}} = \frac{\left. q_\ell \right|_{L^+}}{1 - \varepsilon_{s_0}}$$

or

$$\left. q_s \right|_{L^+} = \frac{\varepsilon_{s_0}}{1 - \varepsilon_{s_0}} \left. q_\ell \right|_{L^+}$$

Substituting the above expression into Equation (2.2.20), $\left. q_\ell \right|_{L^+}$ is found to be

$$\left. q_\ell \right|_{L^+} = \left(1 - \varepsilon_{s_0} \right) \left(-\frac{k}{\mu} \frac{\partial p_\ell}{\partial x} \right)_{x=0} \tag{2.2.22}$$

Substituting Equations (2.1.21c) and (2.1.22) into (2.1.18), the moving boundary condition of the cake/suspension interface is found to be

$$\frac{dL}{dt} = \frac{\varepsilon_s^0}{\varepsilon_s^0 - \varepsilon_{s_0}} \left[\frac{k}{\mu} \frac{\partial p_\ell}{\partial x} \right]_{x=L} + q_{\ell_m} \tag{2.2.23}$$

with the initial condition

$$L = 0, \quad t = 0 \tag{2.2.24}$$

(II) Boundary Conditions
The boundary conditions vary with the mode of operation. Consider the following three specific cases.
a. Constant Pressure Filtration. The boundary conditions are

$$p_\ell = p_0, \quad p_s = 0, \quad \varepsilon_s = \varepsilon_s^0 \qquad \text{at } x = L \tag{2.2.25a}$$

$$-\frac{k}{\mu} \frac{\partial p_\ell}{\partial x} = \frac{-p_\ell}{R_m \mu} \qquad \text{at } x = 0 \tag{2.2.25b}$$

b. Constant Rate Filtration

$$p_s = 0, \qquad \varepsilon_s = \varepsilon_s^0, \qquad \text{at } x = L \tag{2.2.26a}$$

$$-\frac{k}{\mu}\frac{\partial p_\ell}{\partial x} = -\frac{-p_\ell}{R_m \mu} = \text{constant}, \quad \text{at } x = 0 \qquad (2.2.26b)$$

c. Variable Pressure Filtration

$$p_\ell = p_0(t) \quad p_s = 0, \quad \varepsilon_s = \varepsilon_s^0 \quad \text{at } x = L \qquad (2.2.27a)$$

$$-\frac{k}{\mu}\frac{\partial p_\ell}{\partial x} = -\frac{p_\ell}{R_m \mu} \qquad\qquad \text{at } x = 0 \qquad (2.2.27b)$$

For both cases (a) and (c), p_0 is specified. One is interested in obtaining

$$q_{\ell_m}\left(=\left[-\frac{k}{\mu}\frac{\partial p_\ell}{\partial x}\right]_0\right) \text{ as a function of time. The reverse is true for case (b). The}$$

condition imposed on $x = 0$ (namely, Equation (2.2.27b) represented the continuity of filtrate permeation, namely, the rate of permeation on the upstream side and that on the downstream side of the medium are the same. As defined before, R_m is the medium resistance.

■ ■ ■ ▬▬▬▬▬▬▬▬▬▬▬▬▬▬▬▬▬▬▬▬▬▬▬▬▬▬

Illustrative Example 2.1

Show the derivation of Equation (2.2.17a).

Solution

From Equations (2.2.3a), (2.2.10), and (2.2.13b), one has

$$\frac{\partial \varepsilon}{\partial t} = -\frac{\partial q_\ell}{\partial x} \qquad (i)$$

$$\frac{\partial p_\ell}{\partial p_s} = f' \qquad (ii)$$

$$q_\ell = -\varepsilon_s \frac{k}{\mu}\frac{\partial p_\ell}{\partial x} + (1 - \varepsilon_s)\left[-\frac{k}{\mu}\frac{\partial p_\ell}{\partial x}\right]_{x=0} \qquad (iii)$$

From Equation (2.2.12), at $x = 0$, $q_s = 0$, $q_\ell = q_{\ell_m}$ where q_{ℓ_m} is the instantaneous filtration velocity, one has

$$q_{\ell_m} = \left[-\frac{k}{\mu}\frac{\partial p_\ell}{\partial x}\right]_{x=0} \qquad (iv)$$

Combining Equations (i), (iii), and (iv) yields

$$\frac{\partial \varepsilon}{\partial t} = \frac{\partial}{\partial t}(1 - \varepsilon_s) = -\frac{\partial}{\partial x}\left[-\varepsilon_s\frac{k}{\mu}\frac{\partial p_\ell}{\partial x} + (1 - \varepsilon_s)q_{\ell_m}\right]$$

or

$$\frac{\partial \varepsilon_s}{\partial t} = \frac{\partial}{\partial x}\left[-\varepsilon_s\frac{k}{\mu}\frac{\partial p_\ell}{\partial x}\right] - q_{\ell_m}\frac{\partial \varepsilon_s}{\partial x} \qquad (v)$$

$\dfrac{\partial p_\ell}{\partial x}$ may be written as

$$\frac{\partial p_\ell}{\partial x} = \frac{dp_\ell}{dp_s}\frac{\partial p_s}{\partial x} = f'\frac{\partial p_s}{\partial x} \tag{vi}$$

From Equations (2.2.15a) and (2.2.15b), the constitutive relationships of ε_s vs. p_s and k vs. p_s are

$$\varepsilon_s = \varepsilon_s^0\left(1 + \frac{p_s}{p_A}\right)^\beta$$

$$k = k^0\left(1 + \frac{p_s}{p_A}\right)^{-\delta}$$

which may be rewritten as

$$\left(\frac{\varepsilon_s}{\varepsilon_{s_0}}\right) = \left(1 + \frac{p_s}{p_A}\right)^\beta$$

and

$$\frac{k}{k^0} = \left(1 + \frac{p_s}{p_A}\right)^{-\delta} = \left[\left(1 + \frac{p_s}{p_A}\right)^\beta\right]^{-\delta/\beta} = \left(\frac{\varepsilon_s}{\varepsilon_s^0}\right)^{-\delta/\beta} \tag{vii}$$

Substituting Equations (vi) and (vii) into (v), one has

$$\frac{\partial \varepsilon_s}{\partial t} = \frac{\partial}{\partial x}\left[(-f')\varepsilon_s\frac{k_0}{\mu}\left(\frac{\varepsilon_s}{\varepsilon_s^0}\right)^{-\delta/\beta}\frac{\partial p_s}{\partial x}\right] - q_{\ell_m}\frac{\partial \varepsilon_s}{\partial t} \tag{viii}$$

which is Equation (2.2.16b). Since p_s can be related directly to ε_s, ε_s now is the only dependent variable of Equation (2.2.16).

■ ■ ■

■ ■ ■

Illustrative Example 2.2

Calculate p_s as a function of p_ℓ, based on Equations (2.2.10), (2.2.11a)–(2.2.11d), of $CaCO_3$ filter cakes corresponding to the following conditions:

Pressure Drop across the cake 7×10^5 Pa

The constitutive relationship of ε_s vs. p_s is given as

$$\varepsilon_s = \varepsilon_s^0\left(1 + \frac{p_s}{p_a}\right)^\beta$$

$$\beta = 0.13$$

$$\varepsilon_s^0 = 0.20$$

$$p_A = 4.4 \times 10^4 \text{ Pa}$$

Solution

From Equations (2.2.10), (2.2.11a)–(2.2.11d), with the given conditions, the following expressions are obtained:

From Equation (2.2.11a) $p_\ell + p_s = 7 \times 10^5$ (i)

From Equation (2.2.11b) $p_\ell + \int_0^{p_s} \dfrac{dp_s}{(1 - \varepsilon_s)} = 7 \times 10^5$ (ii)

From Equation (2.2.11c) $p_\ell + \int_0^{p_s} \dfrac{\varepsilon_s}{(1 - \varepsilon_s)} \, dp_s = 7 \times 10^5$ (iii)

From Equation (2.2.11d) $p_\ell + \dfrac{\varepsilon_s}{1 - \varepsilon_s} p_s = \dfrac{1 - \varepsilon_s^0}{1 - \varepsilon_s} (7 \times 10^5)$ (iv)

For Equations (i)–(iv), it is more convenient to calculate p_ℓ for specified value of p_s. The results are tabulated as follows:

Values of p_λ vs. p_s

p_s (Pa)	p_ℓ (Pa)			
	Equation (i)	**Equation (ii)**	**Equation (iii)**	**Equation (iv)**
0	7×10^5	7×10^5	7×10^5	7×10^5
5×10^4	6.5×10^5	6.640×10^5	6.866×10^5	7.07×10^5
1×10^5	6×10^5	5.718×10^5	6.718×10^5	6.996×10^5
2×10^5	5×10	4.399×10^5	6.399×10^5	6.803×10^5
3×10^5	4×10^5	3.049×10^5	6.056×10^5	6.552×10^3
4×10^5	3×10^5	1.684×10^5	5.694×10^5	6.192×10^5
5×10^5	2×10^5	0.309×10^5	5.317×10^5	–
5.219×10^5	–	0^*		
6×10^5	1×10^5	–	4.928×10^5	5.442×10^5
7×10^5	0	–	4.527×10^5	–
8×10^5	–	–	4.118×10^5	4.614×10^5
9×10^5	–	–	3.699×10^5	–
1×10^6	–	–	3.272×10^5	3.711×10^5
1.2×10^6	–	–	2.396×10^5	2.742×10^5
1.4×10^6	–	–	1.494×10^5	1.736×10^5
1.6×10^6	–	–	–	0.696×10^5
1.716×10^6	–	–	0^*	–
1.731×10^6	–	–	–	0^*

*Obtained by interpolation.

To show the determination of p_s corresponding to $p_\ell = 0$ from Equations (ii)–(iv), as an example, consider the case with the $p_\ell - p_s$ relationship given by Equation (ii); for $p_s = 5 \times 10^5$ Pa, the corresponding value of p_ℓ is

$$p_\ell = 7 \times 10^5 - \int_0^{5\times10^5} \frac{dp_s}{1 - \varepsilon_s} = 0.307 \times 10^5$$

For $P_s = 6 \times 10^5$ Pa, p_ℓ is found to be

$$p_\ell = 7 \times 10^5 - \int_0^{6\times10^5} \frac{dp_s}{1 - \varepsilon_s} = -1.085 \times 10^5$$

By linear interpolation, the value of p_s which gives $p_\ell = 0$, or p_{s_m}, is

$$p_{s_m} = 5 \times 10^5 + \left(\frac{0.307}{0.307 + 1.085}\right) 10^5 = 5.219 \times 10^5$$

The corresponding solidosity value, ε_{s_m}, is

$$\varepsilon_{s_m} = (0.2)\left[1 + \frac{52.19}{4.4}\right]^{0.13} = 0.279$$

The tabulated results are also shown in Fig. i.

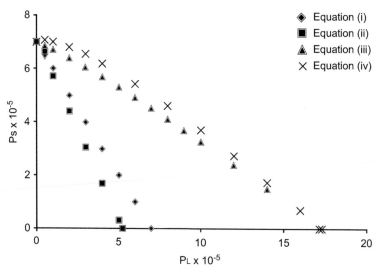

FIGURE I Relation between p_s and p_ℓ.

The results obtained show that for $CaCO_3$ cakes, with $p_\ell - p_s$ relationship given by Equation (i), p_s ranges from 0 to 7×10^5 Pa, the same as p_ℓ. For cases with the relationship given by Equations (ii) and (iii), p_s ranges from 0 to 5.219×10^5 and 0 to 1.716×10^6, respectively. For the last case,

[Equation (iv)], the p_s range is somewhat similar to that of Equation (iii) i.e. from 0 to 1.731×10^6. However, the behavior of the $p_s - p_\ell$ relationship is unusual. This can be seen more clearly by examining the $p_\ell - p_s$ relationship near $p_\ell \cong 7 \times 10^5$. From Equation (iv), we have

p_s (Pa)	p_ℓ (Pa)
4×10^4	7.049×10^5
6×10^4	7.101×10^5
8×10^4	7.036×10^5
1×10^5	6.997×10^5

These calculated results show that near $p_\ell = 7 \times 10^5$, the behavior of the $p_\ell - p_s$ relationship is not monotonic and is physically unrealistic. Therefore, the relationship of Equation (iv) is not valid and should not be used.

■ ■ ■

■ ■ ■ ─────────

Illustrative Example 2.3

As shown in 2.2, cake filtration is described by Equations (2.2.3a), (2.2.3b), and (2.2.12). Show that from these equations and using the void ratio, e, defined as $(1 - \varepsilon_s)/\varepsilon_s$ as the dependent variable and m, the so-called material coordinate and $\bar{t}$ as the independent variables defined below,

$$dm = \varepsilon_s \cdot dx - q_s \cdot dt$$

$$d\bar{t} = dt$$

cake filtration can be shown to be a diffusive process.

Solution

Equations (2.2.3a), (2.2.3b), and (2.2.12) are the volume-averaged continuity equations (mass balance) and the generalized Darcy's law given as

$$\frac{\partial \varepsilon}{\partial t} = -\frac{\partial q_\ell}{\partial x} \tag{i}$$

$$\frac{\partial \varepsilon_s}{\partial t} = -\frac{\partial q_s}{\partial x} \tag{ii}$$

$$\frac{q_{\ell s}}{\varepsilon} = \frac{q_\ell}{\varepsilon} - \frac{q_s}{\varepsilon_s} = -\frac{1}{\varepsilon}\frac{k}{\mu}\left(\frac{\partial p_s}{\partial x}\right) \tag{iii}$$

The void ratio, e, is defined as

$$e = (1 - \varepsilon_s)/\varepsilon_s \tag{iv}$$

From Equation (iv), one has

$$\varepsilon_s = 1/(1+e) \qquad \varepsilon = e/(1+e) \tag{v}$$

From Equations (iii) and (v), one may write

$$q_\ell = q_{\ell s} + e\ q_s \tag{vi}$$

$$q_{\ell s} = -\frac{k}{\mu}\left(\frac{\partial p_\ell}{\partial x}\right) \tag{vii}$$

and

$$q_\ell = e\ q_s - \frac{k}{\mu}\left(\frac{\partial p_\ell}{\partial x}\right) \tag{viii}$$

Substituting Equations (vi) into (i), one has

$$\frac{\partial \varepsilon}{\partial t} = -\frac{\partial}{\partial t}[q_{\ell s} + e\ q_s] = -\frac{\partial q_{\ell s}}{\partial x} - e\ \frac{\partial q_s}{\partial x} - q_s\ \frac{\partial e}{\partial x}$$

or

$$\frac{\partial \varepsilon}{\partial t} + e\left(-\frac{\partial \varepsilon_s}{\partial t}\right) = -\frac{\partial q_{\ell s}}{\partial x} - q_s\ \frac{\partial e}{\partial x} \tag{ix}$$

The left-hand side of the above expression is

$$\frac{\partial \varepsilon}{\partial t} + e\ \frac{\partial \varepsilon}{\partial t} = (1+e)\frac{\partial \varepsilon}{\partial t} = \left(\frac{1}{1+e}\right)\frac{\partial e}{\partial t}$$

Equation (ix) now becomes

$$\frac{de}{dt} + (1+e)\ \frac{\partial q_{\ell s}}{\partial x} + (1+e)q_s\ \frac{\partial e}{\partial x} = 0 \tag{x}$$

In terms of $m - \bar{t}$, the above equation becomes

$$\frac{\partial e}{\partial \bar{t}} - q_s\ \frac{\partial e}{\partial m} + (1+e)\varepsilon_s\ \frac{\partial q_{\ell s}}{\partial m} + (1+e)q_s\ \varepsilon_s\ \frac{\partial e}{\partial m} = 0$$

Since $\varepsilon_s = 1/(1+e)$, one now has

$$\frac{\partial e}{\partial \bar{t}} + \frac{\partial q_{\ell s}}{\partial m} = 0 \tag{xi}$$

From Equation (iii), one may write

$$q_{\ell s} = -\frac{k}{\mu}\frac{\partial p_\ell}{\partial x} = -\frac{k}{\mu}\frac{dp_\ell}{dp_s}\frac{dp_s}{de}\frac{\partial e}{\partial x} = -\frac{k}{\mu}\frac{f'}{(de/dp_s)(1+e)}\frac{\partial e}{\partial m} \tag{xii}$$

where $f' = dp_\ell/dp_s$.

Equation (xi) can now be written as

$$\frac{\partial e}{\partial t} = \frac{\partial}{\partial m}\left[D(e)\frac{\partial e}{\partial m}\right] \tag{xiii}$$

with

$$D(e) = \frac{k(f')}{(1+e)(de/dp_s)\mu} \tag{xiv}$$

Equation (xiii) is the same as the one-dimensional diffusion equation. The quantity $D(e)$ is known as the filtration diffusivity. A more detailed discussion of treating cake filtration as a diffusion problem can be found in the monograph by Tien (2006).

Illustrative Example 2.4

The filtration diffusivity, $D(e)$, is defined as [see Equation (xiv) of Illustrative Example 2.3]:

$$D(e) = \frac{k f'}{(1+e)(de/dp_s)\mu}$$

and e, the void ratio, is given as

$$e = \frac{1-\varepsilon_s}{\varepsilon_s}$$

Obtain values of $D(e)$ vs. e and e vs. p_s for $CaCO_3$ cakes for $0 < p_s < 1 \times 10^6$ Pa with $f' = -1$. The constitutive relationship of ε_s vs. p_s is given in Illustrative Example 2.2. The relationship of k vs. p_s is

$$k = k^0\left(1+\frac{p_s}{p_A}\right)^{-\delta}$$

$$p_A = 4.4 \times 10^4 \text{ Pa}$$

$$\delta = 0.57(-)$$

$$k^0 = 4.87 \times 10^{-14} \text{ m}^2$$

The filtrate viscosity μ may be taken as 10^{-3} Pa s and $f' = -1$.

Solution

By definition, e is defined as

$$e = \frac{1-\varepsilon_s}{\varepsilon_s} = \left(\frac{1}{\varepsilon_s}\right) - 1 = \frac{1}{\varepsilon_s^0}\left(1+\frac{p_s}{p_A}\right)^{-\beta} - 1 \tag{i}$$

Differentiating e with respect to p_s, one has

$$\frac{de}{dp_s} = \frac{-1}{\varepsilon_s^2}\frac{d\varepsilon_s}{dp_s} = -\frac{1}{\varepsilon_s^2}\varepsilon_s^0\beta\left(1+\frac{p_s}{p_A}\right)^{\beta-1}(1/p_A) \tag{ii}$$

Substituting Equations (i) and (ii) into the expression defining $D(e)$, $D(e)$ is now given as

$$D(e) = \frac{k(-f')}{\mu}\frac{1}{(1+e)(-de/dp_s)} = \frac{(-f')k^0\left(1+\dfrac{p_s}{p_A}\right)^{-\delta}}{\mu}(\varepsilon_s)\frac{\varepsilon_s^2}{\varepsilon_s^0(\beta/p_A)\left(1+\dfrac{p_s}{p_A}\right)^{\beta-1}} \tag{iii}$$

$$= \frac{(-f')(k^0)\left(1+\dfrac{p_s}{p_A}\right)^{1-\delta-\beta}\varepsilon_s^3}{\mu\,\varepsilon_s^0(\beta/p_A)} = \left(\frac{-f'}{\mu}\right)\left(\frac{k^0 p_A}{\beta}\right)(\varepsilon_s^0)^2\left(1+\frac{p_s}{p_A}\right)^{1+2\beta-\delta}$$

or $D(e)$ is a function of p_s. The results of e vs. p_s and $D(e)$ vs. p_s are shown below.

Values of e vs. p_s and $D(e)$ vs. p_s can be readily found from Equations (i) and (iii) as shown below.

$p_s(P_a)$	e	D
0	4	6.502×10^{-7}
1×10^3	3.988	6.682×10^{-7}
2×10^3	3.975	6.783×10^{-7}
1×10^4	3.878	7.082×10^{-7}
2×10^4	3.714	8.626×10^{-7}
1×10^5	3.292	1.488×10^{-6}
2×10^5	3.000	2.152×10^{-6}
5×10^5	2.610	3.724×10^{-6}
1×10^6	2.333	5.755×10^{-6}

The results of D vs. p_s and D vs. e are also shown in the following figures.

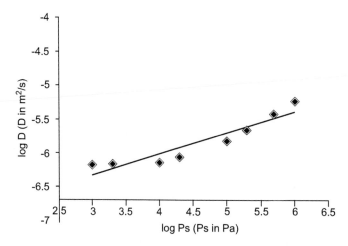

FIGURE I Filtration diffusivity vs. compressive stress.

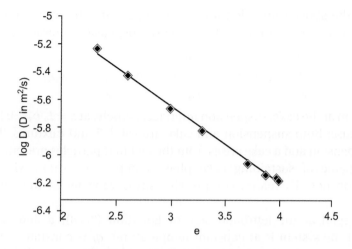

FIGURE II Filtration diffusivity vs. void ratio.

2.3   The Conventional Cake Filtration Theory

As discussed in the preceding section, rigorous analyses of cake filtration can be made from the solution of Equation (2.2.17a) with appropriate initial, boundary, and moving boundary conditions. However, the effort required for such an analysis is not trivial. Equally important, such an approach is not compatible, in many cases, with simple design and scale-up calculations and is also not easily adaptable in evaluating experimental data. In contrast, the conventional cake filtration first proposed by Ruth (1935a,b) and further developed by a number of investigators including Grace, Tiller, and Shirato provide results which can be used easily in practical applications (see Shirato et al. (1987)). A discussion of the conventional theory including its applications and limitations is given below.

The conventional theory of cake filtration can be considered as a simplification of the more exact analysis discussed in 2.2. Beginning with Equations (2.2.3a) and (2.2.12), for the pseudo-steady-state case and with negligible particle velocity, one has

$$q_\ell = -\frac{k}{\mu}\frac{\partial p_\ell}{\partial x} = \text{constant}   0 < x < L(t) \tag{2.3.1}$$

across the entire cake thickness at any instant.

If the relationship between p_ℓ and p_s is given by Equation (2.2.10), Equation (2.3.1) may be written as

$$\mu q_\ell dx = -f' k dp_s \tag{2.3.2a}$$

or

$$\mu q_\ell(\varepsilon_s \rho_s)dx = -f' k \varepsilon_s \rho_s dp_s = -(f'/\alpha)\cdot dp_s \tag{2.3.2b}$$

and α is the specific cake resistance defined by Equation (2.2.15c).

Integrating the above expression for $x = 0$ (or $p_s = p_{s_m}$, the value of p_s at the cake/medium interface) to $x = L$ (or $p_s = 0$ at the cake/suspension interface), one has

$$\int_0^L \mu\, q_\ell(\varepsilon_s \rho_s)\mathrm{d}x = -\int_{p_{s_m}}^0 (f'/\alpha)\mathrm{d}p_s = \int_0^{p_{s_m}} (f'/\alpha)\mathrm{d}p_s \tag{2.3.3}$$

The condition at the cake/suspension interface; namely, at $x = L$, $p_s = 0$ is made on the following premise: Both suspension and cake are solid–liquid mixtures. The difference between a suspension and a cake arises from the fact that particles present in a cake form a network capable of sustaining cake phase compressive stress while particles of a suspension do not. Therefore, compressive stress vanishes at the cake/suspension interface.

For the integral at the left-hand side of Equation (2.3.3), μ and ρ_s are constant (assuming that the system is at constant temperature), q_ℓ is constant across the entire cake thickness as stated before and is, by definition, negative (since filtrate moves along the negative direction of x (see Fig. 2.1). Therefore, one may write

$$\frac{\mathrm{d}V}{\mathrm{d}t} = -q_\ell \tag{2.3.4}$$

where V is the cumulative filtrate volume per unit medium surface ($\mathrm{m^3\ m^{-2}}$).

The left-hand side integral of Equation (2.3.3) may therefore be written as

$$-\mu\rho_s(\mathrm{d}V/\mathrm{d}t)\int_0^L \varepsilon_s\cdot\mathrm{d}x = -\mu(\mathrm{d}V/\mathrm{d}t)(\rho_s\cdot\bar{\bar{\varepsilon}}_s\cdot L) \tag{2.3.5}$$

and $\bar{\bar{\varepsilon}}_s$ is the average cake solidosity, defined as

$$\bar{\bar{\varepsilon}}_s = \frac{1}{L}\int_0^L \varepsilon_s\cdot\mathrm{d}x \tag{2.3.6}$$

The right-hand side integral of Equation (2.3.3) may be written as

$$\int_0^{p_{s_m}} (f'/\alpha)\mathrm{d}p_s = \int_{p_0}^{p_0-\Delta p_c} (1/\alpha)\cdot\mathrm{d}p_\ell = \frac{-\Delta p_c}{(\alpha_{av})_{p_{s_m}}} \tag{2.3.7}$$

where Δp_c is the pressure drop across the cake and the average specific cake resistance, $[\alpha_{av}]_{p_{s_m}}$ is defined as[4]

$$[\alpha_{av}]_{p_{s_m}} = \frac{\Delta p_c}{\int_0^{p_{s_m}} \left(\frac{1}{\alpha}\right)(-f')\mathrm{d}p_s} = \frac{\Delta p_c/p_{s_m}}{\left[\int_0^{p_{s_m}} \left(\frac{1}{\alpha}\right)(-f')\mathrm{d}p_s\right]\Big/p_{s_m}} \tag{2.3.8}$$

---

[4]The subscript, p_{s_m}, refers to the fact that the cake is subject to a compressive stress ranging from zero (at cake surface) to p_{s_m} (at cake base).

Substituting Equations (2.3.5) and (2.3.8) into (2.3.3), $|q_\ell|$ is found to be

$$dV/dt = |q_\ell| = \frac{1}{\mu} \frac{\Delta p_c}{(\rho_s \bar{\bar{\varepsilon}}_s L)[\alpha_{av}]_{p_{sm}}} = \frac{1}{\mu} \frac{\Delta p_c}{w[\alpha_{av}]_{p_{sm}}} \tag{2.3.9}$$

with w being the cake areal mass (kg m^{-2}).

For filtrate permeating across the medium, one may write

$$|q_\ell| = \frac{\Delta p_m}{\mu R_m} \tag{2.3.10}$$

where Δp_m is the pressure drop across the medium.

Combining Equations (2.3.9) and (2.3.10), one has

$$dV/dt = |q_\ell| = \frac{\Delta p_c + \Delta p_m}{\mu \left[w(\alpha_{av})_{p_{sm}} + R_m \right]} = \frac{p_0}{\mu \left[w(\alpha_{av})_{p_{sm}} + R_m \right]} \tag{2.3.11}$$

Equation (2.3.11) is the basic equation of the conventional cake filtration theory and is used widely in estimating cake filtration performance and interpreting and correlating cake filtration results. What is often overlooked is: for a compressible cake, the cake solidosity, permeability, and specific cake resistance vary along the cake thickness. The $(\alpha_{av})_{p_{sm}}$ present in the above expression (defined by Equation (2.3.8)) is a stress-averaged specific cake resistance with p_s ranging from zero to p_{sm}.

A comparison between what is given above and the procedure commonly used in deriving cake filtration equation may be useful as a further examination of the conventional theory. In most liquid–solid separation monographs (including some of the widely used unit operations texts for undergraduate teaching), Equation (2.3.1) is often the starting point in discussing cake filtration. Rewriting Equation (2.3.1) as

$$-\mu \rho_s \varepsilon_s q_\ell dx = k \rho_s \varepsilon_s dp_\ell = (1/\alpha) dp_\ell \tag{2.3.12}$$

and integrating the above expression for $x = 0$ and $x = L$, with the boundary conditions given as

$$x = 0, \quad p_\ell = 0 \tag{2.3.13}$$

$$x = L, \quad p_\ell = p_0$$

one has

$$\frac{dV}{dt} = -q_\ell = \frac{\int_{p_0 - \Delta p_c}^{p_0} \left(\frac{1}{\alpha} \right) \cdot dp_\ell}{\mu \int_0^L \rho_s \varepsilon_s \cdot dx} = \frac{\Delta p_c (\overline{1/\alpha})}{\mu w} \tag{2.3.14}$$

$$\text{with} \quad \left(\frac{\overline{1}}{\alpha} \right) = \frac{\int_{p_0 - \Delta p_c}^{p_0} \left(\frac{1}{\alpha} \right) \cdot dp_\ell}{\Delta p_c} \tag{2.3.15a}$$

and

$$w = \int_0^L \rho_s \varepsilon_s dx \tag{2.3.15b}$$

and $(\overline{1/\alpha})$ may be considered as the average value of $(1/\alpha)$ over p_ℓ ranging from $p_\ell = p_0 - \Delta p_c$ to $p_\ell = p_0$.

At a first glance, the definition given by Equation (2.3.8) and that of Equation (2.3.15a) may appear to be very similar. In fact, for $f' = -1$ (which is commonly assumed), they become identical. However, Equation (2.3.8) gives a physical significance of α_{av} which cannot be seen from Equation (2.3.15a). The structure of a cake, which depends upon the degree of its compactness, is determined by the cake compressive stress. One may expect α to be a function of p_s and this information can be obtained, at least in principle, by forming cakes under different degrees of compression and then determining their porosities and permeabilities. With the knowledge of α vs. p_s and the relationship between p_ℓ and p_s [i.e. Equation (2.2.11)], $[\alpha_{av}]$ can be determined readily from Equation (2.3.8). In other words, $[\alpha_{av}]$ can be estimated based on independent measurements and not only from curve-fitting filtration data. This rather fundamental difference, unfortunately, is often overlooked by workers of cake filtration.

■ ■ ■ ▬▬▬▬▬▬▬▬▬▬▬▬▬▬▬▬▬▬▬▬▬▬▬▬▬▬▬▬▬▬▬▬▬

Illustrative Example 2.5

Outline a procedure of establishing the constitutive relationship of α vs. p_s from results of α_{av} vs. p_s.

Solution

For a cake subject to compression from $p_s = 0$ (at cake surface) to $p_s = p_{s_m}$ (at cake base), the average specific cake resistance α_{av} is defined by Equation (2.3.8), or

$$[\alpha_{av}]_{p_{s_m}} = \frac{\Delta p_c}{\int_0^{p_{s_m}} \left(\frac{1}{\alpha}\right)(-f')dp_s} \tag{i}$$

The pressure drop across the cake, Δp_c, according to Equation (2.3.10) is

$$\Delta p_c = \int_0^{p_{s_m}} (-f')dp_s \tag{ii}$$

From Equations (i) and (ii), one has

$$[\alpha_{av}] = \frac{\int_0^{p_s} (-f')dp_s}{\int_0^{p_s} \left(\frac{1}{\alpha}\right)(-f')dp_s} \tag{iii}$$

The subscript, "p_{s_m}" is dropped for simplification. Equation (iii) may be rewritten as

$$\int_0^{p_s} \left(\frac{1}{\alpha}\right)(-f') \cdot dp_s = \left[\frac{1}{\alpha_{av}}\right]\left[\int_0^{p_s} (-f')dp_s\right]$$

Differentiating the above expression yields

$$\frac{-f'}{\alpha} = \left[\frac{1}{\alpha_{av}}\right][-f'] + \left[\int_0^{p_s} (-f) \cdot dp_s\right] \frac{d}{dp_s}\left[\frac{1}{\alpha_{av}}\right]$$

or

$$\alpha = \left[\frac{1}{\alpha_{av}} + (\Delta p_c)(-1/f')\frac{d}{dp_s}\left(\frac{1}{\alpha_{av}}\right)\right]^{-1} \qquad \text{(iv)}$$

In other words, from the results of α_{av} vs. p_s, the values of $\dfrac{d}{dp_s}\left(\dfrac{1}{\alpha_{av}}\right)$ at various values of p_s can be estimated. The value of Δp_c can be calculated from Equation (ii). This information together with the knowledge of f' can be used to estimate α vs. p_s according to Equation (iv).

For the special case $f' = -1$, Equation (iv) may be simplified to give

$$\alpha = \left[\frac{1}{\alpha_{av}} + p_s\frac{d}{dp_s}\left(\frac{1}{\alpha_{av}}\right)\right]^{-1} \qquad \text{(v)}$$

■ ■ ■

2.4 Expressions of Cake Filtration Performance

The performance of cake filtration may be expressed by the cumulative filtrate volume, V, and/or the solid particles retained as filter cake, w, as functions of time. The relationship between V and w may be obtained from mass conservation considerations as follows. Let $\overline{m}$ be the wet to dry cake mass ratio; for a cake of thickness L, $\overline{m}$ is given as

$$\overline{m} = 1 + \frac{\int_0^L \rho(1-\varepsilon_s)dx}{\int_0^L \rho_s\varepsilon_s dx} = 1 + \frac{\rho(1-\overline{\overline{\varepsilon}}_s)}{\rho_s\overline{\overline{\varepsilon}}_s} \qquad (2.4.1)$$

and $\overline{\overline{\varepsilon}}_s$ is the average cake solidosity defined by Equation (2.3.6).

If s is the particle mass fraction of the feed suspension, by definition, s is given as

$$s = \frac{w}{V\rho + \overline{m}w} \qquad (2.4.2a)$$

and

$$w = \frac{V\rho s}{1 - \overline{m}s} \qquad (2.4.2b)$$

Equation (2.3.11) may be rewritten as

$$\frac{dV}{dt} = \frac{p_0}{\mu\left[(\alpha_{av})_{p_{sm}}\dfrac{V s\rho}{(1-\overline{m}s)} + R_m\right]} \qquad (2.4.3)$$

Equation (2.4.3) is equivalent to (2.3.11), with $\overline{m}$ replacing w as one of the dependent variables. The advantage of Equation (2.4.3) over Equation (2.3.11) is that $\overline{m}$, except for the initial period of filtration, is largely constant. Consequently, under most conditions, integration of Equation (2.4.3) can be easily made as shown below.

The primary operating variable of filtration is the pressure applied. The operating pressure may be kept at constant (constant pressure filtration); kept according to a particular manner, i.e. p_0 is a specified function of time (variable pressure filtration) or so kept that the rate of filtration is constant (constant rate filtration). For the first two cases, filtration performance can be expressed in terms of the cumulative filtrate volume as a function of time. For the last case, one is interested in the operating pressure required to sustain the specified rate of filtration. Both the performance result and the required operating pressure may be obtained from the integration of Equation (2.4.3) as follows:

a. Constant pressure filtration. Equation (2.4.3) may be rewritten as

$$\mu s\rho(1-\overline{m}s)^{-1}[\alpha_{av}]_{p_{sm}}V\frac{dV}{dt}+\mu R_m\frac{dV}{dt}=p_0 \tag{2.4.4}$$

Integrating the above expression with the initial condition, $V=0$, at $t=0$ and the assumption that R_m remains constant, one has

$$\overline{\mu s\rho(1-\overline{m}s)^{-1}[\alpha_{av}]_{p_{sm}}}\frac{V^2}{2}+\mu R_m V=p_0 t \tag{2.4.5}$$

where

$$\overline{(1-\overline{m}s)^{-1}[\alpha_{av}]_{p_{sm}}}=\frac{2}{V^2}\int_0^V(1-\overline{m}s)^{-1}[\alpha_{av}]_{p_{sm}}VdV \tag{2.4.6}$$

b. Variable pressure filtration. If p_0 is a function of time, the relationship between V and t obtained from the integration of Equation (2.4.3) becomes

$$\overline{\mu s\rho(1-\overline{m}s)^{-1}[\alpha_{av}]_{p_{sm}}}\frac{V^2}{2}+\mu R_m V=\int_0^t p_0(t)dt \tag{2.4.7}$$

c. Constant rate filtration. If the filtration rate is kept constant, or $q_\ell=Q=$ constant, one has

$$V=Qt \tag{2.4.8}$$

and the required operating pressure, p_0, is

$$p_0=\mu Q\left\{s\rho(1-ms)^{-1}[\alpha_{av}]_{p_{sm}}Qt+R_m\right\} \tag{2.4.9}$$

The filter cake formed in terms of dry cake mass, w, can be found directly from Equation (2.4.2b) with V known. The filter cake thickness, L, as a function of time can be determined as

$$L=\frac{w}{\rho_s\cdot\overline{\overline{\varepsilon}}_s} \tag{2.4.10}$$

where ρ_s is the cake particle density and $\overline{\overline{\varepsilon}}_s$ the average cake solidosity defined by Equation (2.3.6). From Equation (2.4.1), $\overline{\overline{\varepsilon}}_s$ can be related to $\overline{m}$ as

$$\overline{\overline{\varepsilon}}_s=\left[1+\frac{\rho_s}{\rho}(\overline{m}-1)\right]^{-1} \tag{2.4.11}$$

2.5   Parabolic Law of Constant Pressure Filtration

Equations (2.4.5), (2.4.7), and (2.4.9) are the main results of the conventional cake filtration theory. As constant pressure is often the mode of operation, Equation (2.4.5), in particular, is widely used in design calculations and data interpretations and is known as the parabolic law of constant pressure filtration if the quantity $(1 - \overline{m}s)^{-1}[\alpha_{av}]_{p_{s_m}}$ can be treated as a constant. The assumptions leading to this conclusion are as follows:

(i) During the course of filtration, both the cake solidosity profile and the pressure drop across the cake and, therefore, the cake compressive stress at the cake/medium interface, p_{s_m}, undergo continuous changes. The wet to dry cake mass ratio, $\overline{m}$, as defined by Equation (2.4.1) is a function of the solidosity profile [see Equation (2.4.11)]. As the conventional theory does not give a simple expression of the cake solidosity profile, $\overline{\overline{\varepsilon}}_s$ is often assumed to be the same as the stress-averaged solidosity, $\overline{\varepsilon}_s$, defined as

$$\overline{\varepsilon}_s = \frac{1}{p_{s_m}} \int_0^{p_{s_m}} \varepsilon_s \cdot dp_s \tag{2.5.1}$$

(ii) Evaluating $(1 - \overline{m}s)^{-1}[\alpha_{av}]p_{s_m}$ requires the information of the variations of $\overline{m}$ and $[\alpha_{av}]_{p_{s_m}}$ with V (or time) [see Equation (2.4.6)]. $[\alpha_{av}]_{p_{s_m}}$ is a function of Δp_c or p_{s_m} according to Equation (2.3.8). If $\overline{\varepsilon}_s$ may be used as an approximation of $\overline{\overline{\varepsilon}}_s$, $\overline{m}$ can also be considered as a function of p_{s_m}. Furthermore, with modest medium resistance, R_m, one may assume that both $\overline{m}$ and $[\alpha_{av}]_{p_{s_m}}$ approach to their respective ultimate values (corresponding to the condition, $\Delta p_c \simeq P_0$) rapidly. For the most part of a filtration run, one has

$$\overline{(1 - \overline{m}s)^{-1}[\alpha_{av}]_{p_{s_m}}} = (1 - \overline{m}s)^{-1}\alpha_{av} \tag{2.5.2}$$

and α_{av} is given as

$$\alpha_{av} = \frac{p_0}{\int_0^{p_{s_m}(p_0)} \left(\frac{1}{\alpha}\right)(-f')dp_s} \tag{2.5.3}$$

where $p_{s_m}(p_0)$ is so written to underscore the fact that p_{s_m} is now a function of p_0. With both α_{av} and $\overline{m}$ being evaluated at $\Delta p_c \simeq P_0$, Equation (2.4.5) now becomes

$$\mu s \rho \left[(1 - \overline{m}s)^{-1}_{\Delta p_c = p_0}\right] [\alpha_{av}]_{\Delta p_c = p_0} \frac{V^2}{2} + \mu\, R_m V = p_0 t \tag{2.5.4a}$$

or

$$t/V = \left(\frac{1}{2p_0}\right)\mu s\rho \left[(1 - \overline{m}s)^{-1}_{\Delta p_c = p_0}\right][\alpha_{av}]_{p_{s_m}(p_0)} V + \mu R_m/p_0 \tag{2.5.4b}$$

The above two equations are known as the parabolic law of constant pressure cake filtration. It should be emphasized that these expressions are obtained by simplifying the result of the conventional theory of Equation (2.4.5). The simplifications are made by assuming that the wet to dry cake mass ratio, $\overline{m}$, and the average specific cake resistance are constant and they are evaluated under the condition that the pressure drop across the cake is the same as the operating pressure.

The relationship between Δp_c and p_0 can be found simply from Equations (2.3.9) and (2.3.10), or

$$\frac{\Delta p_c}{p_0} = \frac{\mu s \rho \overline{(1 - \overline{m}s)}^{-1} [\alpha_{av}]_{p_{sm}} V}{\mu s \rho \overline{(1 - \overline{m}s)}^{-1} [\alpha_{av}]_{p_{sm}} V + \mu R_m} \tag{2.5.5}$$

It is clear that one may replace (Δp_c) with p_0 if the first quantity of the denominator outweighs the second quantity. However, since V is zero initially, this means that regardless of the magnitude of the average cake resistance, $[\alpha_{av}]_{p_{sm}}$, there inevitably exists an initial period during which Δp_c is significantly different from p_0. Therefore, Equations (2.5.4a) [or (2.5.4b)] are not valid during the initial period.

An alternate expression of the parabolic law of constant pressure cake filtration can be obtained by considering the medium resistance, R_m, being equivalent to that of a fictitious cake layer corresponding to a cumulative filtrate volume V_m. Equation (2.4.3) may now be written as

$$q_\lambda = \frac{dV}{dX} = \frac{\phi_0}{\mu[\alpha_{av}]_{p_{sm}} \dfrac{\rho s}{1 - \overline{m}} (V + V_m)} \tag{2.5.6}$$

and

$$V_m = \frac{1 - \overline{m}s}{\rho s (\alpha_{av})_{p_{sm}}} R_m \tag{2.5.7}$$

Integrating Equation (2.5.6) subject to the same assumptions used in obtaining Equation (2.5.4a) or (2.5.4b), one has

$$(V + V_m)^2 = A(t + t_m) \tag{2.5.8}$$

and

$$A = \frac{2p_0(1 - \overline{m}s)}{\mu[\alpha_{av}]_{\Delta p_c = p_0} \rho s} \tag{2.5.9a}$$

$$t_m = \frac{V_m^2}{A} = \frac{\mu[\alpha_{av}]_{\Delta p_c = p_0} \rho s}{2p_0(1 - \overline{m}s)} V_m^2 \tag{2.5.9b}$$

One may rewrite Equation (2.5.8) as

$$\frac{t}{V} = \frac{V}{A} + \frac{2V_m}{A} \tag{2.5.10}$$

■ ■ ■ ▬▬▬▬▬▬▬▬▬▬▬▬▬▬▬▬▬▬▬▬▬▬▬▬

Illustrative Example 2.6

Following the procedure described in Section 2.5, obtain the expression of V vs. time of cake filtration if the operating pressure increases incrementally.

♦

Solution

With operating pressure increasing incrementally, or $p_0 = p_1$, $0 < t < t_1$; $p_0 = p_2$, $t_1 < t < t_2$; . . . or $p_0 = p_i$, $t_{i-1} < t < t_i$, for the first period, $0 < t < t_1$, Equation (2.4.5) is readily applicable or

$$\mu s p [(1 - \overline{m}s)^{-1} \alpha_{av}]_{p_1} \frac{V^2}{2} + \mu R_m V = p_1 t \tag{i}$$

and V_1, the cumulative filtrate volume at $t = t_1$, is

$$\mu s p [(1 - \overline{m}s)^{-1} \alpha_{av}]_{p_1} \frac{V_1^2}{2} + \mu R_m V_1 = p_1 t_1 \tag{ii}$$

For the second time period, $t_1 < t < t_2$, with $p_0 = p_2$, the cumulative filtration volume can be found by integrating Equation (2.4.3) with the initial condition $t = t_1$, $V = V_1$, or

$$\mu s \rho [(1 - \overline{m}s)^{-1} \alpha_{av}]_{p_2} \frac{(V - V_1)^2}{2} + \mu R_m (V - V_1) = p_2 (t - t_1) \tag{iii}$$

and the cumulative filtrate volume V_2 at $t = t_2$ is

$$\mu s \rho \left[(1 - \overline{m}s)^{-1} \alpha_{av}\right]_{p_2} \frac{(V_2 - V_1)^2}{2} + \mu R_m (V_2 - V_1) = p_2 (t_2 - t_1) \tag{iv}$$

and for the i-th period, $t_{i-1} < t < t_i$, the V–t relationship is

$$\mu s \rho \left[(1 - \overline{m}s)^{-1} \alpha_{av}\right]_{p_i} \frac{(V - V_{i-1})^2}{2} + \mu R_m (V - V_{i-1}) = p_i (t - t_{i-1}) \tag{v}$$

and the corresponding cumulative filtrate volume at $t = t_i$, V_i, is

$$\mu s \rho \left[(1 - \overline{m}s)^{-1} \alpha_{av}\right]_{p_i} \frac{(V_i - V_{i-1})^2}{2} + \mu R_m (V_i - V_{ij}) = p_i (t_i - t_{i-1}) \tag{vi}$$

From the results obtained above, it is clear that there exists a linear relationship between $(t - t_{i-1})/(V - V_{i-1})$ and $(V - V_{i-1})$ for $t_{i-1} < t < t_i$. Therefore, the results obtained from measurements conducted with incrementally increasing operating pressure, when plotted in the form of $(t - t_{i-1})/(V - V_{i-1})$ vs. $(V - V_{i-1})$, yields a number of linear segments, the slope of which can be used to calculate the average specific cake resistance corresponding to different values of p_s. Murase et al. (1989) first applied this procedure for determining cake properties. A typical set of results obtained by Murase et al. is shown in Fig. i.

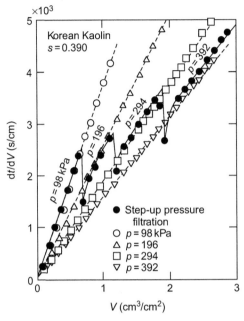

FIGURE I dt/dV vs. V data obtained by Murase et al. (1989). *[Reprinted from T. Murase, E. Iritani, J.H. Cho ad M. Shirato, "Determination of Filtration Characteristics based upon Filtration Tests under Stepped-Up Pressure Conditions", J. Chem. Eng. Japan, 22, 373–377, 1989, with permission of J. of Chem. Eng. Japan].*

■ ■ ■

2.6   Approximate Expressions of Cake Solidosity, Compressive Stress, and Pore Liquid Pressure Profiles

Applying the same assumptions used in developing the conventional cake filtration theory, approximate expressions of cake structure, namely, the cake solidosity, compressive stress, and liquid pressure profiles can be obtained. First, integrating Equation (2.3.2a), one has

$$\int_0^x \mu q_\ell \, dx = -\int_{p_{sm}}^{p_s} (k)(f') \, dp_s \tag{2.6.1}$$

and

$$\int_0^L \mu q_\ell \, dx = -\int_{p_{sm}}^0 (k)(f') \, dp_s \tag{2.6.2}$$

Dividing Equations (2.6.1) by (2.6.2), one has

$$\frac{\int_0^x \mu q_\ell dx}{\int_0^L \mu q_\ell dx} = \frac{\int_{p_s}^{p_{sm}} (k)(f')dp_s}{\int_0^{p_{sm}} (k)(f')dp_s} \tag{2.6.3}$$

The left-hand side quantity of Equation (2.6.3) can be approximated as x/L based on the assumption that $q_s \simeq 0$ and q_ℓ being independent of x. For the right-hand quantity, the permeability, k, is a function of the compressive stress. f' is also a function of the compressive stress (since f' is a function of ε_s), the quantity, $(\int_{p_s}^{p_{sm}} kf'dp_s)/(\int_0^{p_{sm}} kf'dp_s)$, is a function of p_s or ε_s. In other words, Equation (2.6.3) gives an implicit expression of the solidosity profile. This expression can be greatly simplified if the relationship between the cake permeability, k, and the compressive stress p_s is given by Equation (2.2.15b) and $f' = -1$. The result is

$$\frac{x}{L} = \frac{\left(1 + \frac{\Delta p_c}{p_A}\right)^{1-\delta} - \left(1 + \frac{p_s}{p_A}\right)^{1-\delta}}{\left(1 + \frac{\Delta p_c}{p_A}\right)^{1-\delta} - 1} = 1 - \frac{\left(1 + \frac{p_s}{p_A}\right)^{1-\delta} - 1}{\left(1 + \frac{\Delta p_c}{p_A}\right)^{1-\delta} - 1}$$

which, upon rearrangement, gives the expressions of the compressive stress profile

$$\frac{p_s}{p_A} = \left\{1 + \left(1 - \frac{x}{L}\right)\left[\left(1 + \frac{\Delta p_c}{p_A}\right)^{1-\delta} - 1\right]\right\}^{\frac{1}{1-\delta}} - 1 \tag{2.6.4}$$

If the solidosity–compressive stress relationship follows the power-law expression and is given by Equation (2.2.15a), combining Equation (2.6.4) with (2.2.15a), one has

$$\frac{\varepsilon_s}{\varepsilon_s^0} = \left\{\left[\left(1 + \frac{\Delta p_c}{p_A}\right)^{1-\delta} - 1\right]\left(1 - \frac{x}{L}\right) + 1\right\}^{\frac{\beta}{1-\delta}} \tag{2.6.5}$$

which is the same expression given by Tiller et al. (1999).[5]

With $f'(\varepsilon) = -1$, $p_\ell + p_s = \Delta p_c$,[6] the liquid pore pressure profile can be readily obtained from Equation (2.6.4) to be

$$\frac{p_\ell}{p_A} = \frac{\Delta p_c}{p_A} - \left[1 + \left\{\left(1 + \frac{\Delta p_c}{p_A}\right)^{1-\delta} - 1\right\}\left(1 - \frac{x}{L}\right)\right]^{\frac{1}{1-\delta}} + 1 \tag{2.6.6}$$

Once p_ℓ vs. x is known, an estimate of the liquid velocity across the cake, q_ℓ vs. x, can be made as follows: from Equations (2.2.3a) and (2.2.3b) and noting $q_s = 0$ at $x = 0$ (namely, particles do not penetrate into the medium), one has

$$q_\ell + q_s = q_\ell|_m \tag{2.6.7}$$

where the subscript m denotes the cake/medium interface or $x = 0$.

---

[5]Equation (21) of Tiller et al. (1999) after correcting its typographical error is the same as Equation (2.5.5).
[6]If medium resistance is negligible, $\Delta p_c = p_0$.

The generalized Darcy's law [i.e. Equation (2.2.12)] gives

$$\frac{q_\ell}{1-\varepsilon_s} - \frac{q_s}{\varepsilon_s} = \frac{k}{\mu(1-\varepsilon_s)}\left(\frac{\partial p_\ell}{\partial x}\right) \tag{2.6.8}$$

The pore liquid pressure gradient can be obtained for Equation (2.6.6), or

$$\frac{\partial p_\ell}{\partial x} = \frac{p_A}{1-\delta}\frac{C_1}{L}\left[1 + C_1\left(1 - \frac{x}{L}\right)\right]^{\frac{\delta}{1-\delta}} \tag{2.6.9}$$

and

$$C_1 = \left(1 + \frac{\Delta p_c}{p_A}\right)^{1-\delta} - 1 \tag{2.6.10}$$

Substituting Equation (2.6.9) into Equations (2.6.7) and (2.6.8), one has

$$q_\ell + q_s = \frac{k^0\left(1 + \frac{\Delta p_c}{p_A}\right)^\delta}{\mu}\frac{p_A}{1-\delta}\frac{C_1}{L}(1+C_1)^{\frac{\delta}{1-\delta}} \tag{2.6.11}$$

$$q_\ell - q_s = \frac{k^0\left(1 + \frac{\Delta p_c}{p_A}\right)^\delta}{\mu}\frac{p_A}{1-\delta}\frac{C_1}{L}\left[1 + C_1\left(1 - \frac{x}{L}\right)\right]^{\frac{\delta}{1-\delta}} \tag{2.6.12}$$

The above two equations can be solved simultaneously to give the results of q_ℓ vs. x and q_s vs. x. Alternative procedures for estimating internal velocity profiles have also been developed by a number of investigators (Shirato et al., 1969; Tiller et al., 1999; Lee et al., 2000).

■ ■ ■ ▬▬▬▬▬▬▬▬▬▬▬▬▬▬▬▬▬▬▬▬▬▬▬▬▬▬▬▬▬▬▬▬

Illustrative Example 2.7

For $CaCO_3$ cake with the constitutive relationships of ε_s vs. p_s and k vs. p_s given in Illustrative Examples 2.1 and 2.3 and the pressure drop across the cake being 7×10^5 Pa, estimate the solidosity, compressive stress, and filtrate pressure profiles, assuming $dp_\ell + dp_s = 0$.

The constitutive relationships are

$$\varepsilon_s = \varepsilon_s^0\left(1 + \frac{p_s}{p_A}\right)^\beta$$

$$k = k^0\left(1 + \frac{p_s}{p_A}\right)^{-\delta}$$

$$\varepsilon_s^0 = 0.2 \quad k^0 = 4.87 \times 10^{-14} \ m^2 \quad p_A = 4.4 \times 10^4 \ Pa$$

$$\beta = 0.13 \quad \delta = 0.57$$

Solution

The porosity profile is given by Equation (2.6.5) or

$$\frac{\varepsilon_s}{\varepsilon_s^0} = \left\{\left[\left(1+\frac{\Delta p_c}{p_A}\right)^{1-\delta} - 1\right]\left(1-\frac{x}{L}\right)+1\right\}^{\frac{\beta}{1-\delta}} \tag{i}$$

The compressive stress profile is given by Equation (2.6.4) or

$$\frac{p_s}{p_A} = \left\{1+\left(1-\frac{x}{L}\right)\left[\left(1+\frac{\Delta p_c}{p_A}\right)^{1-\delta} - 1\right]\right\}^{\frac{1}{1-\delta}} - 1 \tag{ii}$$

The filtrate pressure profile can be found from the given relationship,

$$dp_\ell + dp_s = 0 \quad \text{or} \quad p_\ell + p_s = \Delta p_c$$

with

$$\beta = 0.13 \quad \delta = 0.57$$

$$1 - \delta = 0.43 \qquad 1/(1-\delta) = 2.3256$$

$$\beta/(1-\delta) = 0.3023$$

$$\Delta p_c/p_A = 700,000/44,000 = 15.9093$$

The porosity profile may be expressed as $\varepsilon_s/\varepsilon_s^0$ vs. x/L [i.e. Equation (i)]. The compressive stress can be more conveniently expressed as $p_s/\Delta p_c$ vs. x/L. From Equation (ii), one may write

$$\frac{p_s}{\Delta p_c} = \left(\frac{p_s}{p_A}\right)\left(\frac{p_A}{\Delta p_c}\right) = (p_A/\Delta p_c)\left\{1+\left(1-\frac{x}{L}\right)\left[\left(1+\frac{\Delta p_c}{p_A}\right)^{1-\delta} - 1\right]\right\}^{\frac{1}{1-\delta}} \tag{iii}$$

The filtrate pressure profile may be expressed as $p_\ell/\Delta p_c = 1 - \dfrac{p_s}{\Delta p_c}$ or

$$\frac{p_\ell}{\Delta p_c} = 1 - \frac{p_s}{\Delta p_c} = 1 - (p_A/\Delta p_c)\left\{1+\left(1-\frac{x}{L}\right)\left[\left(1+\frac{\Delta p_c}{p_A}\right)^{1-\delta} - 1\right]\right\}^{\frac{1}{1-\delta}} \tag{iv}$$

The results from Equations (i), (iii), and (iv) are:

(x/L)	$(\varepsilon_s/\varepsilon_s^0)$	$(p_s/\Delta p_c)$	$p_\ell/\Delta p_c$
0	1.4443	1.0000	0
0.1	1.4127	0.8341	0.1659
0.2	1.3745	0.7470	0.2530
0.3	1.3444	0.5495	0.4505
0.4	1.3311	0.4300	0.5700
0.5	1.6668	0.3249	0.6751
0.6	1.2236	0.2340	0.7660
0.7	1.1615	0.1566	0.8434
0.8	1.1246	0.0947	0.9053
0.9	1.0665	0.0403	0.9597
1.0	1.0000	0.0000	1.0000

The results are also shown in Figs. i and ii.

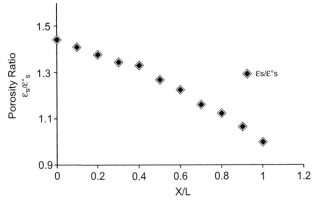

FIGURE I Solidosity profile.

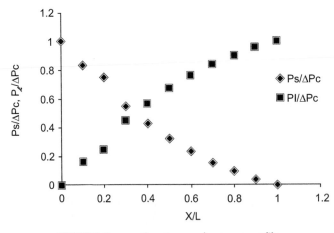

FIGURE II Compression stress and pressure profile.

2.7   Applications of the Conventional Cake Filtration Theory

For the convenience of subsequent discussions, the various relevant expressions given in the previous sections are tabulated in Table 2.1. These expressions are based on the assumption of $dp_\ell + dp_s = 0$.

The two major and most common applications of the conventional cake filtration theory are its use in predicting filtration performance and in determining cake properties from cake filtration data. These two applications will be discussed separately as below.

2.7.1   Prediction of Cake Filtration Performance

For the case of constant pressure filtration, filtration performance is described by the filtrate volume collected vs. time for a given operating pressure. Similarly, for constant rate filtration, performance is described by the required operating pressure, p_0, vs. time for a given filtration rate. To a lesser extent, information of the cake thickness vs. time and the evolution of cake quality ($\overline{m}$ and the solidosity profile) may also be of interest. For predicting filtration performance, different procedures may be used depending upon the degree of simplifications introduced:

a. The simplest (and therefore the least accurate) way of obtaining V vs. t (or p_0 vs. t) is to apply Equation (2.5.4a). The wet to dry cake mass ratio is calculated according to Equation (2.4.1) with $\overline{\overline{\varepsilon}}_s$ approximated as the stress average value $\overline{\varepsilon}_s$ given as

$$\overline{\overline{\varepsilon}}_s \cong \overline{\varepsilon}_s = \frac{1}{p_0} \int_0^{p_0} \varepsilon_s \cdot dp_s \tag{2.7.1}$$

The medium resistance may be taken as its intrinsic values (namely, the value determined using clean filtrate as the test fluid). With the knowledge of V vs. t, the cake thickness L, as a function of time, can be determined from the overall mass balance or

$$\varepsilon_{s_0}(L + V) = \overline{\overline{\varepsilon}}_s \cdot L \tag{2.7.2}$$

where ε_{s_0} is the particle volume fraction of the feed. ε_{s_0} can be expressed in terms of s or

$$\varepsilon_{s_0} = \frac{s/\rho_s}{(s/\rho_s) + [(1-s)/\rho]} \tag{2.7.3}$$

and the solidosity profile is given by Equation (2.6.5).

b. Same as (a) except $\overline{\overline{\varepsilon}}_s$ is estimated according to Equation (2.3.6) with the solidosity profile given by Equation (2.6.5).

c. Same as (b) except the change (increase) of R_m with time is accounted for. As shown by the work of Meeten (2000) and Teoh et al. (2006), the presence of a filter cake over a medium inevitable increases the medium resistance even if there is no particle penetration into and deposition within the medium. Based on the results of Teoh

Table 2.1   Relevant Expressions from the Conventional Cake Filtration Theory (Assuming $dp_\lambda + dp_s = 0$ or $\Delta p_c = p_{s_m}$)

Expression of Instantaneous Filtration Rate
From (2.4.4)

$$\frac{dV}{dt} = \frac{\Delta p_c + \Delta p_m}{\mu\left[(\alpha_{av})_{\Delta p_c}\dfrac{Vs\rho}{(1 - \overline{m}s)} + R_m\right]} = \frac{p_0}{\mu\left[(\alpha_{av})_{\Delta p_c}\dfrac{Vs\rho}{(1 - \overline{m}s)} + R_m\right]} \tag{2.4.3}$$

Expression of the Relationship of V vs. t for the Constant Pressure Case

$$\mu s\rho\left[(1 - \overline{m}s)^{-1}(\alpha_{av})_{\Delta p_c}\right]\frac{V^2}{2} + \mu R_m V = p_0 t \tag{2.5.4a}$$

Expression of the Relationship of p_0 vs. t from the Constant Rate Case

$$p_0 = \mu Q\left\{sp(1 - \overline{m}s)^{-1}(\alpha_{av})_{\Delta p_c}Qt + R_m\right\} \tag{2.4.9}$$

Expression of $(\alpha_{av})_{\Delta p_c}$

$$(\alpha_{av})_{\Delta p_c} = \frac{\Delta p_c}{\int_0^{\Delta p_c}\left(\dfrac{1}{\alpha}\right)dp_\ell} \tag{2.3.8}$$

Expression of $\overline{m}$

$$\overline{m} = 1 + \frac{\rho(1 - \overline{\overline{\varepsilon}}_s)}{\rho_s\overline{\overline{\varepsilon}}_s} \tag{2.4.1}$$

$$\overline{\overline{\varepsilon}}_s = \frac{1}{L}\int_0^L \varepsilon_s \cdot dx \tag{2.3.6}$$

Solidosity Profile

$$\frac{\varepsilon_s}{\varepsilon_s^0} = \left\{\left[\left(1 + \frac{\Delta p_c}{p_a}\right)^{1-\delta} - 1\right]\left(1 - \frac{x}{L}\right) + 1\right\}^{\frac{\beta}{1-\delta}} \tag{2.6.5}$$

Pressure Drop across Cake

$$\frac{\Delta p_c}{p_0} = \frac{\mu s\rho(1 - \overline{m}s)^{-1}[\alpha_{av}]_{p_{s_m}}V}{\mu s\rho(1 - \overline{m}s)^{-1}[\alpha_{av}]_{p_{s_m}}V + \mu R_m} \tag{2.5.5}$$

(2003), R_m may be expressed as a function of Δp_c (i.e. the compressive cake stress at the cake-medium interface) or

$$R_m/R_{m_0} = f(p_s) \tag{2.7.4}$$

with

$$f(p_s) = 1 \quad \text{at } p_s = 0$$

d. The procedures outlined above are based on the assumption that the pressure drop across the filter cake is the same as the operating pressure p_0. This assumption is often justified on the basis that the average specific cake resistance, α_{av}, is high and, therefore, the pressure drop across the medium can be ignored. What is often overlooked is that specific cake resistance values alone do not determine cake resistance. (The cake resistance is equal to the product of α_{av} and the cake mass.) Accordingly, there is always a period during which cake resistance is not dominant and the pressure drop across it is significantly different from the value of p_0.

A proper procedure of predicting cake filtration performance, therefore, should be made through the integration of Equation (2.4.3) with α_{av} evaluated as a function of Δp_c and with Δp_c estimated according to Equation (2.5.5). The change of R_m due to the presence of filter cake can be considered if the relevant experimental data are available (i.e. Equation (2.7.4)).

Tien and Bai (2003) proposed an iterative procedure of predicting filtration performance based on the principle stated above. Briefly speaking, prediction may be made as follows:

(i) A first estimate of V vs. t can be made as described in (a).
(ii) Using the results obtained in (i), Δp_c $\overline{m}$, $(\alpha_{av})_{\Delta p_c}$, and R_m can be estimated as functions of V (or t) using Equations (2.5.5), (2.4.1), (2.3.8), and (2.7.4), respectively.
(iii) The results obtained in (ii) can be used to determine (dV/dt) according to Equation (2.4.3).
(iv) With (dV/dt) vs. V known, a second estimate of V vs. t can be obtained through integration.
(v) The procedure may be repeated until the desired convergence is obtained.

Tien and Bai (2003) found that for cases they considered, a single iteration suffices to yield good results.

■ ■ ■ ▬▬▬▬▬▬▬▬▬▬▬▬▬▬▬▬▬▬▬▬▬▬▬▬▬▬▬▬▬▬▬▬▬

Illustrative Example 2.8

Calculate the cumulative filtrate volume collected from constant pressure filtration (at $p_0 = 5 \times 10^5$ Pa) of 2% (mass) $CaCO_3$ suspension at 10 s, 100 s, 1,000 s, 2,000 s, and 5,000 s. Using procedure (a) given in 2.7.1. The following conditions are given:

$$\varepsilon_s = 0.2\left[1 + \frac{p_s}{4.4 \times 10^3}\right]^{0.13}$$

$$\alpha = 3.85 \times 10^{10}\left[1 + \frac{p_s}{4.4 \times 10^3}\right]^{0.44} \text{ m kg}^{-1}$$

$\rho = 1,000$ kg m^{-3}
$\rho_s = 2655$ kg m^{-3}
$\mu = 1 \times 10^{-3}$ Pa s
$f' = -1$ (i.e. $dp + dp_s = 0$)
$R_m = 2 \times 10^{11}$ m

Solution

The parabolic law of constant pressure filtration (Equation (2.5.4a) may be applied or

$$\mu s \rho [(1 - \overline{m}s)_{p_0}]^{-1} (\alpha_{av})_{p_0} \frac{V^2}{2} + \mu R_m V = p_0 t \tag{i}$$

$s = 0.02$, $p_0 = 5 \times 10^5$ Pa
$R_m = 2 \times 10^{11}$ m
  Assuming that $\Delta p_c = p_0 = 5 \times 10^5$, with $f' = -1$, from Equation (2.5.3)

$$\alpha_{av} = \frac{p_0}{\displaystyle\int_0^{p_0} (1/\alpha) dp_0} = \frac{p_0}{\displaystyle\int_0^{p_0} \left(\frac{1}{3.85 \times 10^{10}}\right) \left[1 + \frac{p_s}{4.4 \times 10^3}\right]^{0.44} dp_s} = 1.852 \times 10^{11} \text{ m kg}^{-1}$$

From Equation (2.4.1), $\overline{m}$ can be estimated as

$$\overline{m} = 1 + \frac{\rho(1 - \overline{\overline{\varepsilon}}_s)}{\rho_s \overline{\overline{\varepsilon}}_s}$$

The average cake solidosity $\overline{\overline{\varepsilon}}_s$ may be approximated by the stress-average solidosity $\overline{\varepsilon}_s$ given as

$$\overline{\overline{\varepsilon}}_s \cong \overline{\varepsilon}_s = \frac{1}{5 \times 10^5} \int_0^{5 \times 10^5} \varepsilon_s^0 \left(1 + \frac{p_s}{4.4 \times 10^4}\right)^{0.13} dp_s$$

$$= \frac{0.2}{5.0 \times 10^5} \frac{4.4 \times 10^4}{1.13} [17.149 - 1] = 0.2514$$

$$\overline{m} = 1 + \frac{(1000)(1 - 02514)}{(0.2514)(2655)} = 2.12$$

Substituting all the values of the relevant quantities into Equation (i),

$$(10^{-3})(0.02)(10^3) \frac{1}{1 - (0.02)(2.12)} (1.852 \times 10^{11}) \frac{V^2}{2} + (10^{-3})(2 \times 10^{11}) V = (5 \times 10^5) \cdot t$$

or

$$3.868 \times 10^3 \, V^2 + 4 \times 10^2 \, V - t = 0$$

$$V = \frac{-4 \times 10^2 + \sqrt{(4 \times 10^2)^2 + (4)(3.868 \times 10^3)t}}{(2)(2.431)(10^3)}$$

The following results are obtained

$t(s)$	$V(m^3/m^2)$
10	0.021
100	0.117
1000	0.459
2000	0.669
5000	1.086

The experimental values of V obtained by Teoh (2003) are 0.030, 0.174, 0.736, 1.10, and 1.85, respectively. It is not surprising that the predictions are less than the experimental values since using $f' = -1$ was found to yield higher values of α_{av}.

■ ■ ■

2.7.2 Determination of Cake Properties from Experimental Filtration Data

Conceptually speaking, the most direct way of determining cake properties, i.e. the constitutive relationships, is to match experimental filtration data with predictions based on different types of constitutive relationships. The problem may be treated as one of search and optimization. If A is a measurable quantity of cake filtration experiment (for example, cumulative filtrate volume), an objective function, ϕ, may be defined as

$$\phi = \sum_{j=1}^{N} \left[A_j^E - A_j^P \right]^2 \tag{2.7.5}$$

where the superscripts E and P denote experimental and predicted values, respectively. The subscript j refers to the j-th measurement (i.e., at $t = t_j$) and there are a total of N measurement values.

Stamatakis and Tien (1991) formulated and successfully applied a search and optimization procedure for treating filtration data of talc suspensions. However, the procedure is computational demanding and may not be practical. For most practical applications, a number of simplified procedures may be applicable, including:

(a) Procedures based on the linearity of t/V vs. V. This is the simplest and most widely used method of determining (α_{av}) from constant pressure filtration data. If the validity of the parabolic law of constant pressure filtration is assumed, from Equation (2.5.4b) one has

$$t/V = \frac{\mu s \rho}{p_0} \left[(1 - \overline{m}s)^{-1} (\alpha_{av}) \right] \frac{V}{2} + (\mu R_m / p_0) \tag{2.7.6}$$

In other words, there exists a linear relationship between t/V vs. V. Therefore, by plotting experimental data in the form of t/V vs. V, a straight line can be expected.

From the intercept and the slope, α_{av}(at $p_s = p_0$) and the medium resistance can be estimated by assuming that the value of $\bar{m}$ is known.

(b) Procedures based on the linear relationship of $(dV/dt)^{-1}$ vs. V. From Equation (2.4.3), one has

$$\left(\frac{dV}{dt}\right)^{-1} = \frac{\mu s\rho}{p_0}(\alpha_{av})(1 - \bar{m}s)^{-1}V + (\mu R_m/p_0) \tag{2.7.7}$$

Therefore, a straight line results from a plot of $(dV/dt)^{-1}$ vs. V. From the knowledge of the slope and intercept of the line, as before (α_{av}) and R_m may be determined.

(c) Procedures based on the linear plot of $(dV/dt)^{-2}$ vs. t. From Equation (2.4.3), the initial filtrate rate, $(dV/dt)_0$, is

$$\left(\frac{dV}{dt}\right)_0 = p_0/(\mu R_m) \tag{2.7.8}$$

Also, the square of Equation (2.7.7) is

$$\left(\frac{dV}{dt}\right)^{-2} = \left[\frac{\mu s\rho}{p_0}(\alpha_{av})(1 - \bar{m}s)^{-1}\right]^2 V^2$$

$$+ 2\left[\frac{\mu s\rho}{p_0}(\alpha_{av})(1 - \bar{m}s)^{-1}V\right](\mu R_m/p_0) + (\mu R_m/p_0)^2 \tag{2.7.9}$$

Combining the above two equations yields

$$\left(\frac{dV}{dt}\right)^{-2} - \left[\left(\frac{dV}{dt}\right)_0\right]^{-2} = 2\left[\frac{\mu s\rho}{p_0}(\alpha_{av})(1 - \bar{m}s)^{-1}\right]\left\{\mu s\rho(\alpha_{av})(1 - \bar{m}s)^{-1}(V^2/2) + \mu R_m V\right\}\Big/p_0$$

$$= 2\left[\frac{\mu s\rho}{p_0}(\alpha_{av})(1 - \bar{m}s)^{-1}\right]t \tag{2.7.10}$$

According to Equation (2.7.10), there is a linear relationship between $(dV/dt)^{-2}$ vs. t. This procedure can be used to evaluate α_{av} and $(dV/dt)_0$ which give the values of R_m.

(d) Procedures based on the linear relationship of V vs. $t^{1/2}$. If the medium resistance can be ignored, from Equation (2.5.4a), one has

$$V = \left[\frac{2p_0}{\mu s\rho}\frac{(1 - \bar{m}s)}{\alpha_{av}}\right]^{1/2}t^{1/2} \tag{2.7.11}$$

Therefore by fitting experimental data to the above expression, the value of α_{av} may be obtained. The method advanced by Landman et al. (1999) can be shown to be identical to this procedure.

(e) Procedures based on fitting data to Equation (2.4.5). The procedure of (d) can be generalized by fitting filtration data of V vs. t to Equation (2.4.5). This procedure can be viewed as a special case of the optimization/search procedure based on Equation (2.7.5) with predictions obtained using the conventional theory of cake filtration.

In addition to the above-mentioned procedure, cake solidosity and specific cake resistance (or permeability) as functions of p_s can also be determined by independent measurements using the so-called compression–permeability (C–P) cell. Corroborations of cake property values obtained from filtration data with C–P measurements have been established, therefore giving credence to the physical significance of $(\alpha)_{av}$ and ε_s used in the conventional cake filtration theory. For a more complete discussion of cake property determination, see Tien (2006).

An example to demonstrate the use of some of the procedures given above in the determination of cake properties is given below.

■ ■ ■ ▬▬▬▬▬▬▬▬▬▬▬▬▬▬▬▬▬▬▬▬▬▬▬▬▬▬▬▬▬▬▬▬▬▬▬

Illustrative Example 2.9

Teoh (2003) obtained constant pressure filtration data of 2.0% $CaCO_3$ suspension at $p_0 = 5$ bars as follows.

Time (s)	Cake Thickness (m)	Cumulative Filtrate Volume (m^{-3} m^{-2})
12	–	3.67×10^{-2}
42	–	9.82×10^{-2}
92	–	1.69×10^{-1}
192	–	2.74×10^{-1}
292	–	3.58×10^{-1}
378	0.01	–
392	–	4.28×10^{-1}
592	–	5.48×10^{-1}
792	–	6.52×10^{-1}
992	–	7.45×10^{-1}
1038	0.02	–
1992	–	1.12
2268	0.03	–
2992	–	1.41
3208	0.04	–
3992	–	1.66
4992	–	1.87
5620	0.05	–

Other data: $\rho = 1,000$ kg m^{-3}; $\rho_s = 2,665$ kg m^{-3}; and $\mu = 10^{-3}$ Pa s.

Determine:

(1) The wet to dry cake mass ratio.
(2) Determine the average specific cake resistance on the basis of
  (i) t/V vs. V plot
  (ii) fitting V vs. t data according to the parabolic law of constant pressure filtration
  (iii) V being linearly proportional to $t^{1/2}$
  (iv) fitting $(dV/dt)^{-1}$ as a linear function of V
(3) What are the values of the medium resistance for each of the four cases of (2)?

Solution

To calculate $\bar{m}$ and $\bar{\bar{\varepsilon}}_s$ from Equations (2.4.10), (2.4.1), and (2.4.2b), one has

$$L = \frac{w}{\rho_s \bar{\bar{\varepsilon}}_s} \tag{i}$$

$$\bar{m} = 1 + \frac{\rho(1 - \bar{\bar{\varepsilon}}_s)}{\rho_s \bar{\bar{\varepsilon}}_s} \tag{ii}$$

$$w = \frac{V \rho s}{1 - \bar{m}s} \tag{iii}$$

Combining the above three equations yields

$$L = \frac{1}{\rho_s \bar{\bar{\varepsilon}}_s} \frac{V \rho_s}{1 - \bar{m}s} = \frac{V \rho_s}{\rho_s \bar{\bar{\varepsilon}}_s} \frac{1}{1 - s\left[1 + \frac{\rho(1 - \bar{\bar{\varepsilon}}_s)}{\rho_s \bar{\bar{\varepsilon}}_s}\right]} \tag{iv}$$

Solving for $\bar{\bar{\varepsilon}}_s$, one has

$$\bar{\bar{\varepsilon}}_s = \frac{s(\rho/\rho_s)[1 + (V/L)]}{1 - s(\rho/\rho_s)} \tag{v}$$

Accordingly, with V vs. t and L vs. t known, the average cake solidosity at various times can be calculated, according to the above expression. Once $\bar{\bar{\varepsilon}}_s$ is known, $\bar{m}$ can be readily determined from Equation (v).

From the data given, values of V corresponding to the time when specified cake thickness was reached may be estimated by interpolating the given V vs. t data. Based on the V values, the following results are obtained.

t(s)	L(m)	V(m³/m²)		$\bar{\bar{\varepsilon}}_s$		$\bar{m}$
378	0.01	0.418		0.325		1.78
1038	0.02	0.716		0.297		1.85
2268	0.03	1.200		0.311		1.83
3208	0.04	1.467		0.285		1.94
5672	0.05	2.002*		0.311		1.83
			av.	0.306	av.	1.85

*Obtained by extrapolation based on value of V at $t = 3992$ and 4992 s.

From the given *V–t* data, values of $t^{1/2}$ and t/V can be readily calculated. Using the central difference approximation, values of (dV/dt) and $(dV/dt)^{-1}$ at $t = (t_{i+1} + t_i)/2$, can be estimated. These results are given as follows.

t	$t^{1/2}$	V	t/V	(dV/dt)	$(dV/dt)^{-1}$
12	3.464	0.0367	326		
				2.05×10^{-3}	4.878×10^2
42	6.481	0.0982	428		
				1.416×10^{-3}	7.062×10^2
92	9592	0.169	544		
				1.05×10^{-3}	9.524×10^2
192	13.856	0.274	701		
				8.4×10^{-4}	1.19×10^3
292	17.088	0.358	816		
				7×10^{-4}	1.429×10^3
392	19.799	0.428	916		
				6×10^{-4}	1.667×10^3
592	24.331	0.548	1080		
				5.2×10^{-4}	1.923×10^3
792	28.192	0.652	1215		
				4.65×10^{-4}	2.151×10^3
992	31.496	0.745	1332		
				3.75×10^{-4}	2.667×10^3
1992	44.632	1.12	1779		
				2.9×10^{-4}	3.448×10^3
2992	54.699	1.41	2132		
				2.5×10^{-4}	4×10^3
3992	63.180	1.66	2405		
				2.1×10^{-4}	4.76×10^3
4992	70.654	1.89	3670		

From Equations (2.5.4b), one has

$$t/V = \left(\frac{1}{2p_0}\right)\mu\rho_s(1 - \overline{m}s)^{-1}(\alpha_{aV})V + \mu R_m/p_0 \tag{vi}$$

The above may be rewritten as

$$t = \left(\frac{1}{2p_0}\right)\mu\rho_s(1 - \overline{m}s)^{-1}(\alpha_V)V^2 + [\mu R_m/p_0] \cdot V \tag{vii}$$

Alternatively for the case of negligible medium resistance, one has

$$V = \left[\frac{2p_0(1 - \overline{m}s)}{\mu\rho_s\,\alpha_{av}}\right]^{1/2}t^{1/2} \tag{viii}$$

Equation (2.4.3) may be rewritten as

$$\left(\frac{dV}{dt}\right)^{-1} = (\mu/p_0)\left[(\alpha_V)Vs\rho(1-\overline{m}s)^{-1}V + R_m\right] \tag{ix}$$

The results of t/V vs. V, t vs. V, V vs. $t^{1/2}$, and $(dV/dt)^{-1}$ vs. V are shown in Figs. (i)–(iv). The expressions of t/V as a linear function V, t as a 2nd order polynomial of V, V as a linear function of $t^{1/2}$, and $(dV/dt)^{-1}$ as a linear function of V were established by linear regressions. The results are

$$t/V = 1.258 \times 10^3\ V + 3.497 \times 10^2 \tag{x}$$

$$t = 1.182 \times 10^3\ V^2 + 4.22 \times 10^2 \tag{xi}$$

$$\left(\frac{dV}{dt}\right)^{-1} = 2.409 \times 10^3\ V + 424 \tag{xii}$$

$$V = 2.534 \times 10^{-2}\ t^{1/2} \tag{xiii}$$

By comparing Equations (x)–(xiii) with their respective counterparts of Equations (vi)–(ix), the values of α_{av} and R_m can be readily determined using $\overline{m}$ obtained from part (1). For example, from Equations (x) and (vi), one has

$$t/V = 1.258 \times 10^3\ V + 3.5 \times 10^2$$

$$\frac{\mu\rho s(\alpha_{av})}{2p_0(1-\overline{m}s)} = 1.258 \times 10^3$$

$$\frac{\mu R_m}{p_0} = 350$$

with $\mu = 10^{-3}$ Pa s, $p_0 = 5 \times 10^5$ Pa

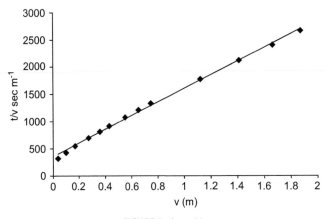

FIGURE I t/v vs. V

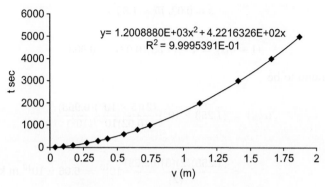

FIGURE II *t* vs.*V* according *to* $t = a_1v + a_2v^2$

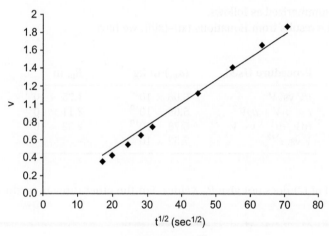

FIGURE III *V* vs. $t^{1/2}$

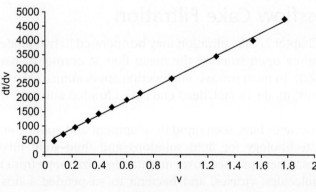

FIGURE IV *at/av* vs. *V*

$$s = 0.02, \overline{m} = 1.85$$

$$(1 - \overline{m}s) = 1 - (1.85)(0.02) = 0.963$$

α_{av} and R_m are found to be

$$(\alpha_{av}) = (1.258 \times 10^3) \frac{(2)(5 \times 10^5)(0.963)}{(0.02)10^{-3}(10^3)}$$

$$= \frac{(1.258)(2)(5)(0.963)}{(2)} 10^{10} = 6.06 \times 10^{10} \text{ m kg}^{-1}$$

$$R_m = (350)(5 \times 10^5)/(10^{-3}) = 1.75 \times 10^{11} \text{ m}^{-1}$$

The results are summarized as follows.
Together with the results from Equations (xi)–(xiii), we have

Procedure Used	(α_{av}) m kg^{-1}	R_m m^{-1}
t/V vs. V	6.06×10^{10}	1.75×10^{11}
$t = a_1V + a_2V^2$	5.69×10^{10}	2.11×10^{11}
$(dV/dt)^{-1}$ vs. V	5.78×10^{10}	2.32×10^{11}
V vs. $t^{1/2}$	7.52×10^{10}	–

Procedure (iii) of (2) does not give R_m values. For the other three procedures, the results are given above.

■ ■ ■

2.8   Application of the Conventional Theory to Crossflow Cake Filtration

As mentioned in Chapter 1, cake filtration may be operated in two modes – dead end and crossflow – depending upon whether the mean flow is normal or parallel to the filter medium (see Fig. 2.3). In most process engineering applications, cake filtration is carried out in the "dead end" mode. In fact, dead end cake filtration and cake filtration are often synonymous.

The past three decades have seen rapid development and applications of membrane-based separation technology for fluid solutions and fluid–solid mixtures. For liquid systems, the species to be separated cover an extended size range from monovalent ions, proteins, macromolecules, viruses, and bacteria to suspended solids of supramicron sizes. Separation is accomplished using membranes of different types of materials with

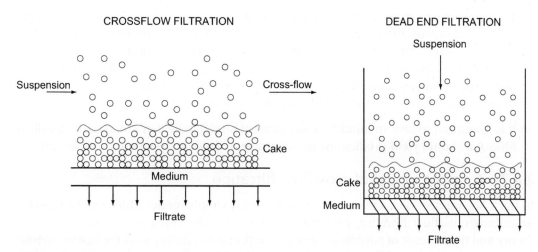

FIGURE 2.3 Crossflow Filtration vs. Dead End Filtration.

a variety of morphological structures. If filters are considered as devices which effect separation of a feed-through selective passage of the constituents of the feed, any separation process involving the use of membranes is a filtration operation although it is obvious that the nature and the mechanism by which ionic species are retained in reverse osmosis are totally different from those of supramicron particles in micro-filtration. In discussing possible extensions of the conventional cake filtration theory to membrane filtration, the problems considered are those of micro-filtration and possibly ultrafiltration and the purpose is to remove particulates present in feed streams.

A large number of studies on the analysis of crossflow membrane filtration have been reported in the literature in recent years. Although a degree of success on the understanding of crossflow filtration has been achieved, the results have not yet yielded any rational base necessary for design calculations and scale-up. In the following sections, a general discussion of the problems involved is given and the difficulties present in the analysis of crossflow filtration are discussed.

2.8.1   Features of Crossflow Filtration

Before presenting a simple yet rational analysis of crossflow filtration, it may be useful to recall the salient features of the process. Based on a number of experimental investigations, these features may be summarized as follows:

1. There exists a threshold filtration velocity (critical flux) below which flux declination with time is not observed. Experimentally, it was shown that for constant rate filtration, at sufficiently low rate, the operating pressure required to maintain a constant rate does not change with time. However, with high filtration rate, the required pressure is found to increase with time. One may therefore conclude that cake formation does not take place in crossflow filtration if the filtration velocity is below the critical flux value.

2. For constant pressure crossflow filtration, if the operating pressure is sufficiently high, the instantaneous filtrate rate decreases with time but the rate of decrease diminishes with time, suggesting a possible achievement of steady-state filtration.

3. There is a preferential deposition of smaller particles in crossflow filtration. For polydisperse suspensions, filter cakes formed in crossflow filtration are composed of particles smaller than those in dead end operation.

It is important to bear in mind that any rational model to be constructed for crossflow filtration should yield predictions not contrary to these above-mentioned features.

2.8.2   A simple model of crossflow filtration

In the following sections, a plausible and sufficiently simple model which can be used for describing crossflow filtration performance based on the conventional cake filtration theory and the concept of particle adhesion is outlined. Beginning with the instantaneous filtrate rate expression of Equation (2.3.11), one has

$$\frac{dV}{dt} = \frac{p_0}{\mu\left[w(\alpha_{av})_{\Delta p_c} + R_m\right]} \tag{2.8.1}$$

with the assumption of $dp_\ell + dp_s = 0$. The subscript of α_{av} is now written as Δp_c.

The major difference between the dead end and the crossflow mode resides in the fact that in the former case, all particles transported to membrane surface associated with filtrate flow become deposited, while for the crossflow case only some of these particles are deposited. To obtain the V–w relationship for the crossflow case, for a time interval of δt, the increase of V and that of w are δV and δw, respectively. If β is the fraction of particles transported to the membrane surface, which become deposited, the increase of the particles transported to the membrane surface associated with filtrate flow is $\delta w/\beta$. Similarly, the increase of the filtrate entrained with the cake over δt is $\delta[(\overline{m} - 1)w]$, where $\overline{m}$ is the wet to dry cake mass ratio. Therefore, the feed mass treated over δt is the sum of $\rho\delta V$, $\delta w/\beta$, and $\delta[(\overline{m} - 1)w]$. If s is the particle mass friction of the feed, s can be expressed as

$$s = \frac{(dw/\beta)}{(dw/\beta) + d[(\overline{m} - 1)w] + \rho dV} \tag{2.8.2}$$

After rearrangement, Equation (2.8.2) becomes

$$dV = \frac{1 - s[(\overline{m} - 1)\beta + 1]}{\beta s\rho}dw + \frac{w}{\beta s\rho}d(\overline{m} - 1) \tag{2.8.3}$$

where $\overline{m}$, as shown before, is nearly constant except initially. Therefore, the last term of the above expression may be omitted. From integration, w is found to be

$$w = s\rho\int_0^t \frac{\beta(dV/dt)}{1 - s[(\overline{m} - 1)\beta + 1]}dt \tag{2.8.4}$$

Substituting Equation (2.8.4) into (2.8.1), one has

$$\frac{dV}{dt} = \frac{p_0}{\mu s \rho (\alpha_{av})_{\Delta p_c} \int_0^t \frac{\beta (dV/dt) dt}{1 - s[\overline{m} - 1)\beta + 1]} + \mu R_m} \tag{2.8.5}$$

Note that the above expression reduces to Equation (2.4.3) with

$$\beta = 1.$$

Differentiating Equation (2.8.5) with time, the following expression is obtained:

$$\frac{d}{dt}\left(\frac{dV}{dt}\right) = \frac{-(\alpha_{av})_{\Delta p_c} \mu \rho s \, \beta \left(\frac{dV}{dt}\right)^3}{p_0 [1 - s\{(\overline{m} - 1)\beta + 1\}]} \tag{2.8.6a}$$

$$= \frac{\left(\frac{dV}{dt}\right)^2}{1 - s\{(\overline{m} - 1)\beta + 1\}} \frac{1}{\frac{1}{\beta} \int_0^1 \frac{\beta (dV/dt) dt}{1 - s[(\overline{m} - 1)\beta + 1]} + \frac{R_m}{(\alpha_{av})_{\Delta p_c} s \rho \beta}} \tag{2.8.6b}$$

$$= \frac{\beta \left(\frac{dV}{dt}\right)^2}{\frac{1}{\beta} \int_0^t \beta \left(\frac{dV}{dt}\right) dt + \frac{R_m}{(\alpha_{av})_{\Delta p_c} s \rho}} \quad \text{if } s\{(\overline{m} - 1)\beta + 1\} \ll 1 \tag{2.8.6c}$$

A few remarks about Equations (2.8.5)–(2.8.6c) may be in order. These equations can be seen as a generalization of the results of some of the previous studies. For example, with β being constant, Equation (2.8.6c) becomes the same as the expression given by Bowen et al. (2001). Equation (2.8.6a) is the same as the result of Hong et al. (1997) if β is assumed to be unity and $(\alpha_{av})_{\Delta p_c}$ is taken to be the value of α_{av} corresponding to $\varepsilon_s = (\varepsilon_s)_{max}$, $\Delta p_c = p_0$, and permeability given by Happel's model.

For the results given above to be consistent with the features of crossflow filtration mentioned above, certain constraints on the value of β should be recognized. During the course of filtration, β, generally speaking, cannot be a constant, but varies with time unless filtration proceeds with a filtration velocity less than the critical flux. β can be expected to increase with the decrease of particle size in accordance with the observation of preferred deposition of smaller particles. To achieve the steady-state for $t > t_c$, β should vanish for $t \geq t_c$. The steady-state filtration velocity $(dV/dt)_{steady}$ is

$$\left(\frac{dV}{dt}\right)_{study} = \frac{p_0}{\mu s \rho (\alpha_{av}) \int_0^{t_c} \frac{\beta (dV/dt) dt}{1 - s[(\overline{m} - 1)\beta + 1]} + \mu R_m} \quad \text{for} \quad t > t_c \tag{2.8.7}$$

If the above derivation is used as a basis of describing crossflow filtration, the information required for predicting filtration performance for a given set of conditions includes the constitutive relationships as well as the knowledge about β, the significance of which will be discussed in the following section.

2.8.3   Evaluation of β and Prediction of Filtration Performance

As mentioned before, β is the fraction of particles transported to the cake/membrane surface being deposited. One may therefore inquire, what are the reasons for some of the transported particle not being deposited? In earlier studies, it was speculated that, similar to the well-known concentration polarization phenomenon found in reverse osmosis, accumulation of particles at the cake/suspension interface leads to particle movement away from the interface due to the Brownian diffusion effect, for example. Simple calculations, however, show that the extent of back-diffusion of this kind is rather insignificant and not consistent with experimental observation. Moreover, this hypothesis is not in accord with preferred deposition of smaller particles observed experimentally. Later suggestions that transport of particles away from the interface may be caused by the so-called shear induced diffusion seems equally non-viable. The magnitude of the shear induced diffusivity necessary for the hypothesis to be valid was found to be unreasonably large. More important, there are uncertainties about the determination of the shear induced diffusivity or even the concept of the shear induced diffusion.

A logical and practical way to account for the failure of particles to be deposited can be made based on particle adhesion considerations. The cake/suspension interface can be expected to be not smooth and may be described in terms of the presence of protrusions of different heights at the interface. A particle which reaches the interface and makes contact with a protrusion is subject to a hydrodynamic drag along the direction of the main flow and another force along the direction of the filtrate flow, or normal to the main flow. The interplay of these forces and a simplified assumption about the protrusion height distribution results in an expression of the particle adhesion probability, γ, given as (Stamatakis and Tien, 1993)

$$\gamma = 1 - \left[1 - \frac{1}{\sqrt{(F_p/F_q)^2 + 1}}\right] \frac{d_p}{2h_{max}} \tag{2.8.8}$$

where F_p is the hydrodynamic drag experienced by the contacting particle along the direction of the main flow, F_q the hydrodynamic drag along the direction of the filtrate flow, and h_{max} the maximum protrusion height. A schematic diagram demonstrating the interplay is given in Fig. 2.4.

If the extent of particle transport due to either the Brownian and/or induced shear diffusion is insignificant, γ can be taken as β. To evaluate γ, the values of F_p, F_q, and h_{max} must be known. For their estimations, Stamatakis and Tien (1993) suggested that

$$F_p = 3\pi\mu d_p(\tau_w/\mu)(d_p/2)f_1 \qquad (2.8.9)^7$$

$$F_q = F_{q_1} + F_{q_2} \qquad (2.8.10)$$

$$F_{q_1} = 2\pi\mu d_p(dV/dt)f_2 \qquad (2.8.11)^7$$

$$F_{q_2} = (\pi/6)d_p^3(\rho_s - \rho) \qquad (2.8.12)$$

and

$$h_{max} = d_p/2 \qquad (2.8.13)$$

where τ_w is the shear stress at the fluid/medium interface. f_1 and f_2 are the hydrodynamic retardation factors.

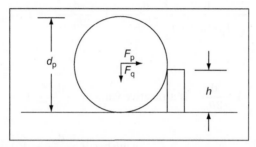

FIGURE 2.4 Deposition according to Adhesion Hypothesis.

With the knowledge of the flow conditions within a membrane module, for a given set of operating conditions β can be determined as a function of the instantaneous filtration rate, (dV/dt). This information, together with Equation (2.7.5), can be used to predict filtration performance of V vs. t for constant pressure filtration. An example demonstrating the estimation of β is given below.

■ ■ ■ ▬▬▬▬▬▬▬▬▬▬▬▬▬▬▬▬▬▬▬▬▬▬▬▬

Illustrative Example 2.10

Calculate γ as a function of (F_p/F_q) and d_p/h_{max} according to Equation (2.8.8). Comment on the results obtained.

Solution

According to Equation (2.8.8), γ is given as

$$\gamma = 1 - \left[1 - \frac{1}{\sqrt{(F_p/F_q)^2 + 1}}\right]\frac{d_p}{2h_{max}}$$

It is clear that the quantity in the bracket decreases with the decrease of (F_p/F_q). In particular,

$$1 - \frac{1}{\sqrt{(F_p/F_q)^2 + 1}} \to 0 \quad \text{as } F_p/F_q \to 0$$

---

[7] F_p and F_{q_1} are given by the Stokes law corrected for the hydrodynamic retardation effect.

$$1 - \frac{1}{\sqrt{(F_p/F_q)^2 + 1}} \to 1 \quad \text{as } F_p/F_q \to \infty$$

Accordingly,

$$1 - \left[1 - \frac{1}{\sqrt{(F_p/F_q)^2 + 1}}\right] \frac{d_p}{2h_{max}} \to 1 \quad \text{as } F_p/F_q \to 0$$

$$1 - \left[1 - \frac{1}{\sqrt{(F_p/F_q)^2 + 1}}\right] \frac{d_p}{2h_{max}} \to 1 - \frac{d_p}{2h_{max}} \quad \text{as } F_p/F_q \to \infty$$

Also,

$$1 - \frac{d_p}{2h_m} \quad \text{is a positive value if} \quad \frac{d_p}{2h_m} < 1$$

$$\text{vanishes if} \quad \frac{d_p}{2h_m} = 1$$

$$\text{is a negative value if} \quad \frac{d_p}{2h_m} > 1$$

In order to account for the fact that the extent of particle adhesion diminishes as the crossflow velocity increases, $d_p/2h_m$ is taken to be 1 (see Stamatakis and Tien, 1993). The results are

F_p/F_q	γ
0.1	0.995
0.5	0.894
0.8	0.781
1.0	0.707
1.5	0.555
2.0	0.447
4.0	0.242
10.0	0.098
20	0.050
30	0.033
50	0.020

■ ■ ■

■ ■ ■ ▬▬▬▬▬▬▬▬▬▬▬▬▬▬▬▬▬▬▬▬▬▬▬▬▬▬▬▬▬

Illustrative Example 2.11

Estimate the performance of crossflow filtration of suspensions of calcium carbonate particles based on Equation (2.8.5) under the following conditions

CaCO$_3$ Suspensions	Particle Diameter	$d_p = 5 \ \mu m$
	Suspension Concentration	$s = 0.001$
Operating Condition	Constant Pressure	$p_0 = 30 \ kPa$
Medium Resistance	$R_m = 6 \times 10^9 \ m$	

Specific Cake Resistance
$$\alpha = \alpha^0 \left[1 + \frac{p_s}{p_A}\right]^{0.44} \ m \ kg^{-1}$$

$$\alpha^0 = 3.85 \times 10^{10} \ m \ kg^{-1}$$

$$p_A = 4.4 \times 10^3 \ Pa$$

Filtration module may be approximated by the parallel plates geometry with the distance between the plates, $b = 1$ mm.

The flow is laminar with an average flow velocity of 2.0 m s^{-1}. The other information are:

$$\mu = 10^{-3} \ Pa \ s, \quad \rho = 1,000 \ Kg \ m^{-3}$$

Solution

Equation (2.8.5) is given as

$$q_{\ell_m} = \frac{dV}{dt} = \frac{p_0}{\mu s \rho (\alpha_{av})_{\Delta p_c} \int \frac{\beta q_{\ell m} dt}{1 - s[(\overline{m} - 1)\beta + 1]} + \mu R_m} \tag{i}$$

Equation (i) may be simplified as follows: Since $\overline{m}$ is of the order of unity and β is always less than unity $(\overline{m} - 1)\beta + 1 \cong 1$. Equation (i) now becomes

$$q_{\ell_m} = \frac{dV}{dt} = \frac{p_0}{\frac{\mu s \rho}{1 - s}(\alpha_{av})\Delta p_c \int_0^t \beta q_{\ell_m} dt + \mu R_m} \tag{ii}$$

There are two parameters present in Equation (ii), $(\alpha_{av})_{\Delta P_c}$ and β. The average specific cake resistance (α_{av}) is calculated with p_s ranging from $p_s = 0$ to $p_s = \Delta p_c$, the pressure drop across filter cakes. Assuming $f' = -1$, $p_{s_m} = \Delta p_c$, from Equation (2.3.8), one has

$$(\alpha_{av})_{\Delta P_c} = \frac{\Delta p_c}{\int_0^{\Delta p_c} \left(\frac{1}{\alpha}\right) dp_s} \tag{iii}$$

The pressure drop across the cake may be written as

$$\Delta p_c = p_0 - \Delta p_m = p_0 - q_{\ell_m} \mu R_m \tag{iv}$$

where Δp_m is the pressure drop across the medium. In other words, $(\alpha_{av})_{\Delta p_c}$ can be directly related to q_{ℓ_m} (and as expected varies with time).

As mentioned in 2.8.3, β may be taken, as an approximation, to be the same as γ, the adhesion probability, given by Equation (2.8.8). At the cake/suspension interface, the protrusion height can be expected to be of the order of d_p. (The presence of particle dendrites is not likely because of the relatively large crossflow velocity.) If one assumes $h_{max} = d_p/2$, from Equation (2.8.8), one has

$$\beta = \gamma = \frac{1}{\sqrt{(F_p/F_q)^2 + 1}} \tag{v}$$

There are a variety of forces acting on a particle at (or nearly at) the cake/medium surfaces. If one assumes that the dominant forces in the case are the hydrodynamic drag force, for F_q, the force acting on the direction normal to the medium, according to the Stokes law, is

$$F_q = 3\pi\mu d_p q_{\ell_m} \tag{vi}$$

For F_p, the force acting parallel to the cake/medium interface, according to O'Neill (1968), is

$$F_p = 1.7009[3\pi\mu d_p](u)\Big|_{\frac{d_p}{2}} \tag{vii}$$

where the quantity, 1.7009, accounts for the hydrodynamic retardation effect. $(u)|_{(d_p/2)}$ is the fluid velocity (main flow) at a point with a distance, $d_p/2$, away from the cake/medium interface.

For laminar flow, the velocity profile between two parallel plates is given as

$$\frac{u}{u_{max}} = 1 - \left(\frac{y_c}{b/2}\right)^2 \tag{viii}$$

where y_c is the distance measured from the mid-point between the two plates. In terms of the distance from either plates, $y = (b/2) - y_c$, the profile is

$$\frac{u}{u_{max}} = \frac{4y}{b} - \frac{4y^2}{b^2}$$

and

$$u\Big|_{\frac{d_p}{2}} = u_{max}\left[\left(\frac{4}{b}\right)(d_p/2) - (d_p/b)^2\right] \cong u_{max}(2d_p/b) \tag{ix}$$

$$d_p/b \ll 1$$

Substituting Equations (vi) and (ix) into (v), one has

$$\beta = \left\{\left[\frac{(1.7009)(3\pi\mu d_p)u_{max}(2d_p/b)}{3\pi\mu d_p q_{\ell_m}}\right]^2 + 1\right\}^{-1/2} \tag{x}$$

The maximum velocity, $u_{max} = 1.5\,\bar{u}$ with $\bar{u}$ being the average velocity. $\bar{u}$ may be taken as the crossflow velocity. Equation (x) shows that β, similar to $(\alpha_{av})_{\Delta p_c}$, is a function of q_{ℓ_m}.

In Equation (ii) [or Equation (i)], physically speaking, q_{ℓ_m} is the dependent variable and t the independent variable. However, since both β and $(\alpha_{av})_{\Delta P_c}$ are directly related to q_{ℓ_m} through Equations (x) and (iii) and (iv), for the numerical solution of Equation (ii), it is more convenient to obtain the results of t as a function of (q_{ℓ_m}). The specific procedure used is described below.

Considering a set of values of $(q_{\ell_m})_i$, $i = 1, 2, \ldots$, with $(q_{\ell_m})_i$ being the values at $t = t_i$, the integral present in the denominator of Equation (ii) may be approximated as

$$\int_0^{t_i} \beta q_{\ell_m} \, dt = [\beta_0(q_{\ell_m})_0 + \beta_1(q_{\ell_m})_1]\frac{\Delta t_1}{2} + \left[\beta_1(q_{\ell_m})_1 + \beta_2(q_{\ell_m})_2\frac{\Delta t_2}{2}\right]$$

$$+ \cdots [\beta_{i-1}(q_{\ell_m})_{i-1} + \beta_i(q_{\ell_m})_i]\frac{\Delta t_i}{2} \tag{xi}$$

$$= \sum_{k=1}^{i-1}\left[\beta_{k-1}(q_{\ell_m})_{k-1} + \beta_k(q_{\ell_m})_k\right]\frac{\Delta t_k}{2} + \left[\beta_{i-1}(q_{\ell_m})_{i-1} + \beta_i(q_{\ell_m})_i\right]\frac{\Delta t_i}{2}$$

where β_k is the values of β obtained from Equation (x) with $q_{\ell_m} = (q_{\ell_m})_k$ and

$$\Delta t_k = t_k - t_{k-1}, \quad (q_{\ell_m})_k = (q_{\ell_m})_{k-1} - \Delta(q_{\ell_m}) \quad k = 1, 2, \ldots i \tag{xii}$$

Substituting Equation (xi) into (ii), after rearrangement, one has

$$\frac{p_0}{(q_{\lambda_m})_i} - \mu R_m = \frac{\mu s \rho}{1 - s}(\alpha_{av})_i \left[\sum_{k=1}^{i-1}\left\{\beta_{k-1}(q_{\ell_m})_{k-1} + \beta_k(q_{\ell_m})_k\right\}\frac{\Delta t_s}{2} + \frac{\Delta t_i}{2}\right.$$

$$\left. \times \left\{\beta_{i-1}(q_{\ell_m})_{i-1} + \beta_i(q_{\ell_m})_i\right\}\right] \tag{xiii}$$

and $(\alpha_{av})_i$ is the value of α_{av} calculated from Equation (iii) with ΔP_c from Equation (iv) with $q_{\ell_m} = (q_{\ell_m})_i$.

From the above expressions, one has

$$t_i - t_{i-1} = \frac{2(1-s)}{\mu s \rho(\alpha_{av})_i} \frac{\left\{p_0/(q_{\ell_m})_i\right\} - \mu R_m}{\beta_{i-1}(q_{\ell_m})_{i-1} + \beta_i(q_{\ell_m})_i} - \sum_{k=1}^{i-1}\frac{\beta_{k-1}(q_{\ell_m})_{k-1} + \beta_k(q_{\ell_m})_k}{\beta_{i-1}(q_{\ell_m})_{i-1} + \beta_i(q_{\ell_m})_i}\Delta t_k \tag{xiv}$$

To begin the calculation, the following equations may be used:

$$t_1 = \frac{2(1-s)}{\mu s \rho(\alpha_{av})_1} \frac{\left\{p_0/(q_{\ell_m})_1\right\} - \mu R_m}{\beta_0(q_{\ell_m})_0 + \beta_1(q_{\ell_0})_1} \tag{xv}$$

Equations (iii), (x), (xiv), and (xv), with given conditions, may be used to obtain the performance results. First, the initial filtration rate, $(q_{\ell_m})_0$, is simply

$$(q_{\ell_0})_0 = \frac{p_0}{\mu R_m} = \frac{30,000}{(10^{-3})(6 \times 10^9)} = 0.005 \text{ m s}^{-1}$$

The numerical results obtained using $\Delta q_{\ell_m} = 10^{-4}$ m s^{-1} are given below.

q_{ℓ_m}(m s^{-1})	Time(min)	β	ΔP_c (Pa)	α (m kg^{-1})
0.005	0	0.4400	–	–
0.0045	0.0478	0.4035	3,000	4.35×10^{10}
0.004	0.2913	0.3649	6,000	4.75×10^{10}
0.003	0.3940	0.2820	12,000	5.40×10^{10}
0.002	3.2863	0.1923	18,000	5.93×10^{10}
0.0015	7.6744	0.1454	21,000	6.16×10^{10}
0.001	25.2968	0.0975	24,000	6.39×10^{10}
0.0005	195.829	0.0489	27,000	6.60×10^{10}

The results show that there is a rapid decline of the filtration rate initially followed by a more moderate and then very slow decrease. The filtration rate does not reach a steady-state since β vanishes only at $t = \infty$. The results also show that the medium resistance remains significant even when the filtration rate reduces to less than half of its initial value. The omission of the medium resistance is not justified in this case.

Problems

2.1. A slurry containing 4% by weight of insoluble solids (with density $\rho_s = 1500$ kg m^{-3}) in water is filtered through a plate-and-frame press with a total area of 0.25 m^2. The operating pressure is 180 kPa. For one test run, the following data were obtained:

Time	Filtrate Volume (m^3)
0	0
7	0.05
27	0.10
118	0.20

For the cake formed, the wet to dry cake mass ratio is 1.85. The filtrate viscosity is 0.9 cp and density 1,000 kg m^{-3}.

  a. Calculate the mass fraction of the slurry
  b. Calculate the average cake solidosity, $\bar{\varepsilon}_s$
  c. Based on the filtration data given and using the t/V vs. V plot, determine the average specific resistance and the medium resistance

d. Your coworker suggests that as the specific cake resistance, α, is defined as $\alpha = (k\rho_s\varepsilon_s)^{-1}$; therefore, one may calculate k_{av} to be $(\alpha_{av}\bar{\bar{\varepsilon}}\rho_s)^{-1}$ with α_{av} from (c) and $\bar{\bar{\varepsilon}}_s$ from (b). Is this suggestion valid?

2.2. An aqueous suspension is filtered using a leaf filter with area of 0.02 m² at 180 kPa. The following data were obtained:

Time	Filtration Volume (liters)
79	0.566
651	1.70
1145	2.27

Filtrate viscosity: 1.2 cp, Density of solids 3050 kg m^{-3}
Filtrate Density: 1,000 kg m^{-3}
Mass fraction of solids of test suspension: 0.08
Average cake solidosity $\bar{\bar{\varepsilon}}_s = 0.32$
Calculate the following
a. Volume of filtrate per unit dry cake mass collected
b. Final cake thickness
c. Average cake specific resistance
d. Medium resistance
e. An expression relating filtration volume with time

2.3. For the same suspension of Problem 2.2, if the filtration rate is to be kept at 0.06 liters/s, what is the required operating pressure?

2.4. Laboratory filtration experiments conducted with a certain aqueous suspension yielded the following data:

Volume of Filtrate (liters)	Test 1	Test 2	Test 3	Test 4
0.5	17.3	6.8	6.3	5.0
1.0	41.3	19.0	17.0	11.5
1.5	720	34.6	24.2	19.8
2.0	108.3	53.4	37.0	30.1
2.5	152.1	76.0	51.7	42.5
3.0	201.7	102.0	69.0	56.8
3.5		131.2	88.8	73.0
4.0		163.0	110.0	91.2
5.0			164.0	133.0
6.0				1825

$p_0 = 6.7, 16.2, 28.2$ and 363 lb/in² for tests 1, 2, 3, and 4.

The filter area is 440 cm^2, the mass of cake solid per unit filtrate is 23.5 gram/liter, and filtration is carried out at 25 °C. The solid density is 2600 kg m^{-3}.

(a) Determine the average specific cake resistance and media resisted for each test based on the linear relationship of $(dV/dt)^{-1}$ vs. V.

(b) Assuming $f' = -1$, based on the results of (a), develop empirical expressions of α_{av} vs. p_s and R_m vs. p_s. (Hint: For simplicity, the empirical expression may be assumed to be of the form $y = ax^b$.)

2.5. A rotary drum filter with 30% submergence is to be used for the filtration of a $CaCO_3$ slurry containing 240 kg $CaCO_3$ per m^3. The pressure drop is kept at 7×10^4 Pa. Assuming that the moisture content of the cake is 50% (volume basis), estimate the filter area required to filter 0.4 m^3/min slurry under the following conditions assuming that the medium resistance is negligible.

1. The specific cake resistance of $CaCO_3$ cakes is given as

$$\alpha = \alpha^0 \left(1 + \frac{p_s}{p_A}\right)^{0.44}$$

$$\alpha^0 = 3.85 \times 10^{10} \text{ m kg}^{-1}$$

$$p_A = 4.4 \times 10^4 \text{ Pa}$$

2. Density of $CaCO_3$, $\rho_s = 2665$ kg m^{-3}

2.6. The constitutive relationships of ε_s vs. ρ_s, and α vs. p_s for Kaolin cakes are

$$\varepsilon_s = 0.34 \left[1 + \frac{p_s}{p_A}\right]^{0.17}$$

$$\alpha = 5.47 \times 10^{11} \left[1 + \frac{p_s}{p_A}\right]^{0.85}$$

$$p_A = 8.7 \times 10^4 \text{ Pa}$$

$$\alpha \text{ in m kg}^{-1}$$

Calculate the average specific cake resistance over the range $0 < p_s < p_{sm}$ with $p_{sm} = 10^5$, 5×10^5 and 10^6 Pa with $f' = -\varepsilon_s/(1 - \varepsilon_s)$.

2.7. From a series of constant pressure filtration experiments, the following average specific cake resistance results were obtained:

P_0, Operating Pressure (kPa)	$(\alpha)_{av}$, Average Specific Cake Resistance (m/kg)
70	1.4×10^{11}
104	1.8×10^{11}
140	2.1×10^{11}
210	2.7×10^{11}
400	4.0×10^{11}
800	5.6×10^{11}

Use the procedure outlined in Illustrative Example 2.5, obtain the relationship of α vs. p_s from the data given above assuming that $f' = -1$, and the compressive stress at the cake/medium surface may be assumed to be P_0 (namely, the medium resistance is negligible).

2.8. Assuming that the constitutive relationships of ε_s vs. p_s, k vs. p, and α vs. p_s can be represented by the power-law equation of Equations (2.2.15a), (2.2.15b), and (2.2.15c), the coefficient and experimental values for carbonyl ion, kaolin fluid D, and activated sludge are

Cake Material	ε_s^0	$(\alpha^0 \rho_s)(\mathrm{m}^{-2})$	β	n	δ	$p_A(\mathrm{kPa})$
Carbonyl Iron (Grade E)	0.575	2.34×10^{13}	30	$\simeq 0$	$\simeq 0$	–
Kaolin Flat D	0.14	2.98×10^{13}	0.12	0.4	0.52	11
Activated Sludge	0.05	3.62×10^{14}	0.26	1.4	1.66	1.90

Calculate filtration rate as a function of the operating pressure ($p_0 = 10$, 20, 50, 100, 150 kPa) for filtration of carbonyl ion, kaolin suspension, and activated sludge assuming $f' = -1$. What kind of conclusions you may draw based on your calculations?

Hint: Consider the following in making your conclusion.

1. If the exponent of the power-law expression vanishes, the material may be viewed as incompressible.
2. The filtration rate can be seen to be proportional to $\Delta p_c/(w\alpha_{av})$ [see Equation (2.3.11)].

2.9. Based on the results obtained of Problems 2.4, calculate filtration performance of $CaCO_3$ suspensions (V vs. t, L vs. t) with $p_0 = 200$ kPa according to procedure (d) given in 2.7.1.

2.10. Calculate crossflow filtration performance according to Equation (2.8.6c) for the following conditions

1. β may be taken to be the same as the adhesion probability given by Equation (2.8.8) with $d_p/(2h_{max}) = 1$.
2. The constitutive relationship of α vs. p_s is

$$\alpha(\text{in m kg}^{-1}) = 3.58 \times 10^{10}\left[1 + \frac{p_s}{p_A}\right]^{0.44}$$

3. The medium resistance is $R_m = 1.6 \times 10^{11}\ \mathrm{m}^{-1}$
4. Suspension crossflow velocity is $0.5\ \mathrm{m\ s}^{-1}$
5. Filtrate viscosity is 10^{-3} Pa s; filtrate density 1,000 kg m^{-3}, and solid density 2,500 kg m^{-3}.

   Hint: You should first specify the initial conditions used to integrate Equation (2.8.6c).

2.11. Same as problem 2.10, calculate the performance of V vs. t if β is assumed to be unity. Compare the results with those of problem 2.10.

References

Bowen, W.R., Yousef, H.H.S, Colvo, J.F., 2001. Sep. Pur. Tech. 24, 297.

Hong, S., Faibish, R.S., Elimelech, H., 1997. J. Colloid Interface Sci. 196, 267.

Landman, K.A., Stankovich, J.M., White, L.R., 1999. AIChE J. 45, 1875.

Lee, D.J., Ju, S.P., Kwon, J., Tiller, F.M., 2000. AIChE J. 46, 110.

Meeten, G.H., 2000. Chem. Eng. Sci. 55, 1755.

Murase, T., Iritani, E., Cho, J.H., Shirato, M., 1989. J. Chem. Eng. Jpn. 22, 373.

O'Neill, M.E., 1968. Chem. Eng. Sci. 23, 1387.

Rietema, K., 1982. Chem. Eng. Sci. 37, 1125.

Ruth, B.F., 1935a. Ind. Eng. Chem. 27, 108.

Ruth, B.F., 1935b. Ind. Eng. Chem. 27, 806.

Shirato, M., Sambuichi, M., Kato, H., Aragaki, T., 1969. AIChE J. 15, 405.

Shirato, M., Murase, M., Iritani, E., Tiller, F.M., Alciatone, A.F., 1987. Filtration in the chemical process industry. In: Matteson, M.J., Orr, C. (Eds.), Filtration: Principles and Practices, second ed. Marcel Dekker, New York.

Smiles, D.E., 1970. Chem. Eng. Sci. 25, 985.

Stamatakis, K., Tien, C., 1993. AIChE J. 39, 1292.

Teoh, S.K., 2003. Studies in Filter Cake Characterization and Modeling, PhD Thesis, National University of Singapore .

Teoh, S.K., Tan, R.B.H., Tien, C., 2006. AIChE J. 52, 3427.

Tien, C., Teoh, S.K., Tan, R.B.H., 2001. Chem. Eng. Sci. 56, 5361.

Tien, C., Bai, Renbi, 2003. Chem. Eng. Sci. 58, 1323.

Tien, C., 2006. Introduction to Cake Filtration: Analyses, Experiments and Applications. Elsevier.

Tiller, F.M., Lu, R., Kwon, K.W., Lee, D.J., 1999. Water. Res. 31, 15.

Wakeman, R.J., Tarleton, E.S., 1999. Filtration: Equipment Selection, Modeling and Process Simulation. Elsevier Advanced Technology, Oxford.

3

Post-Treatment Processes of Cake Filtration

Notation

A	parameter of Equation (3.1.6)
a_i	coefficient of Equation (3.2.9)
B	parameter of Equation (3.1.21) $(-)$
c	concentration
c_i	solute concentration of the i-th stage of washing
c_0	initial value of c
c_w	concentration of wash liquor
$(c_w)_i$	inlet value of c_w
$(c_w)_e$	exit value of c_w
$(c_w^*)_e$	defined by Equation (3.3.2a)
$D(e)$	filtration diffusivity $(\mathrm{m^2\,s^{-1}})$
D_{av}	average (effective) filtration diffusivity $(\mathrm{m^2\,s^{-1}})$
D	diffusivity tensor $(\mathrm{m^2\,s^{-1}})$
D_L	axial dispension coefficient $(\mathrm{m^2\,s^{-1}})$
D_m	molecular diffusivity $(\mathrm{m^2\,s^{-1}})$
D'_m	D_m corrected by turtuosity factor (see Equation (3.3.22a)) $(\mathrm{m^2\,s^{-1}})$
d_e	equivalent particle diameter (m)
e	void ratio $(-)$
e_{av}	average void ratio $(-)$
e_0	initial value of e $(-)$
e_∞	ultimate (equilibrium) value of e $(-)$
$\mathrm{erf}(x)$	error function of argument x
$\mathrm{erfc}(x)$	complementary error function of argument x
E	defined by Equation (3.3.27) $(-)$
F	fraction of solute washed away, defined by Equation (3.3.1) or a correction factor defined by Equation (3.2.13) $(-)$
f_i	fraction of wash water de-watered in the i-th stage $(-)$
k_{av}	average permeability defined by Equation (3.2.11) $(\mathrm{m^2})$
k_i	coefficient of Equation (3.3.19a)
L	suspension height or cake thickness (m)
L_0	initial value of L (m)
L_1	transition value of L (m)
L_∞	equilibrium value of L (m)
M	quantity of solids $(\mathrm{m^3\,m^{-2}})$ present in a solid–liquid mixture, or moisture content $(-)$
M'	moisture content based on dry cake basis
m	spatial coordinate of filtration diffusion equation based on solids' values $(\mathrm{m^3\,m^{-2}})$
$\overline{m}$	wet to dry cake mass ratio $(-)$
N_{cap}	capillary number defined by Equation (3.2.5) $(-)$

Principles of Filtration, DOI: 10.1016/B978-0-444-56366-8.00003-7

N_{Pe_m}	Peclet number defined by Equation (3.3.22b) (−)
$N_{Pe'_m}$	Peclet number defined by Equation (3.3.22c) (−)
N_{Re}	Reynolds number defined by Equation (3.3.22e) (−)
N_{Sc}	Schmidt number defined by Equation (3.3.22d) (−)
p	pressure (Pa)
p^*	dimensionless pressure defined as p/p_b (−)
p_b	threshold pressure given by Equation (3.2.1) (Pa)
p_s	compressive stress (Pa)
R	defined as $1-F$ (−)
R_m	medium resistance (m^{-2})
S	cake saturation (−)
S_∞	irreducible saturation (−)
S_R	relative saturation defined by Equation (3.2.6) (−)
s	mass fraction of solids in a solid–liquid mixture (−)
t	time (s)
t_c	starting time of consolidation (s)
t_m	a fictitious time corresponding to V_m [see Equation (2.5.9b) (s)
t^*	dimensionless time defined by Equation (3.1.20) (−)
u_c	extent of consolidation defined by Equation (3.1.13)
u	velocity vector (m s^{-1})
u_g	gas velocity (m s^{-1})
u_g^*	dimensionless gas velocity defined by Equation (3.2.10b) (−)
u_s	superficial velocity (m s^{-1})
u_w	superficial velocity of wash water (m s^{-1})
V_m	equivalent filtrate volume defined by Equation (2.5.7) (m^3 m^{-2})
v_v	equal to $L\varepsilon_{av}$ (m)
v_w	given as $u_w t$ (m)
$(V_f)_i$	cake liquid of the i-th stage cake washing (m^3)
$(V_w)_i$	wash liquid of the i-th stage (m^3)
w	wash ratio defined by Equation (3.3.2b) (−)
w_0	solids mass given by Equation (3.1.5) (kg m^{-2})
x	an empirical factor of Equation (3.3.11)

Greek Letters

α	specific cake resistance (m kg^{-1}) or parameter of Equation (3.3.6)
α_{av}	average specific cake resistance (m kg^{-1})
γ	parameter of Equation (3.3.19b)
Δp	imposed pressure (Pa)
Δp^*	dimensionless ΔP, defined by Equation (3.2.10c) (−)
ΔL	increment of column height (m)
Δt	time increment (s)
ε_{av}	average cake porosity (−)
ε_s	cake solidosity (−)
$\bar{\bar{\varepsilon}}_s$	average cake solidosity (−)
ε_s^0	value of ε_s at the zero-stress state (−)
ε_{s_0}	initial value of ε_s

η	creep constant (see Equation (3.1.21)) (s^{-1})
θ	defined by Equation (3.2.10a) (−)
μ	viscosity (Pa s)
ν	parameter of Equation (3.1.19)
ρ	filtrate density ($kg\,m^{-3}$)
ρ_s	solids density ($kg\,m^{-3}$)
σ	surface tension ($N\,m^{-1}$)
τ	turtuosity factor (−)

As a separation process, cake filtration is applied to separate solid from liquid of solid/liquid suspensions. However, the degree of separation achieved may not meet the required specifications. Consequently, additional processes may be used to further enhance the degree of separation.

By post-treatment processes, we refer to a collection of technologies that can be used to reduce the residue filtrate present in filter cakes or to remove certain substances present in filter cakes. As an introductory text, it is impractical to make a thorough and complete presentation of the subject. Instead, our attention will be restricted to three of the most commonly applied processes: deliquoring of filtrate by mechanical force, deliquoring by airflow, and cake washing.

3.1 Deliquoring by Mechanical Force: Expression and Consolidation

The term "expression" was used by Shirato et al. (1967) to describe the removal of liquid from solid/liquid mixtures. A schematic diagram demonstrating the operation of expression is shown in Fig. 3.1. The mixture to be treated is first placed into a cell equipped with a permeable septum at the bottom of the cell and a movable piston at the top. Through the action of mechanical force, the piston moves downward as liquid is progressively squeezed out. The liquid removed passes through the bottom septum or through both the septum and piston if the piston is also permeable to liquid flow.

If the solid concentration of the solid–liquid mixture to be treated is sufficiently low, deliquoring by expression proceeds in two stages. During the first stage, liquid is removed in a manner similar to what takes place in cake filtration (dead-end mode). Filter cake is formed at the medium surface until the cake embraces all the particles present in the mixture initially. Further compression of the cake by the piston constitutes the consolidation stage (second stage). Consolidation ceases when mechanical equilibrium is reached. On the other hand, if the solid concentration of the solid–liquid mixture is sufficiently high (or the solid/liquid mixture is in the "cake" or "semi-solid" state), expression coincides with consolidation. If one

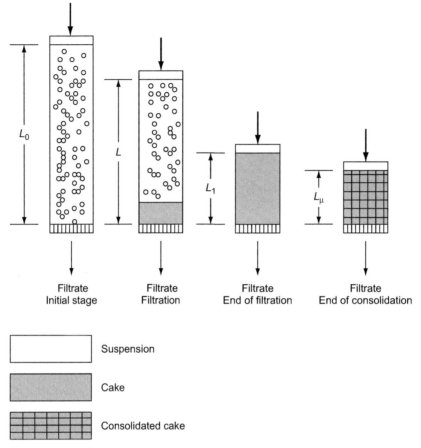

FIGURE 3.1 Expression of a solid/liquid mixture.

considers filter cake as a state in which cake particles form a network capable of sustaining finite compression stress, a solid–liquid mixture which is in the "cake" state requires that its solid concentration (in terms of ε_s) should be (see Equation (2.7.3))

$$\varepsilon_s = \frac{s}{s + (1 - s)(\rho_s/\rho)} > \varepsilon_s^{0\,[1]} \tag{3.1.1}$$

where ε_s^0 is the solidosity at the zero-stress state.

[1]The meanings of the symbols remain the same as defined before unless stated otherwise.

3.1.1   Onset of Consolidation

As stated before, two different situations may be encountered in expression. First, for a solid–liquid mixture with $\varepsilon_{s_0} > \varepsilon_s^0$, assuming that the constitutive relationship of ε_s vs. p_s is given as

$$\varepsilon_s = \varepsilon_s(p_s)$$

or

$$p_s = p_s(\varepsilon_s) \tag{3.1.2}$$

consolidation begins if the mixture is subject to a compressive stress, p_s, with

$$p_s > p_s(\varepsilon_{s_0}) \tag{3.1.3}$$

Secondly, if the solid concentration of the mixture, ε_{s_0}, is less than ε_s^0, expression of the mixture proceeds in two stages as stated before and consolidation begins upon the completion of the cake filtration stage. If the height of the solid–liquid mixture is denoted by L with $L = L_0$ initially, $L_0 - L$ is the volume of water removed per unit cross-sectional area of medium, or V. If L_1 is the height at the end of the filtrate stage, the cumulative filtrate volume is $L_0 - L_1$. Since the filter cake now consists of all particles present in the mixture initially, one has

$$\frac{L_0 - L_1}{w_0} = \frac{1 - \overline{m}s}{s\rho} \tag{3.1.4}$$

and

$$w_0 = (L_0)(\varepsilon_{s_0})\rho_s \tag{3.1.5}$$

where w_0 is the solid mass present in the mixture.

Equation (3.1.4) is obtained from (2.4.2b) with w_0 replacing w and $L_0 - L_1$ replacing V. It can be used to calculate the amount of liquid removed at the end of the filtration stage. Based on this information, the time duration of the filtration stage can be calculated from Equation (2.5.4b) or its equivalent [Equation (2.5.10)].

Onset of consolidation can be identified from experimental data in the following manner: One may re-write Equation (2.5.8) by replacing V with $L_0 - L$ to give

$$(L - L_0) + V_m = A^{1/2}(A + t_m)^{1/2} \tag{3.1.6}$$

or

$$\frac{\Delta L}{\Delta\sqrt{t + t_m}} = A^{1/2} \tag{3.1.7}$$

with V_m and t_m defined by Equations (2.5.9a) and (2.5.9b).

As stated before, $\overline{m}$ remains essentially constant and $(\alpha)_{av}$ can be approximated as the average specific cake resistance, with p_s ranging from zero to p_0 (assuming $dp + dp_2 = 0$). A plot of $\Delta L/\Delta t^{1/2}$ vs. t should yield a straight line parallel to the time-coordinate. Onset of consolidation is shown when this behavior is violated (see Fig. 3.2).

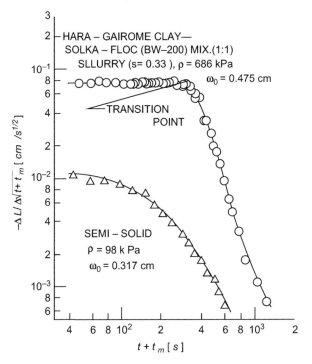

FIGURE 3.2 Transition from filtration to consolidation. *Reprinted from Shirato, M., Murase, T., Kato, H., Fukaya, S., 1970. Fundamental analysis for expression under constant pressure. Filtration and Separation, May/June 1970, 277–282, with permission of Elsevier.*

■ ■ ■ ━━

Illustrative Example 3.1

A column of $CaCO_3$ suspension (2% by mass) of height 0.5 m undergoes expression with an applied pressure of 5×10^5 Pa.

(a) Determine whether or not the process proceeds as filtration initially.

(b) If the answer to (a) is affirmative, determine the time when the filtration stage is complete. What is the cake thickness at that time?

The following data are given:
Cake constitutive relationships

$$\varepsilon_s = 0.02\left[1 + \frac{p_s}{4.4 \times 10^4}\right]^{0.13}$$

$$\alpha = 3.85 \times 10^{10}\left[1 + \frac{p_s}{4.4 \times 10^4}\right]^{0.44} \text{ m kg}^{-1}$$

p_s in Pa

Medium resistance: $R_m = 2 \times 10^{11}$ m^{-1}

The relevant physical properties are

$$\rho_s = 2665 \text{ kg m}^{-3}$$

$$\rho = 1000 \text{ kg m}^{-3}$$

$$\mu = 10^{-3} \text{ Pa s}$$

Solution

(a) The initial particle volume fraction of the suspension, ε_{s_0} is

$$\varepsilon_{s_0} = \frac{(s/2665)}{(s/2665) + (1-s)/1000} = 0.0076 < 0.02$$

Therefore, expression starts as cake filtration.

(b) Let the suspension height be L_1 upon completing the first stage. From Equation (3.1.4), one has

$$\frac{L_0 - L_1}{w_0} = \frac{1 - \overline{m}s}{s\rho}$$

The total mass present in the suspension initially, w_0 is

$$w_0 = (L_0)(\varepsilon_{s_0})(\rho_s) = (0.5)(0.0076)(2665) = 10.13 \text{ kg}$$

The wet to dry cake mass ratio, $\overline{m}$ is given as

$$\overline{m} = \frac{\overline{\overline{\varepsilon}}_s \rho_s + (1 - \overline{\overline{\varepsilon}}_s)\rho}{\overline{\overline{\varepsilon}}_s \rho_s}$$

The average cake solidosity, $\overline{\overline{\varepsilon}}_s$, can be found from the cake solidosity profiles as discussed in Chapter 2. As an approximation, the average spatial solidosity may be taken as the stress average. At the end of the filtration stage, the medium resistance is insignificant and can be ignored. One may consider the filter cake subject to a compressive stress ranging from $p_s = 0$ to $p_s = 5 \times 10^5$ Pa (since $f' = -1$). The stress average solidosity, $\overline{\varepsilon}_s$ is (see Illustrative Example 2.8)

$$\overline{\varepsilon}_s = \frac{(0.2) \int_0^{5 \times 10^5} \left(1 + \frac{p_s}{4.4 \times 10^4}\right) dp_s}{5 \times 10^5} = 0.2514$$

and $\overline{m} = 2.12$

The cumulative filtrate collected at the end of the first stage, $L_0 - L_1$ is given as

$$L_0 - L_1 = w_0 \frac{1 - \overline{m}s}{s\rho} = (10.13)\frac{1 - (0.02)(2.12)}{(0.06)(1000)} = 0.485 \text{ m}$$

$$L_1 = 0.5 - 0.485 = 0.015 \text{ m}$$

The filtration performance results, V vs. t is given by Equation (2.5.4a) or

$$\mu s \rho [1 - \overline{m}s]^{-1} (\alpha_{av}) \frac{V^2}{2} + \mu R_m V = p_0 t \qquad \text{(i)}$$

The average specific cake resistance can be taken to be the value between $p_0 = 0$ and $p_s = 5 \times 10^5$ Pa. $(\alpha)_{av}$ is found to be (see Illustrative Example 2.8)

$$\alpha_{av} = 1.852 \times 10^{11} \text{m kg}^{-1}$$

With the given conditions from Equation (i), one has

$$3.865 \times 10^3 (L_0 - L_1)^2 + 4 \times 10^2 (L_0 - L_1) = t$$

The time when the first stage ends or the beginning of the consolidation is

$$t = (3.865 \times 10^3)(0.485)^2 + (4 \times 10^2)(0.485) = 909 + 194 = 1073 \text{ s}$$

■ ■ ■

3.1.2   Consolidation Calculation

As shown in Example 2.2, de-watering of solid–liquid mixtures can be described by the diffusion equation of filtration or

$$\frac{\partial}{\partial m} \left[D(e) \frac{\partial e}{\partial m} \right] = \frac{\partial e}{\partial t} \qquad \text{(3.1.8)}$$

with the void ratio, e, being the dependent variable.

With the spatial domain, m, the volume of solids confined for consolidation, extends from zero to M which is given as

$$M = (L_0)(\varepsilon_{s_0}) \qquad \text{(3.1.9)}$$

where L_0 is the initial mixture height ($\text{m}^3 \text{m}^{-2}$) and ε_{s_0} is the initial solid volume fraction.

The boundary conditions are

$$\text{at} \quad m = 0, \quad e = e_\infty \qquad \text{(3.1.10a)}$$

$$m = M \quad \partial e / \partial m = 0 \qquad \text{(3.1.10b)}$$

Equation (3.1.10a) is based on the assumption that the medium resistance is negligible and e_∞ is the value of e at the completion of consolidation (i.e., when mechanical equilibrium is reached). Equation (3.1.10b) states that at the piston surface, the relative solid/liquid motion vanishes. In other words, the liquid and solid velocities are the same.

If consolidation is preceded by a filtration stage, the initial condition is

$$\text{At} \quad t = 0, \quad e = f(m) \quad 0 < m < M \qquad \text{(3.1.11a)}$$

where $f(m)$ corresponds to the e-profile at the end of the filtration stage. On the other hand, if the particle concentration of the solid–liquid mixture is sufficiently high, consolidation begins with a uniform mixture or

$$e = e_0 = \text{constant} \quad \text{for } 0 < m < M \tag{3.1.11b}$$

The solution of Equation (3.1.8) gives the void ratio profile (e vs. m) as a function of time. To express the extent of de-watering, the amount of water retained within the system is

$$\int_0^M \frac{1 - \varepsilon_s}{\varepsilon_s} \, dm = \int_0^M e \, dm = M \, e_{av} \tag{3.1.12a}$$

with

$$e_{av} = \frac{\int_0^M e \, dm}{M} \tag{3.1.12b}$$

The extent of de-watering may be expressed by u_c defined as

$$u_c = \frac{\int_0^M e_0 \, dm - \int_0^M e \, dm}{\int_0^M e_0 \, dm - \int_0^M e_\infty \, dm} = \frac{(e_{av})_0 - e_{av}}{(e_{av})_0 - e_\infty} \tag{3.1.13}^2$$

where the superscripts "0" and "∞" denote the initial and the ultimate states, respectively. e_∞ is the void ratio value when the cake under compression reaches its mechanical equilibrium, i.e., e being the value of e corresponding to $p_s = p_0$. The initial void ratio, e_0, may be either constant or a function of m.

The filtration diffusivity, D, in general, is a function of e (or p_s) (see Illustrative Example 2.3). Therefore, the solution of Equation (3.1.8) can only be obtained numerically. For more information, see Ramarao et al. (2002).

3.1.3   Approximation Solution of Consolidation

Approximate solution of Equation (3.1.8) may be obtained if one replaces the filtration diffusivity by an average (effective) value, D_{av}. Equation (3.1.8) becomes

$$D_{av} \frac{\partial^2 e}{\partial m^2} = \frac{\partial e}{\partial t} \tag{3.1.14}$$

which is the consolidation equation used by soil scientists. The following two solutions are frequently found in the filtration/consolidation literature:

If the initial e-profile is uniform, $e_0 = \text{constant}$, for constant pressure consolidation with negligible medium resistances, the extent of de-watering is

$$u_c = 1 - \frac{8}{\pi^2} \sum_{n=1}^{\infty} \frac{1}{(2n-1)^2} \exp\left[-\frac{(2n-1)^2 \pi^2}{4M^2} D_{av} t \right] \tag{3.1.15}$$

[2]It is simple to show that this definition is equivalent to $(L_0 - L)/(L_0 - L_\infty)$ when L_0 is the initial solid/liquid mixture height (or $m^3 m^{-2}$), L, the height at a given time and L_∞ the ultimate height.

If the initial e-profile is sinusoidal, u_c is given as

$$u_c = 1 - \exp\left[-\frac{\pi^2 D_{av}}{4M^2}t\right]$$

(3.1.16)

Leclerc and Rebouillat (1985) argued that Equation (3.1.16) can be applied in general, although Equation (3.1.15) seems to be a more logical choice if consolidation begins with a solid–liquid mixture with sufficiently high solid concentration.

An alternate expression of Equation (3.1.16) is (Carslaw and Jaeger, 1959, p. 97)

$$u_c = 2\left(\frac{D_{av}t}{M^2}\right)^{1/2}\left[\pi^{-1/2} + 2\sum_{n=1}^{\infty}(-1)^n i\,\mathrm{erfc}\left(\frac{nM}{\sqrt{D_{av}t}}\right)\right]$$

(3.1.17)

where

$$i\,\mathrm{erfc}(x) = \int_x^{\infty} \mathrm{erfc}(\xi)\,d\xi$$

(3.1.18a)

and $\mathrm{erfc}(x)$ is the complementary error function defined as

$$\mathrm{erfc}(x) = \frac{2}{\sqrt{\pi}}\int_x^{\infty} e^{-\xi^2}\,d\xi$$

(3.1.18b)

For small values of t, Equation (3.1.17) is more convenient to use as the infinite series of Equation (3.1.15) is slow in convergence for small exponents.

3.1.4 Empirical Equations Describing Consolidation/De-watering Performance

As discussed before, de-watering performance can be estimated through the solution of Equation (3.1.8) with appropriate initial and boundary conditions or using the approximate solutions given by Equation (3.1.15), (3.1.16), or (3.1.17). The solution of Equation (3.1.8), however, can only be obtained numerically, and numerical solution is cumbersome to use in scale-up and preliminary design. Numerical solution is also not suitable in data interpretation. To a degree, this is also true regarding the approximate solutions of Equations (3.1.15)–(3.1.17).

Equally important, Equation (3.1.8) is formulated with certain simplification including the assumption that the state of a cake structure (i.e. cake solidosity ε_s or void ratio) depends only on the compressive stress. With this assumption, the secondary effect of consolidation (or the creeping effect) is ignored. This simplification becomes a major problem in many cases. With a relatively large operating time, secondary effect often becomes important and cannot be ignored.

To account for such deficiencies, empirical expressions of consolidation performance have been proposed. Sivaram and Swamee (1977) proposed the following expression of consolidation de-watering:

$$u_{\mathrm{c}} = \frac{\sqrt{\dfrac{4t^*}{\pi}}}{\left[1 + \left(\dfrac{4t^*}{\pi}\right)^{\nu}\right]^{1/2\nu}} \qquad (3.1.19)$$

where

$$t^* = \frac{D_{\mathrm{av}}t}{M^2} \qquad (3.1.20)$$

where ν is the consolidation behavior index.[3]

A few remarks about Equation (3.1.19) may be in order. First, Equation (3.1.19) is identical to the leading term of Equation (3.1.17) for small values of t^*. Second, since the value of ν can only be determined from relevant data, the use of Equation (3.1.19) is largely for data interpretation. Its value as a predictive tool is limited.

A more general expression of consolidation de-watering was obtained with the premise that the change of the void ratio experienced in cake consolidation may be considered to be due to the combined effect of the primary (described by Equation (3.1.8)) and the secondary (creeping) consolidation. Based on the simple Voight model, u_{c} can then be expressed as (Shirato et al., 1974)

$$u_{\mathrm{c}} = (1 - B)\left[1 - \exp\left(-\frac{\pi^2}{4}\frac{D_{\mathrm{av}}}{M^2}t\right)\right] + B[1 - \exp(-\eta\, t)] \qquad (3.1.21)$$

if the initial e-profile is sinusoidal.

Similarly, for the case of uniform e-profile initially,

$$u_{\mathrm{c}} = (1 - B)\left[1 - \frac{8}{\pi^2}\sum_{n=1}^{\infty}\frac{1}{(2n-1)^2}\exp\left(-\frac{(2n-1)^2\pi^2}{4M^2}D_{\mathrm{av}}t\right)\right] + B[1 - \exp(-\eta\, t)] \qquad (3.1.22)$$

where η is the creep constant (s). B is the weighting factor for the contribution due to creep and $1-B$ due to the primary consolidation.

The model parameters of the various expressions given above [i.e. Equations (3.1.15), (3.1.16), (3.1.19), (3.1.21), and (3.1.22], namely, D_{av}, ν, B and η, generally speaking, can be determined by matching de-watering data with these expressions. In principle, the determination can be carried out using the optimization-search procedure described in 2.7.2. However, the computation effort required for such an approach may not be justifiable for preliminary design calculations. In the following, we discussed some simple ad hoc methods suggested by Shirato et al. (1970) for the determination of these parameters.

For the determination of D_{av}, (3.1.16), Equation (3.1.16) may be rewritten as

$$\frac{\pi^2}{4}\frac{D_{\mathrm{av}}}{M^2}t = \ln\left(\frac{1}{1 - u_{\mathrm{c}}}\right) \qquad (3.1.23)$$

[3]In the original work of Sivaram and Swamee (1977), ν was taken to be a constant and equal to 2.85. This condition was relaxed later by Shirato et al. (1979) in order to better account for the secondary effect.

If the time corresponding to 90% de-watering (or $u_c = 0.9$) is $(t)_{0.9}$, the above expression becomes

$$D_{av} = \frac{4}{\pi^2} \frac{M^2}{(t)_{0.9}} \ln\left[\frac{1}{0.1}\right] = 0.933 \, M^2/(t)_{0.9} \tag{3.1.24}$$

In other words, D_{av} is obtained by matching one consolidation data point with Equation (3.1.11).

Using the same procedure, D_{av} of Equation (3.1.15) can be estimated according to the following expression

$$D_{av} = 0.848 \, M^2/(t)_{0.9} \tag{3.1.25}$$

For the determination of the parameters present in Equation (3.1.19), for small values of $4t^*/\pi$, u_c becomes

$$u_c = \sqrt{\frac{4t^*}{\pi}} = \left(\frac{4 \, D_{av} \, t}{\pi \, M^2}\right)^{1/2} \tag{3.1.26}$$

From the above expression, u_c vs. $\sqrt{t}$ gives a straight line with a slope of $\left(\frac{4 \, D_{av}}{\pi \, M^2}\right)^{1/2}$.

Accordingly, from a plot of u_c vs. $t^{1/2}$, the value of D_{av} can be readily determined.

Once D_{av} is known, the experimental data can be converted into the form of u_c vs. t^*. On the other hand, from Equation (3.1.19), a graph of u_c vs. t^* with various values of ν can be easily constructed. By superimposing experimental data of u_c vs. t^* on this graph, the question of whether Equation (3.1.19) may be used to describe the de-watering process and, if so, what is the value of ν, can be answered. An example of applying the procedure for estimating ν is shown in Fig. 3.3.

The parameters B and η of Equation (3.1.21) [or (3.1.22)], can be obtained from the latter part of the de-watering data. With completion of the primary consolidation, one has

$$1 - \exp\left(-\frac{\pi^2}{4} \frac{D_{av}}{M^2} t\right) \cong 1$$

Equation (3.1.21) becomes

$$\begin{array}{r} u_c = 1 - B \exp(-\eta t) \\ \text{or} \quad 1 - u_c = B \exp(-\eta t) \end{array} \tag{3.1.27}$$

Therefore, a plot of $\ln(1-u_c)$ vs. t yields a straight line and B and η can be found from the intercept and the slope of the line. Fig. 3.4 illustrates the application of the procedure.

With B and η known, the creeping effect may be included in the determination of the filtration diffusivity, D_{av}. For this purpose, we may define a corrected u_c, $(u_c)_{cor}$ to be

$$(u_c)_{cor} = \frac{(L_0 - L) - B(L_0 - L_\infty)[1 - \exp(-\eta t)]}{(1 - B)(L_0 - L_\infty)} \tag{3.1.28}$$

The denominator gives the maximum amount of de-watering by primary consolidation. The numerator is the difference between the de-watering achieved minus that due to creep, or de-watering due to primary consolidation.

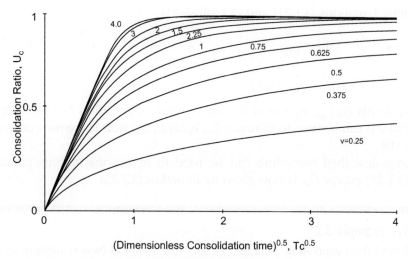

FIGURE 3.3 U_c as a function of t* and *v* according to Equation (3.1.19).

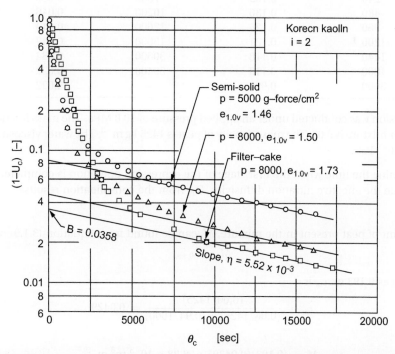

FIGURE 3.4 Determination of 2nd step consolidation parameters of Equation (3.1.27). *Reprinted from Shirato, M., Murase, T., Atsumu, K., Nakami T., Suzuki, H., 1978. Creep constants in expression of compressible solid–liquid mixtures. J. Chem. Eng. Japan 11, 334–336, with permission of J. Chem. Eng. Japan.*

Combining Equations (3.1.21) and (3.1.28) yields

$$(u_c)_{cor} = \left[1 - \exp\left(-\frac{\pi^2}{4M^2}D_{av}t\right)\right] + \frac{B}{1-B}[1 - \exp(-\eta t)] - \frac{B}{1-B}[1 - \exp(-\eta t)]$$

$$= 1 - \exp\left(-\frac{\pi^2}{4M^2}D_{av}t\right)$$

In other words $(u_c)_{cor}$ vs. t is of the same form as Equation (3.1.16). Accordingly, Equation (3.1.24) can be used to estimate D_{av} except $(t)_{0.9}$ now is the time corresponding to $(u_c)_{cor} = 0.9$.

The above-described procedure can be used to obtain the various parameters of Equation (3.1.22) except D_{av} is now given by Equation (3.1.25).

■ ■ ■ ▬▬▬▬▬▬▬▬▬▬▬▬▬▬▬▬▬▬▬▬▬▬▬▬▬▬▬▬

Illustrative Example 3.2[4]

Data collected from expressing a column of solid (peat)–liquid (water) mixture as column height vs. time are as follows:

Time (s)	Column Height (m)	Time (s)	Column Height
0	0.193	5400	0.073
240	0.163	7800	0.056
480	0.136	10200	0.046
960	0.125	12600	0.043
1200	0.121	15000	0.042
1440	0.118	30600	0.034
1800	0.113	59400	0.032
3000	0.099	∞	0.022

Expression was conducted under an applied pressure of 10.8 MPa. The mass fraction of the peat–water mixture is 0.06. The mass density of peat is 1425 $kg\,m^{-3}$. The water viscosity may be taken as 10^{-3} Pa s.

1. Determine the time when the first stage of the expression operation is complete.
2. Evaluate the effective filtration diffusivity, D_{av} from the consolidation results.

Solution

The amount of peat present in the mixture, M, can be found from Equation (3.1.9) or

$$M = L_0\,\varepsilon_{s_0}$$

L_0 is given as 0.193 m. ε_{s_0} is

$$\varepsilon_{s_0} = \frac{(0.06/1425)}{(0.06/1425) + (0.94/1000)} = 0.0429$$

and

$$M = (0.193)(0.0429) = 8.28 \times 10^{-3}\ m^3\,m^{-2}$$

[4]Problem taken from Example 6.1, Wakeman and Tarleton (1999) but with different solution.

1. Determination of the time when the first stage of expression is complete.

   From the column height results, the cumulative filtrate collected ($m^3 \, m^{-2}$) may be expressed as L_0-L.

   For the first stage of expression, Equation (2.5.4a) gives the relationship between the cumulative filtration collected vs. time, or

$$\mu s p\left[(1 - \overline{m}s)^{-1}(\alpha_{av})\right] \frac{(L_0 - L)^2}{2} + \mu R_m(L_0 - L) = p_0 t \tag{i}$$

The instantaneous filtration rate, $d(L_0 - L)/dt$ is given by Equation (2.4.3) or

$$-\frac{d(L_0 - L)}{dt} = \frac{p_0}{\mu\left[\alpha_{av}\dfrac{Vs\rho}{1 - \overline{m}s} + R_m\right]} \tag{ii}$$

with sufficient cake thickness, the medium resistance may be ignored. The above two expressions become

$$\mu s \rho\left[(1 - \overline{m}s)^{-1}(\alpha_{av})\right] \frac{(L_0 - L)^2}{2} \cong p_0 t \tag{iii}$$

and

$$-\frac{dL}{dt} = \frac{p_0}{\mu(\alpha_{av})(1 - \overline{m}s)^{-1}s\rho(L_0 - L)} = \frac{p_0}{\left[2\,\mu(\alpha_{av})(1 - \overline{m}s)^{-1}s\rho\right]^{1/2}} t^{-1/2} \tag{iii}$$

The above expression may be written as

$$-\frac{dL}{d(t^{1/2})} = \frac{2p_0}{\left[2\,\mu(\alpha_{av})(1 - \overline{m}s)^{-1}s\rho\right]^{1/2}} \tag{iv}$$

For the consolidation stage, the extent of consolidation U_c is given by Equation (3.1.19).

$$u_c = \frac{L_0 - L}{L - L_\infty} = \frac{\sqrt{4t^*/\pi}}{\left[1 + \left(\dfrac{4t^*}{\pi}\right)^v\right]^{1/2v}} \tag{v}$$

where L_0' is the column height with the completion at the first stage and t^* is given as

$$t^* = \frac{D_{av}(t - t_c)}{M^2} \tag{vi}$$

and t_c is the time when the first stage is complete.

   During the initial period of consolidation, the denominator of the expression of u_c may be taken as unity. u_c is given as

$$u_c = \frac{L_0 - L}{L_0 - L_\infty} \cong \sqrt{\frac{4D_{av}}{\pi M^2}(t - t_0)} \tag{vii}$$

and

$$-\frac{dL}{d(t^{1/2})} = (L'_0 - L_\infty)\sqrt{\frac{4D_{av}}{\pi M}}\sqrt{\frac{t}{t - t_c}} \qquad \text{(viii)}$$

Equation (iv) suggests that during the first stage of expression, $dL/d(t^{1/2})$ remains essentially constant. However, for consolidation operation, $dL/d(t^{1/2})$ is a function of time as shown by Equation (viii). This difference may be used to determine the transition from filtration to consolidation.

From the given data, one has

t(s)	$t^{1/2}(s^{1/2})$	L(m)	$-\Delta L/\Delta(t^{1/2})(ms^{-1/2})$
0	0	0.193	
			8.09×10^{-4}
240	15.49	0.163	
			4.21×10^{-3}
480	21.91	0.136	
			1.21×10^{-3}
960	30.48	0.125	
			1.19×10^{-3}
1200	34.64	0.121	
			9.06×10^{-4}
1400	37.95	0.118	
			1.12×10^{-3}
1800	42.43	0.113	
			1.13×10^{-3}
3000	54.77	0.090	
			1.39×10^{-3}
5400	23.48	0.073	
			1.15×10^{-3}
7800	88.32	0.056	
			7.89×10^{-4}
10200	101	0.046	
			2.67×10^{-4}
12600	112.24	0.043	
			8.9×10^{-5}
15000	122.47	0.042	
			1.88×10^{-4}
30600	174.93	0.034	
			2.91×10^{-5}
5940	243.72	0.032	

The above results shows that $-\Delta L/\Delta t^{1/2}$ is essentially constant (with the exception of the first two entries) until $t \cong 5400\text{–}7800$.

To identify the transition, consider the values of $\Delta L/(\Delta t^{1/2})$ from row 3 to row 9 (the values of the first two rows were ignored since they are likely to be in error). These 7 values vary from 9.06×10^{-4} to 1.39×10^{-3} with the majority within 1.12–1.21. The average of the seven values is 1.16 which may be taken as the value for $\Delta L/\Delta t^{1/2}$ during the first stage of expansion.

   To identify the point corresponding to the transition from filtration to consolidation, the transition point is given by the intersection point between the horizontal line $-\Delta L/\Delta t^{1/2} = 1.16 \times 10^{-3}$ and the line obtained by connecting the two points $[t = (7800 + 10200)/2 = 9000, \ -\Delta L/\Delta t^{1/2} = 7.89 \times 10^{-4}]$ and $[t = (10200 + 12600)/2 = 11400, \ -\Delta L/\Delta t^{1/2} = 2.67 \times 10^{-4}]$. The intersection point is found to be

$$-\Delta L/\Delta(t^{1/2}) = 1.16 \times 10^{-3}$$

$$t = 9000 - (11400 - 9000)\frac{1.16 \times 10^{-3} - 7.89 \times 10^{-4}}{(7.89 - 2.67) \times 10^{-4}}$$

$$= 9000 - 1719 \cong 7280$$

The following figure demonstrates the procedure used.

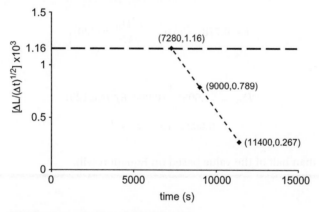

FIGURE I Determination of transition time.

The corresponding column height can be estimated by interpolating the data given

$$L_0 = 0.056 + \frac{(7800 - 7280)}{2400}(0.073 - 0.056)$$

$$= 0.056 + 0.004 = 0.06 \text{ m}$$

2. Evaluation of the effective (average) filtration diffusivity.
   The above results indicate that consolidation begins $t = t_c = 7280$ s with a column height, $L_0' = 0.06$ m. Accordingly, U_c during the consolidation stage are:

Time	$t' = t - t_c$	$t'^{1/2}$	L	$U_c = (L_0' - L)/(L_0' - L_\infty)$
7280	0	0	0.06	0
7800	520	22.80	0.056	0.105
9000	1720	41.47	0.050	0.263
10200	2920	54.04	0.046	0.368
15000	7720	87.86	0.042	0.474
30600	23320	152.71	0.034	0.684
59400	52120	228.30	0.032	0.778
∞	∞		0.022	

According to Equation (vii), during the initial period of the consolidation stage, the results of u_c vs. $t^{*1/2}$ yield a straight line with a slope of $\sqrt{\dfrac{4D_{av}}{\pi}}$. Using the first three data points of the consolidation step, the slope is estimated to be $0.18/30 = 0.006$. D_{av} is estimated to be

$$D_{av} = \frac{\pi}{4}M^2(\text{slope})^2$$
$$= \left(\frac{\pi}{4}\right)(0.00828)^2(0.006)^2 = 1.938 \times 10^{-9} \text{ m}^2 \text{ s}^{-1}$$

Alternatively, if one matches the consolidation data with Equation (3.1.16) for the highest value of u_c, one has

$$1 - 0.778 = \exp\left[-\frac{\pi^2 D_{av}}{4M^2}(52120)\right]$$

and

$$D_{av} = (1.505)\frac{4}{\pi^2}(0.00828)^2/(52120)$$
$$= 0.802 \times 10^{-9} \text{ m}^2 \text{ s}^{-1}$$

which is less than half of the value based on Equation (vii).

■ ■ ■

■ ■ ■

Illustrative Example 3.3

Calculate consolidation performance (u_c vs. time) according to Equations (3.1.16), (3.1.19), and (3.1.21) for the following conditions.

$M = 5.18 \times 10^{-3}$ m,　　$D_{av} = 1.298 \times 10^{-8}$ m^2 s^{-1}　　$i = 2$　　i: number of draining surface,
$B = 0.082$　　　　　　$\eta = 4.76 \times 10^{-5}$ s　　　　　　　　$\nu = 1.4$

These data were given by Shirato et al. (1970, 1978, 1987) for Clay-Solka Floc mixtures.

Solution

As the number of draining surface, $i = 2$, Equations (3.1.16), (3.1.21), and (3.1.19) are modified to be

$$u_c = 1 - \exp\left[-\frac{\pi^2}{4}\frac{D_{av}}{(M/2)^2}t\right]$$

$$u_c = (1-B)\left[1 - \exp\left(-\frac{\pi^2}{4}\frac{D_{av}}{(M/2)^2}t\right)\right] + B[1 - \exp(-\eta t)]$$

$$u_c = \left\{\left[\frac{4}{\pi}\frac{D_{av}t}{(M/i)^2}\right]\bigg/\left[1 + \left(\frac{4}{\pi}\frac{D_{av}}{(M/i)^2}t\right)\right]^{1/\nu}\right\}^{1/2}$$

The numerical results obtained are as follows:

Time (s)	Equation (3.1.16)	u_s (Equation (3.1.19))	Equation (3.1.21)
50	0.213	0.194	0.3446
100	0.379	0.343	0.4736
200	0.605	0.562	0.6271
300	0.761	0.695	0.7182
400	0.852	0.779	0.778
500	0.908	0.830	0.8194
600	0.943	0.862	0.8496
800	0.978	0.895	0.8898
1000	0.992	0.909	0.9149
1500	0.999	0.917	0.9483

From the results shown above, it is clear that these three expressions [i.e., Equations (3.1.16), (3.1.19), and (3.1.21)] give significantly different predictions. Furthermore, the differences do not display any fixed pattern as the respective u_c vs. t curves cross each other at certain points. This is not surprising since all these equations are either empirical or solutions from the filtration diffusion equations with different simplifying assumptions. More important, the model parameter of these equations, in spite of their supposedly physical significance, are, in many ways, just fitting parameters. One may expect that parameters obtained by fitting experimental data to a particular u_c expression based on certain specific assumptions, when used in predicting consolidation performance, may yield results with significant uncertainties. One should keep this limitation in mind in estimating consolidation performance. ■ ■ ■

3.2   Deliquoring by Suction or Blowing

For certain applications, filtrate left in filter cakes needs to be reduced to a minimum. For this purpose, filtrate may be drained off by gravity, centrifugal, or other force, air suction, or air blowing. Calculations of the extent of deliquoring by the last two processes are discussed in the following.

For cakes saturated with filtrate, airflow through the cake occurs only if the pressure difference imposed exceeds a threshold value, p_b, given as (Wakeman, 1976)

$$p_b = \frac{4.6(1 - \varepsilon_{av})\sigma}{\varepsilon_{av}d_e} \tag{3.2.1}$$

where σ is the interfacial tension of the filtrate and ε_{av} is the average cake porosity. (Note: filter cake formed at the end of filtration is not homogeneous for compressible cakes.) d_e is the equivalent particle diameter which may be estimated from the Kozeny–Carmen equation based on average cake properties (i.e., specific cake resistance and porosity) or

$$d_e = 13.42\sqrt{\frac{1 - \varepsilon_{av}}{\alpha_{av}\rho_s\varepsilon_{av}^3}} \tag{3.2.2}^{[5]}$$

If one defines cake saturation, S, as

$$S = \frac{\text{volume of liquid}}{\text{volume of cake void}} \tag{3.2.3}$$

The degree of saturation of a filter cake depends on the pressure difference imposed on the cake as shown in Fig. 3.5 which is commonly referred to as the capillary pressure–saturation curve. As shown in this figure, there is a threshold pressure below which the cake remains saturated. Furthermore, there is a certain amount of liquid within the cake pore space which cannot be removed by suction or blowing. This saturation value is known as the irreducible saturation, S_∞. For air suction or blowing

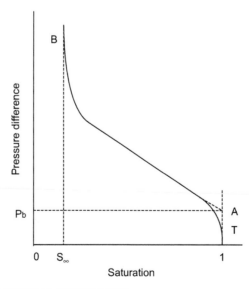

FIGURE 3.5 A typical capillary pressure vs. saturation curve.

[5]Obtained from the Kozeny–Carman Equations given by Equation (3.2.11).

drained filter cakes, S_∞ may be estimated according to the following empirical expression (Wakeman, 1976):

$$S_\infty = 0.155\left[1 + 0.031\, N_{\text{cap}}^{-0.49}\right] \tag{3.2.4}$$

where the capillary number, N_{cap} is defined as

$$N_{\text{cap}} = \frac{\varepsilon_{\text{av}}^3 d_e^2 (\rho_\ell g L + \Delta p)}{(1 - \varepsilon_{\text{av}})^2 L \, \sigma} \tag{3.2.5}$$

where Δp is the imposed pressure drop for gas flow.

The reduced saturation, S_R, defined as

$$S_R = \frac{S - S_\infty}{1 - S_\infty} \tag{3.2.6}$$

is often used to express the liquid (moisture) content of deliquored cake. Alternatively, the moisture content can also be given as M', defined as mass of liquid per unit mass of dry cake. M' and mass fraction, M can be expressed as

$$M' = \frac{(\varepsilon_{\text{av}})(S)(\rho)}{(1 - \varepsilon_{\text{av}})\rho_s} \tag{3.2.7a}$$

$$M = \frac{M'}{1 + M'} \tag{3.2.7b}$$

A number of empirical correlations of deliquoring by flowing air or suction have been developed in the past. The utility of these correlations, in general, is rather restricted. Nelson and Dahlstrom (1957) suggested a method of correlating deliquoring data and the correlation obtained can then be used for prediction if appropriate experimental data are available.

In principle, prediction of deliquoring performance by air suction (or flow) can be made by the solution of the relevant volume-averaged continuity equations. A number of studies along this line were made by Wakeman (1979a–c) which yield results of average reduced cake saturation and average gas (air) flow velocity as functions of time and the gas-phase pressure drop imposed. The results are shown in Fig. 3.6 (reduced S_R) and Fig. 3.7 (average u_g^*).

To facilitate the use of these results, the following empirical expressions based on the numerical results were obtained (Wakeman, and Tarleton, 1999).

For average reduced cake saturation

$$S_R = \left[1 + 1.08\,\{\theta(\Delta p^*)\}^{0.88}\right]^{-1} \quad \text{for } 0.096 \leq \theta(\Delta p^*) \leq 1.915 \tag{3.2.8a}$$

$$= \left[1 + 1.46\{\theta(\Delta p^*)\}^{0.48}\right]^{-1} \quad \text{for } 1.915 \leq \theta(\Delta p^*) \leq 204 \tag{3.2.8b}$$

For average gas (air) flow rate at the outlet

$$\log u_g^* = \sum_{i=0}^{4} a_i (\log \theta)^i \tag{3.2.9}$$

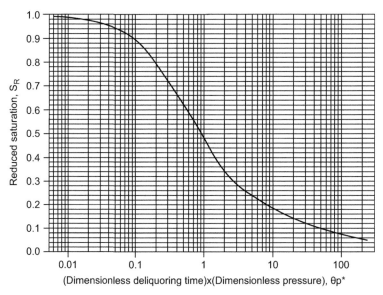

FIGURE 3.6 Reduced saturation vs. $\theta(\Delta p^*)$. *Reprinted from Wakeman, R.J., Tarleton, E.S., 1999. Filtration: Equipment Selection, Modeling and Process Simulation, first ed. Elsevier Advanced Technology, with permission of Elsevier.*

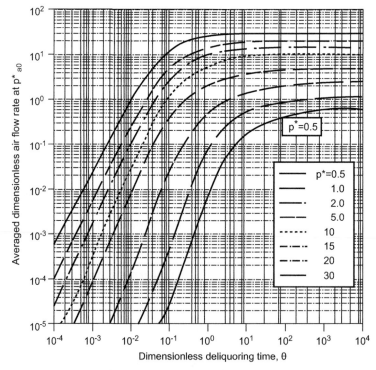

FIGURE 3.7 Averaged dimensionless airflow rate vs. dimensionless deliquoring times at various dimensionless pressure. *Reprinted from Wakeman, R.J., Tarleton, E.S., 1999. Filtration: Equipment Selection, Modeling and Process Simulation, first ed. Elsevier Advanced Technology, with permission of Elsevier.*

The values of the coefficient a_i are given in Table 3.1.
The various dimensionless quantities are defined as

$$\theta = \frac{k_{av} p_b t}{\mu\, \varepsilon_{av}(1 - S_\infty)L^2} \qquad (3.2.10a)$$

$$u_g^* = \frac{u_g\, \mu_g\, L}{k_{av}\, p_b} \qquad (3.2.10b)$$

$$\Delta p^* = \frac{\Delta p}{p_b} \qquad (3.2.10c)$$

The cake average permeability k_{av} present in the above expression is given as

$$k_{av} = \frac{\varepsilon_{av}^3 d_e^2}{180(1 - \varepsilon_{av})^2} \qquad (3.2.11)$$

The results shown in Figs. 3.6 and 3.7 and Equations (3.2.8a)–(3.2.9) were obtained subject to certain restrictions, namely, the inlet gas-phase pressure $(p_a^*)_i$ is assumed to be

Table 3.1 Values of Coefficients of Equations (3.2.9) (Wakeman and Tarleton, 1999)

p^*	a_0	a_1	a_2	a_3	a_4	Region of Validity
0.5	−2.2828	1.9809	−0.3649	−0.0786	−0.0049	$10^{-4} \le \theta < 20$
	−1.6166	1.1793	−0.3787	0.0560	−0.0031	$20 \le \theta \le 10^6$
	0.1487	0	0	0	0	$\theta > 10^6$
1	−1.2305	1.6891	−0.6050	−0.2142	−0.0262	$10^{-4} \le \theta < 5$
	−0.8702	0.7782	−0.2668	0.0435	−0.0027	$5 \le \theta \le 10^6$
	0.1367	0	0	0	0	$\theta > 10^6$
2	−0.3275	0.5687	−1.1410	−0.3354	−0.0359	$10^{-4} \le \theta < 2$
	−0.2570	0.5046	−0.1602	0.0253	−0.0016	$2 \le \theta \le 10^6$
	0.4456	0	0	0	0	$\theta > 10^6$
5	0.3310	0.4431	−0.4369	−0.0126	−0.0058	$10^{-4} \le \theta < 0.8$
	0.2765	0.4395	−0.2264	0.0606	−0.0062	$0.8 \le \theta \le 10^4$
	0.7202	0	0	0	0	$\theta > 10^4$
10	0.4085	−0.6765	−1.3865	−0.3329	−0.0304	$10^{-4} \le \theta < 0.3$
	0.7071	0.4015	−0.2104	0.0503	−0.0046	$0.3 \le \theta \le 10^3$
	1.0039	0	0	0	0	$\theta > 10^3$
15	0.8288	−0.4022	−1.0965	−0.2551	−0.0227	$10^{-4} \le \theta < 0.18$
	0.9755	0.2690	−0.2139	0.1139	−0.0256	$0.18 \le \theta \le 10^2$
	1.1614	0	0	0	0	$\theta > 10^2$
20	0.7981	−0.8823	−1.3651	−0.3367	−0.0314	$10^{-4} \le \theta < 0.1$
	1.1400	0.2554	−0.1888	0.0724	−0.0108	$0.1 \le \theta \le 10^2$
	1.3010	0	0	0	0	$\theta > 10^2$
30	0.9539	−1.1129	−1.2848	−0.2757	−0.0225	$10^{-4} \le \theta < 0.05$
	1.4050	0.1420	−0.1481	0.0762	−0.0143	$0.05 \le \theta \le 10^2$
	1.4771	0	0	0	0	$\theta > 10^2$

100 and the gas (air) to liquid (water) viscosity ratio, $\mu_g/\mu = 0.0185$. In order to extend the results to cases with different gas-phase inlet pressure, the average gas velocity given by Fig. 3.6 or Equation (3.2.9) should be corrected according to

$$u_g^* = (u_g^*)_{\text{from Equation (3.2.9)}} \, F \qquad (3.2.12)$$

and the correction factor, F, is given as

$$F = \frac{(p_{g_0}^*)_{\text{ref}}}{p_{g_0}^*} \left[\frac{(p_{g_0}^*)^2 - (p_{g_i}^*)^2}{(p_{g_0}^*)^2 - (p_{g_i}^*)_{\text{ref}}^2} \right] \qquad (3.2.13)$$

where $p_{g_0}^*$ and $p_{g_i}^*$ are the dimensionless gas-phase outlet pressure and inlet pressure (with $p^* = p/p_b$). The subscript "ref" refers to the reference state and is defined as

$$(p_{g_i}^*)_{\text{ref}} = 100 \qquad (3.2.14a)$$

$$(p_{g_0}^*)_{\text{ref}} = 100 - \Delta p^* = 100 - \left[(p_{g_i}^*) - (p_{g_0}^*) \right] \qquad (3.2.14b)$$

Comparisons of predictions based on Equations (3.2.8a)–(3.2.9) with experimental data were made by Condie et al. (1996) and Carleton and Salway (1993). Reasonably good agreement was observed in most cases. However, caution must be exercised in using these results. The concept of the relative permeability was used in describing the liquid and gas flow during air suction (or blowing) and the relative permeability was assumed to be a function of S_R. The accuracy of the numerical results, therefore, depends upon the validity of the concept used in developing correlations, such as Equation (3.2.9). As our knowledge of two phase flow-through porous media is far from being complete, the question posed therefore cannot be answered unequivocally.

An example is given below to illustrate the procedure which may be used to predict cake deliquoring using the numerical results of Wakeman's.

■ ■ ■ ▬▬▬▬▬▬▬▬▬▬▬▬▬▬▬▬▬▬▬▬▬▬▬▬▬▬▬▬

Illustrative Example 3.4

A saturated filter cake is subject to air suction for 100 seconds with the following conditions:

Vacuum	60 kPa
Average Specific Cake Resistance	10^9 m kg^{-1}
ε_{av}	0.43
ρ_s	1900 kg m^{-3}
Cake Thickness	0.08 m
Filtration Viscosity, μ	0.001 Pa s
Filtrate Density	1000 kg m^{-3}
Filtrate Surface Tension	0.07 N m^{-1}

Determine the cake moisture content at the end of air suction. What is the airflow rate?

Solution

1. Calculate the equivalent particle size, d_e (Equation (3.2.2))

$$d_e = 13.42 \left[\frac{0.57}{(10^9)(1900)(0.43)^3} \right]^{1/2} = 2.607 \times 10^{-5}\text{ m}$$

2. Calculate N_{cap} (Equation (3.2.5))

$$N_{cap} = \frac{(0.43)^3(2.607 \times 10^{-5})\left[(10^3)(9.8)(0.08) + 6 \times 10^4\right]}{(0.57)^2(0.07)(0.08)}$$

$$= 1.806 \times 10^3$$

3. Calculate S_∞ (Equation (3.2.4))

$$S_\infty = 0.155 \left\{ 1 + 0.031 \left[\frac{10^3}{1.806} \right]^{0.49} \right\} = 0.261$$

4. Calculate the threshold pressure p_b [Equation (3.2.1)]

$$p_b = \frac{(4.6)(0.57)(0.07)}{(0.43)(2.607 \times 10^{-5})} = 1.639 \times 10^4\text{ N m}^{-1}$$

5. Calculate k_{av}. By definition, one has

$$k_{av} = \frac{1}{\alpha_{av}\rho_s(1 - \varepsilon_{av})} = \frac{1}{(10^9)(1900)(0.57)} = 9.234 \times 10^{-13}\text{ m}^2$$

6. Calculate the dimensionless time, θ (Equation (3.2.10a))

$$\theta = \frac{(9.243 \times 10^{-13})(1.637 \times 10^4)(100)}{(0.001)(0.43)(1 - 0.261)(0.08)^2} = 0.745$$

7. Calculate pressure drop

Inlet gas (air) pressure	1.013×10^5 Pa (Atmosphere Pressure)
Outlet gas pressure	$(1.013–0.6)\ 10^5$ Pa $= 4.13 \times 10^4$ Pa

$$\Delta p^* = \frac{6 \times 10^4}{1.637 \times 10^4} = 3.66$$

and

$$\theta(\Delta p^*) = (0.745)(3.66) = 2.731$$

8. Estimate Average Cake Saturation after suction using Equation (3.2.8b)

$$S_R = \left[1 + 1.46(2.731)^{0.48}\right]^{-1} = 0.297$$

and

$$S = S_R(1 - S_\infty) + S_\infty = 0.297(1 - 0.261) + 0.261 = 0.481$$

From Equation (3.2.7a)

$$M' = \frac{(0.43)(0.481)(10^3)}{(0.5)(1900)} = 0.191 \text{ kg } H_2O/\text{kg drag cake}$$

Moisture mass fraction at the end of deliquoring

$$= \frac{0.191}{1 + 0.191} = 0.16 \text{ or } 16\%$$

9. To estimate the average gas (air) flow rate, first calculate u_g^* at $\theta = 0.745$ and $\Delta p^* = 3.66$. Equation (3.2.9) gives u_g^* as a function of θ for specified values of Δp^*. $\Delta p^* = 3.66$ being between $\Delta p^* = 2$ and 5.
At $\Delta p^* = 2.0$, $a_0 = -0.3275$, $a_1 = 0.5687$, $a_2 = 1.141$, $a_3 = 0.3354$, $a_4 = 0.0369$, $\theta = 0.745$, $\log \theta = -0.1278$, $(\log \theta)^2 = 0.0163$, $(\log \theta)^3 = -0.0021$, $(\log \theta)^4 = 0.0003$

$$\sum_{i=0}^{4} a_i (\log \theta)^i = -0.4185$$

and

$$u_g^* = 10^{-0.4185} = 0.3815$$

At $\Delta p^* = 5.0$, $a_0 = 0.3310$, $a_1 = 0.4431$, $a_2 = -0.4369$, $a_3 = -0.0126$, $a_4 = -0.0058$

$$\sum_{i=0}^{4} a_i (\log \theta)^i = 0.2744$$

$$u_g^* = 10^{0.2744} = 1.881$$

By linear interpolation, for $\Delta p^* = 3.66$,

$$u_g^* = 0.3815 + (1.881 - 0.3815)\frac{1.66}{3} = 1.2112$$

To correct the airflow rate for $p_{g_{in}}^*$ being different from the reference value of 100, the correction factor, F is found according to Equation (3.2.13):

$$F = \frac{(p_{g_0}^*)_{\text{ref}}}{p_{g_0}^*} \frac{(p_{g_0}^*)^2 - (p_{g_i}^*)^2}{(p_{g_0}^*)^2_{\text{ref}} - (p_{g_i}^*)^2_{\text{ref}}}$$

$$p_{g_0}^* = \frac{4.13 \times 10^4}{1.637 \times 10^4} = 2.523$$

$$p_{g_i}^* = \frac{1.013 \times 10^5}{1.637 \times 10^4} = 6.118$$

$$(p_{g_i}^*)_{\text{ref}} = 100$$

$$(p_{g_0}^*)_{\text{ref}} = 100 - 3.66 = 96.34$$

The corrected u_g^* is

$$(u_g^*)_{\text{cor}} = 1.2112 \frac{96.34}{2.523} \frac{(2.523)^2 - (6.188)^2}{(96.34)^2 - (100)^2} = 2.0544$$

From Equation (3.2.10b), the average given (air) flow rate is

$$u_g = 2.0544 \frac{(9.234 \times 10^{-13})(1.637 \times 10^4)}{(0.0815)(0.001)(0.08)} = 2.0901 \times 10^{-2} \text{ m s}^{-1} \text{ at 60 kPa vacuum}$$

■ ■ ■

3.3 Washing of Filter Cakes

If the recovered solid of cake filtration (i.e. filter cake particles) is the main product, and product purity is important, removing filtrate retained in and impurities adhered to filter cakes by washing may be necessary. Cake washing can be accomplished by passing a wash liquid (usually water) through filter cakes (saturated or deliquored) or by re-slurrying filter cakes using wash liquid and then refiltering the resulting suspensions. For both cases, the flow rate and the amount of wash liquid required for a given degree of solute removal are the important factors to be considered.

3.3.1 Representation of Cake-Washing Results

The physical process of cake washing is depicted in Fig. 3.8. As wash liquid enters into a cake, it, at first, displaces the filtrate present in the cake and washing is accomplished by filtrate displacement. As washing progresses, solute present in the filtrate in the immediate neighborhood of cake particles may diffuse into the wash liquid and there is also solute transport due to axial dispersion. This combined diffusion–dispersion process constitutes the second phase of cake washing.

As an indication of the progress of cake washing, consider a filter cake of unit cross-sectional area. The solute concentration of the filtrate present within the cake initially

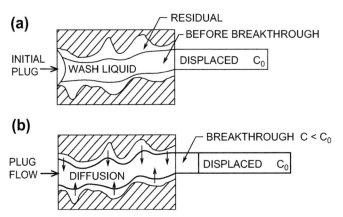

FIGURE 3.8 Schematic representation of cake washing.

is c_0. If the solute concentration of the liquid used for washing is $(c_w)_i$ and that of the effluent is $(c_w)_e$, the amount of the solute removed up to time t is

$$\int_0^t u_w \left[(c_w)_e - (c_w)_i \right] dt$$

where u_w is the superficial velocity of the wash liquid.

A complete washing of a filter cake is achieved when the cake becomes saturated with the wash liquid with solute concentration of $(c_w)_i$. Accordingly, the maximum solute removable by washing is

$$\left[c_0 - (c_w)_i \right] L \, \varepsilon_{av}$$

The fraction of the removable solute removed by washing, F, is

$$F = \frac{\int_0^t u_w \left[(c_w)_e - (c_w)_i \right] dt}{\left[c_0 - (c_w)_i \right] L \, \varepsilon_{av}}$$

$$= \int_0^w (c_w^*)_e \, dw \tag{3.3.1}$$

where

$$(c_w^*)_e = \frac{(c_w)_e - (c_w)_i}{c_0 - (c_w)_i} \tag{3.3.2a}$$

$$w = \frac{u_w t}{L \, \varepsilon_{av}} = \frac{v_w}{v_v} \tag{3.3.2b}$$

where $v_w = u_w t$ is the cumulative wash liquid (in $m^3 \, m^{-2}$) passing through the cake and $v_w = L \, \varepsilon_{av}$ is the cake void volume. w, the wash ratio, expresses the cumulative wash liquid volume in terms of multiples of the void volume.

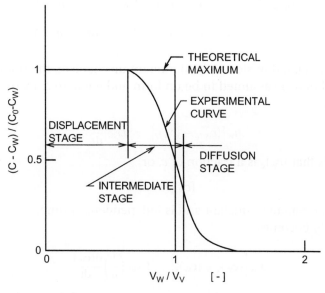

FIGURE 3.9 A typical wash curve of effluent concentration vs. wash ratio.

Wash experiments yield data of c_w vs. t which can be easily transformed into c_w^* vs. w, commonly referred to as the wash curve. A typical wash curve is shown in Fig. 3.9.

The efficiency of cake washing can be seen from the relationship of F vs. w. The washing efficiency can also be expressed by the fraction of solutes remaining in the washed cake, R, given as

$$R = 1 - F \qquad (3.3.3)$$

3.3.2 Empirical Expression of F (or R) vs. w

From Equation (3.3.1) it is clear that the wash efficiency can be readily obtained from the wash curve. Physically speaking, the effectiveness of washing depends on the degree of the displacement of the resident liquid (filtrate) by the invading fluid (wash liquid) and the solute diffusion into and through the wash liquid. As a starting point, it may be instructional to obtain the expression of c_w or F with one of the mechanisms being dominant.

Consider the case with filtrate displacement being the dominant mechanism. With this assumption, the effluent solute concentration is the same as the initial filtrate solute concentration until all the filtration present in the cake void initially is replaced or

$$(c_w)_e = c_0 \qquad \text{if} \qquad u_w t < L\varepsilon_0 \qquad \text{or} \qquad w < 1$$

$$(c_w)_e = (c_w)_i \qquad \text{if} \qquad u_w t > L\varepsilon_0 \qquad \text{or} \qquad w > 1 \qquad (3.3.4)$$

On the other hand, if diffusion is the dominant mechanism, the solute concentration of the pore liquid may be assumed to be uniform and equal to c. By overall solute mass balance, one has

$$u_w \left[(c_w)_e - (c_w)_i \right] = L\varepsilon_{av} \frac{dc}{dt} \qquad (3.3.5)$$

If one assumes that $(c_w)_e$ is a fraction of c, or

$$(c_w)_e = \alpha c \qquad (3.3.6)$$

And furthermore, α remains constant and is independent of time,
    Equation (3.3.5) becomes

$$u_w \left[(c_w)_\lambda - (c_w)_i \right] = L\,\varepsilon_{av} \left(\frac{1}{\alpha} \right) \frac{d(c_w)_s}{dt} \qquad (3.3.7)$$

Integrating the above equation with the initial condition of $(c_w)_e = c_0$ at $t = 0$ yields

$$\ell n \frac{(c_w)_e - (c_w)_i}{c_0 - (c_w)_i} = -\frac{u_w}{L\varepsilon_{av}} \alpha t \qquad (3.3.7)$$

or

$$c_w^* = \frac{(c_w)_e - (c_w)_i}{c_0 - (c_w)_i} = \exp \left[-\alpha \frac{u_w}{L\varepsilon_{av}} t \right] = \exp[-\alpha w] \qquad (3.3.8)$$

The above equation was first given by Rhodes (1934).
    Substituting Equation (3.3.8) into Equation (3.3.1) yields

$$R = 1 - F = \left(1 - \frac{1}{\alpha} \right) + \frac{1}{\alpha} \exp(-\alpha w) \qquad (3.3.9)$$

The above expression suggests an approximate linear relationship between ℓn R vs. w. Choudhury and Dahlstrom (1957) showed that the relationship was obeyed in the case of aluminum trihydrate cakes with $w < 2.0$ (see Fig. 3.10).
    A more commonly used empirical expression of wash efficiency based on a large body of experimental data proposed by Choudhury and Dahlstrom (1957) is

$$R = 1 - F = \left[1 - (F)_{w=1} \right]^w \qquad (3.3.10)$$

which predicts R (or F) based on values of F at $w = 1$. For simple calculations, Holdich (1996) suggested the use of 0.7 for $(F)_{w=1}$ although experimentally determined value of $(F)_{w=1}$ ranged rather widely from a minimum of 0.35 to a maximum of 0.85 (Choudhury and Dahlstrom, 1957).

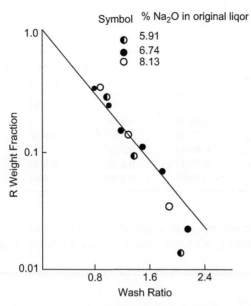

FIGURE 3.10 Experimental data demonstrating the nearby linear relationship of log R vs. wash ratio.

A more elaborate expression of R was given by Moncrieff (1965) for small values of w. R is given as

$$R = 1 - F = (1 - x) \exp\left[-\frac{w - x}{1 - x}\right] \qquad (3.3.11)$$

where x is an empirical factor.

Values of R vs. w for various values of x according to Equation (3.3.11) are shown in Fig. 3.11. Some of Choudhury and Dahlstrom's data are superimposed on the figure, suggesting the use of $x \simeq 0.3 \sim 0.4$. Shirato et al. (1987) stated that the simplicity of Equation (3.3.10) justifies its use for $w > 1.0$. However, for $w < 1$, Equation (3.3.11) may be applied.

■ ■ ■ ▬▬▬▬▬▬▬▬▬▬▬▬▬▬▬▬▬▬▬▬

Illustrative Example 3.5

Calculate and compare the predicted values of F according to Equation (3.3.10) with $(F)_{w=1} = 0.7$ and 0.825 and according to Equation (3.3.11) with $x = 0.3$, 0.4 and 0.5.

Solution

Equations (3.3.10) and (3.3.11) may be written as

$$F = 1 - \left[1 - (F)_{w=1}\right]^{w}$$

and

$$F = 1 - (1 - x) \exp\left[-\frac{w - x}{1 - x}\right]$$

The results are

$$F$$

	Equation (3.3.10)		Equation (3.3.11)		
w	$(F_w)_{w-1} = 0.7$	0.825	$x = 0.3$	$x = 0.4$	$x = 0.5$
0.4	0.3822	0.4908	0.393	0.400	0.389
0.6	0.5144	0.6486	0.544	0.570	0.591
0.8	0.6183	0.7520	0.657	0.692	0.726
1.0	0.7	0.8250	0.742	0.779	0.816
2.0	0.91	0.9694	0.938	0.958	0.975
4.0	0.992	0.9991	0.9965	0.9987	0.9995

For Equation (3.3.10), increasing $F_{w=1}$ increases the predicted value of F. The numerical results validate what can be seen in Fig. 3.11, namely, for $w < 0.5$. Equation (3.3.11) is insensitive to the value of x. ■ ■ ■

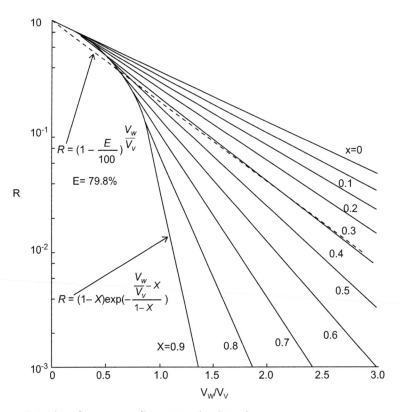

FIGURE 3.11 Relationship of R vs w according to Equation (3.3.11).

3.3.3   Diffusion–Dispersion Model of Cake Washing

Generically speaking, cake washing can be considered as a mass transfer process taking place in porous media. It is similar to a number of fixed-bed operations such as adsorption, chromatography, and catalytic reactions and has been studied by a number of investigators (Sherman, 1964; Han and Bixler, 1967; Han, 1967; Wakeman and Attwood, 1990). In particular, the work of Sherman and that of Wakeman and Attwood were based on the fixed-bed adsorption study of Lapidus and Amundson (1952). A discussion of the Lapidus–Amundson work is given below.

Consider a cake initially saturated with filtrate, the solute concentration of which is c_0. A wash liquid with solute concentration $(c_w)_i$ enters into the cake at time $t = 0$. The solute concentration of the cake pore liquid, c, is given by the following equation:

$$\nabla \cdot [D \cdot \nabla c] - \nabla \cdot [u\,c] = \frac{\partial c}{\partial t} \tag{3.3.12}$$

where D is the dispersion–diffusion coefficient tensor and u the pore liquid velocity vector.

For the case of homogeneous medium, with the pore liquid flow being uniform and one dimensional and the dispersion–diffusion effect limited to the axial (flow) direction, Equation (3.3.12) reduces to

$$D_L \frac{\partial^2 c}{\partial x^2} - \frac{u_s}{\varepsilon} \frac{\partial c}{\partial x} = \frac{\partial c}{\partial t} \tag{3.3.13}$$

where x is the coordinate along the wash liquid flow direction and u_s the wash liquid superficial velocity. ε is the cake porosity and is assumed to be the same as the average cake porosity at the end of filtration. D_L is the axial dispersion coefficient.

In specifying the spatial domain over which Equation (3.3.13) applies, it is logical to consider x extends from $x = 0$ (cake/medium interface) to $x = L$ (L, cake length). However, because of the difficulty of imposing appropriate boundary conditions, Lapidus and Amundson considered the cake to be of infinite thickness or $0 < x - \infty$. The wash liquid solute concentration at exit $(c_w)_e$ is taken to be the value of c at $x = L$.

The initial condition of Equation (3.3.13) is

$$c = c_0 \qquad x > 0 \qquad t < 0 \tag{3.3.14}$$

The boundary conditions are

$$-\varepsilon D_L \frac{\partial c}{\partial x} = u_s \big[(c_w)_i - c\big] \quad \text{at} \quad x = 0 \tag{3.3.15a}$$

$$\text{and} \quad \frac{\partial c}{\partial x} = 0 \quad \text{at} \quad x \to \infty \tag{3.3.15b}$$

If the axial dispersion effect is ignored, Equation (3.3.15a) becomes

$$c = (c_w)_i \quad \text{at } x = 0 \tag{3.3.15c}$$

The solution of Equation (3.3.13) corresponding to the initial and boundary conditions of Equations (3.3.14), (3.3.15b) and (3.3.15c) is

$$\frac{c - c_0}{(c_w)_i - c_0} = \frac{1}{2}\left\{ \text{erfc}\left[\frac{x - \frac{u_s t}{\varepsilon}}{\sqrt{4D_L t}}\right]\right\} + \exp\left(\frac{x u_s}{\varepsilon D_L}\right)\text{erfc}\left[\frac{x + \frac{u_s t}{\varepsilon}}{\sqrt{4D_L t}}\right] \tag{3.3.16}$$

where erfc(y) is the complementary error function with argument y.

The effluent wash liquid solute concentration is taken to be c at x = L, or

$$\frac{(c_w)_e - c_0}{(c_w)_i - c_0} = \frac{1}{2}\left\{ \text{erfc}\left[\frac{L - \frac{u_s t}{\varepsilon}}{\sqrt{4D_L t}}\right] + \exp\left(\frac{L u_s}{\varepsilon D_L}\right)\text{erfc}\left[\frac{L + \frac{u_s t}{\varepsilon}}{\sqrt{4D_L t}}\right]\right\} \tag{3.3.17}$$

The wash ratio, w by Equation (3.3.2b), may be used to replace t as the independent variable. The above expression becomes

$$(c_w)_e^* = \frac{(c_w)_e - (c_w)_i}{c_0 - (c_w)_i} = 1 - \frac{(c_w)_e - (c_0)}{(c_w)_i - c_0} = 1 - \frac{1}{2}\left[\text{erfc}\left\{\frac{1 - w}{2\sqrt{w}}\sqrt{\frac{u_s L}{\varepsilon D_L}}\right\}\right.$$
$$\left. + \exp\left(\frac{L u_s}{\varepsilon D_L}\right)\text{erfc}\left\{\frac{1 + w}{2\sqrt{w}}\sqrt{\frac{u_s L}{\varepsilon D_L}}\right\}\right] \tag{3.3.18}$$

Cake washing is often carried out in such a manner that a pool of liquid is formed above cake surface and solute back diffusion to the liquid pool may occur. Consequently, the boundary condition given by Equation (3.3.15a) may not be valid. To account for this back-diffusion effect, Sherman (1964) suggested that instead of Equation (3.3.15a), the boundary condition at cake surface should be

$$\text{at } x = 0, \quad c = (c_w)_i \sum_{i=0}^{4} k_i t^i \tag{3.3.19a}$$

where k_i's are empirical constants to be determined by matching experiments and predictions based on using Equation (3.3.19a). Based on the same argument, Wakeman and Attwood (1990) proposed the use of the following boundary conditions

$$\text{at } x = 0, \quad c = k_0 \exp\left[-\gamma \frac{u_s}{\varepsilon L} t\right] \tag{3.3.19b}$$

With Equation (3.3.19b) replacing Equation (3.3.15a), the solution of Equation (3.3.13) was found to be (Eriksson et al., 1996)

$$\frac{(c_w)_e - (c_w)_i}{c_0 - (c_w)_i} = 1 - \frac{1}{2}\left[\text{erfc}\left\{\frac{1 - w}{2w}\sqrt{\frac{u_s L}{\varepsilon D_L}}\right\} + \exp\left(\frac{L u_s}{\varepsilon D_L}\right)\text{erfc}\left\{\frac{1 + w}{2w}\sqrt{\frac{u_s L}{\varepsilon D_L}}\right\}\right]$$

$$+ k_0\sqrt{\frac{u_s L}{\pi \varepsilon W D_L}}\int_1^\infty \exp\left[\frac{u_s L}{2\varepsilon D_L} - \gamma\left(1 - \frac{1}{w^2}\right)w - \frac{w u_s L}{(\varepsilon)4 D_L}\frac{1}{w^2} - \frac{u_s L w^2}{4\varepsilon w D_L}\right]dw \tag{3.3.20}$$

A comparison of the above expression with Equation (3.3.18) shows that using Equations (3.3.19b) instead of (3.3.13) results in the presence of an additional term in the expression of the wash liquid effluent concentration expression.

For analyzing cake washing and calculating cake-washing performance, Equation (3.3.18) [or Equation (3.3.20)] may be used in two ways. Sherman (1964) showed that by matching selected cake-washing data with the solution of Equation (3.3.13), one may obtain the value of the axial dispersion coefficient.[6] With D_L known, cake-washing performance for various conditions may be predicted. Alternatively, cake-washing results may be predicted directly from Equations (3.3.18) or (3.3.20). In this connection, one should bear in mind the limitation of these two equations. First, the result of Lapidus and Amundson (1952) was based on the assumption of homogeneous medium. For filter cakes, they are likely to be non-homogeneous with significant porosity variation across the cake length. Another problem is the estimation of the axial dispersion coefficient. In spite of the large number of studies on the subject in the past, accurate estimation of D_L remains difficult if not impossible. The most up-to-date correlation of D_L of liquid systems was proposed by Delgado (2007) based on experimental data reported in the past five decades. His results are

$$\frac{D_L}{D'_m} = 1 \qquad\qquad N_{Pe_m} < 0.1 \qquad\qquad\qquad (3.3.21a)$$

$$\frac{D_L}{D'_m} = \frac{N_{Pe'_m}}{(0.8/N_{Pe'_m}) + 0.4} \qquad 0.1 < N_{Pe_m} < 4 \qquad\qquad (3.3.21b)$$

$$\frac{D_L}{D_m} = \frac{N_{Pe_m}}{\left[18 N_{Pe'_m}^{-1.2} + 2.35 N_{Sc}^{-0.38}\right]^{1/2}} \qquad N_{Pe_m} > 4$$
$$N_{Re} < 10 \qquad\qquad (3.3.21c)$$

$$\frac{D_L}{D'_m} = \frac{N_{Pe'_m}}{(25 N_{Sc}^{1.14}/N_{Pe'_m}) + 0.5} \qquad N_{Pe_m} < 10^6$$
$$\qquad\qquad (3.3.21d)$$
$$N_{Re} > 10$$

$$\frac{D_L}{D'_m} = \frac{N_{Pe'_m}}{2} \qquad N_{Pe_m} > 10^6 \qquad\qquad\qquad (3.3.21e)$$

where

$$D'_m = D_m/\tau \qquad\qquad\qquad (3.3.22a)$$

$$N_{Pe_m} = (u_s/\varepsilon)d_p/D_m \qquad\qquad (3.3.22b)$$

---

[6]Specifically, Sherman reported his results in terms of the value of D_L/u_s.

$$N_{Pe'_m} = (u_s/\varepsilon)d_p/D'_m \qquad (3.3.22c)$$

$$N_{Sc} = \mu/(\rho\,D_m) \qquad (3.3.22d)$$

$$N_{Re} = (d_p\,\rho\,u_s)/(\mu\varepsilon) \qquad (3.3.22e)$$

where D_m is the binary (solute-filtrate or wash liquid) diffusivity and D'_m the solute effective diffusivity (equal to D_m corrected by the tortuosity factor, τ). N_{Sc} and N_{Re} are the Schmidt and Reynolds numbers. N_{Pe_m} and $N_{Pe'_m}$ are the Peclet numbers based on D_m and D'_m, respectively,

■ ■ ■ ▬▬▬▬▬▬▬▬▬▬▬▬▬▬▬▬▬▬▬▬▬▬▬▬▬▬▬▬▬▬▬▬▬▬▬

Illustrative Example 3.6

Sherman (1964) determined the axial dispersion coefficient D by matching experimental washing data with predictions from the solution Equation (3.3.13) with the initial conditions given by Equation (3.3.19a). The conditions he used in one series of experiments are as follows:

$$d_p = 3 \times 10^{-3} \text{ m}$$

$$\varepsilon = 0.45$$

$$L = 5.29 \times 10^{-2} \text{ m}$$

$$u_s = 4.3 \times 10^{-4} - 3.74 \times 10^{-3} \text{ m s}^{-1}$$

The axial dispersion coefficient results obtained may be expressed as

$$\frac{D_L}{u_s} = 0.233 \text{ cm} = 2.33 \times 10^{-3} \text{ m} \qquad (i)$$

(a) Estimate D_L based on Delgado's correlation [Equations (3.3.21a)–(3.3.21e) at $u_s = 5 \times 10^{-4}$, 1×10^{-3}, 2×10^{-3} and 3×10^{-3} m s^{-1}. Compare them with Sherman's results.
(b) Wakeman and Tarleton (1999) gave the following expressions for D_L:

$$D_L/D_m = 0.707 + 1.75\frac{d_p(u_s)/\tau}{D_m} \qquad L > 10^{-1} \text{ m} \qquad (ii)$$

$$= 0.707 + 5.55\left[\frac{d_p(u_s/\tau)}{D_m}\right]^{0.96} \qquad L < 10^{-1} \text{ m} \qquad (iii)$$

Estimate D_L according to the above expressions and compare the results with Equation (i).
(c) What kind of conclusions can you make based on results of (a) and (b)?

Solution

(a) In addition to the conditions given, additional information used are:

$$\rho = 1000 \text{ kg m}^{-3} \qquad \mu = 10^{-3} \text{ Pa s}$$

$$D_m = 10^{-9} \text{ m}^2 \text{ s}^{-1} \qquad \tau = \sqrt{2} \qquad D'_m = \frac{10^{-9}}{\sqrt{2}} = 7.07 \times 10^{-10} \text{ m}^2 \text{ s}^{-1}$$

The various dimensionless groups may be calculated as

$$N_{Sc} = \frac{\mu}{\rho D_m} = \frac{10^{-3}}{10^3 \times 10^{-9}} = 10^3$$

$$N_{Re} = \frac{(d_p)(u_s/\varepsilon)(\rho)}{\mu} = (3 \times 10^{-3})(u_s/0.45)(10^3)/(10^{-3}) = (6.67 \times 10^3)u_s$$

$$N_{Pe_m} = (N_{Re})(N_{Sc})$$

$$N_{Pe'_m} = (N_{Pe_m})(\tau)$$

The numerical values of the various dimensionless groups vs. u_s are (round off to 3 digits)

u_s	N_{Re}	N_{Pe_m}	$N_{Pe'_m}$
5×10^{-4}	3.33	3.33×10^3	4.71×10^3
1×10^3	6.67	6.67×10^3	9.43×10^3
2×10^3	13.34	1.33×10^4	1.88×10^4
3×10^3	20.01	2.00×10^4	2.82×10^4

For $u_s = 5 \times 10^{-4} \text{ m s}^{-1}$, $N_{Re} = 3.33$, Equation (3.3.21c) may be used

$$\frac{D_L}{D_m} = \frac{4.71 \times 10^3}{\left[\dfrac{1.81}{(4.7 \times 10^3)^{1.12}} + \dfrac{2.35}{(10^3)^{0.38}} \right]^{1/2}} = 1.14 \times 10^4$$

$$D_L = (1.14 \times 10^4)(7.07 \times 10^{-10}) = 8.06 \times 10^{-6} \text{ m}^2 \text{ s}^{-1}$$

and

$$D_L/u_s = (8.06 \times 10^{-6})/(5 \times 10^{-4}) = 1.61 \times 10^{-2} \text{ m}$$

For $u_s = 1 \times 10^{-3}$, $N_{Re} = 6.67$, from Equation (3.3.21c), one has

$$\frac{D_L}{D'_m} = 2.29 \times 10^4$$

$$D_{\mathrm{m}} = 1.62 \times 10^{-5} \ \mathrm{m^2 \ s^{-1}}$$

$$D_{\mathrm{L}}/u_{\mathrm{s}} = 1.62 \times 10^{-2} \ \mathrm{m}$$

For $u_{\mathrm{s}} = 2 \times 10^3$, or greater, the corresponding N_{Re} exceeds 10. Accordingly, Equation (3.3.21d) is used
For $u_{\mathrm{s}} = 2 \times 10^{-3} \ \mathrm{m \ s^{-1}}$

$$\frac{D_{\mathrm{L}}}{D'_{\mathrm{m}}} = \frac{1.88 \times 10^4}{\dfrac{2.5(10^3)^{1.14}}{1.88 \times 10^4} + 0.5} = 4.7 \times 10^3$$

$$D_{\mathrm{L}} = (4.7 \times 10^3)(7.07 \times 10^{-10}) = 3.32 \times 10^{-6} \ \mathrm{m^2 \ s^{-1}}$$

$$D_{\mathrm{L}}/u_{\mathrm{s}} = 1.66 \times 10^{-3} \ \mathrm{m}$$

and

$$\text{For } u_{\mathrm{s}} = 3 \times 10^{\,3} \ \mathrm{m \ s^{-1}}$$

$$\frac{D_{\mathrm{L}}}{D'_{\mathrm{m}}} = \frac{2.82 \times 10^4}{\dfrac{2.5(10^3)^{1.14}}{2.82 \times 10^4} + 0.5} = 9.96 \times 10^3$$

$$D_{\mathrm{m}} = (9.96 \times 10^3)(7.07 \times 10^{-10}) = 7.05 \times 10^{-6} \ \mathrm{m^2 \ s^{-1}}$$

and

$$D_{\mathrm{L}}/u_{\mathrm{s}} = 2.35 \times 10^{-3} \ \mathrm{m}$$

The difference between the estimation based on Delgada's correlation and the value obtained by Sherman is insignificant at $u_{\mathrm{s}} = 3 \times 10^{-3} \ \mathrm{m \ s^{-1}}$ but becomes more significant as u_{s} decreases.

(b) As all Sherman's experiments were made with cake thickness less than 10 cm, the following expression may be used

$$D_{\mathrm{L}}/D_{\mathrm{m}} = 0.707 + 55.5 \left[\frac{d_{\mathrm{p}}(u_{\mathrm{s}}/\varepsilon)}{D_{\mathrm{m}}} \right]^{0.96}$$

The results are

$U_{\mathrm{s}}(\mathrm{m \ s^{-1}})$	$N_{\mathrm{Pe_m}} = d_{\mathrm{p}}(u_{\mathrm{s}}/\varepsilon)/D_{\mathrm{m}}$	$(D_{\mathrm{L}}/D_{\mathrm{m}})$	D_{L}	$D_{\mathrm{L}}/u_{\mathrm{s}}$
5×10^{-4}	3.33×10^3	1.34×10^5	1.34×10^{-4}	0.268
1×10^{-3}	6.67×10^3	2.60×10^5	2.6×10^4	0.26
2×10^{-3}	1.33×10^3	5.05×10^5	5.5×10^4	0.253
3×10^{-3}	2.00×10^3	7.47×10^5	7.47×10^4	0.249

The estimated results according to Equation (ii) are two orders of magnitude greater than that of Sherman's.

(c) In examining the various estimated D_L values and comparing them with Sherman's result, one should bear in mind, that in all likelihood, Sherman matched his experiments with the solution of Equation (3.3.13) in an ad hoc manner. It is therefore difficult to assess the accuracy of the results he reported. Delgado's correlation claimed an average deviation of less than 20% based on data collected with $L \geq 150\, d_p$, a condition significantly different for that used by Sherman in his experiments. Taking both factors into account, the order-of-magnitude agreement observed is rather surprising.

The significant difference between the correlation mentioned in Wakeman and Attwood's work and Delgado's is difficult to explain since Wakeman and Attwood gave no details about the expression (whether it was their own work or correlation proposed by others) given in the monograph of Wakeman and Tarleton (1999). It may be pertinent to point out that a correlation for transverse dispersion coefficient D_T by Fetter (1999) is of the form

$$\frac{D_T}{D'_m} = 1 + 0.055\, N_{Pe'_m}$$

The coefficient of the $N_{Pe'_m}$ term in the above expression is one-thousandth of that of Equation (iii). Could there be a mistake?

■ ■ ■

Illustrative Example 3.7

Calculate $(c_w)_e^*$ as a function of w according to Equation (3.3.18) with $\dfrac{u_s L}{\varepsilon D_L} = 25$.

Solution

Equation (3.3.18) is written as

$$(c_w)_e^* = 1 - \frac{1}{2}\left[\mathrm{erfc}\left\{ \frac{1-w}{2\sqrt{w}}\sqrt{\frac{u_s L}{\varepsilon D_L}} \right\} + \left[\exp\left(\frac{u_s L}{\varepsilon D_m} \right) \right] \mathrm{erfc}\left\{ \frac{1+w}{2\sqrt{w}}\sqrt{\frac{u_s L}{\varepsilon D_L}} \right\} \right] \qquad \text{(i)}$$

The complementary error function is defined as

$$\mathrm{erf}(c) = 1 - \mathrm{erf}(x) = \frac{2}{\sqrt{\pi}} \int_x^\infty e^{-\zeta^2}\, d\zeta \qquad \text{(ii)}$$

and

$$\frac{2}{\sqrt{\pi}} \int_0^\infty e^{-\zeta^2}\, d\zeta = 1 \qquad \text{(iii)}$$

With $\dfrac{u_s L}{\varepsilon D_L}$ given, $(c_w)_e^*$ can be readily calculated as a function of w. $\mathrm{erfc}(x)$ can be seen to decrease rapidly with the increase of x. Since $\mathrm{erfc}(x)$ decreases rapidly with the increase of x, therefore, as $(u_sL)/(\varepsilon D_L)$ increases, the two complementary error functions of Equation (i) decrease. On the other hand, $\exp\left(\frac{u_sL}{\varepsilon D_L}\right)$ increases with the increase of $(u_sL)/(\varepsilon D_L)$. The second term of Equation (i) is a product of two quantities with contradictory dependence on $(u_sL)/(\varepsilon D_L)$. A more explicit expression indicating the dependence of this term on $\frac{u_sL}{\varepsilon D_L}$ is desired.

To obtain such an expression which is applicable under the conditions in cake washing, first, the cake Peclet number, $(u_sL)/(\varepsilon D_L)$, may be assumed to be moderately large, say of the order 10^2. Next, consider the magnitude of $(1+w)/(2\sqrt{w})$. It is simple to show that this quantity is always greater than unity since

$$\frac{(1+w)^2}{4w} - 1 = \frac{(1-w)^2}{4w} > 0$$

$$\text{Therefore, } \frac{1+w}{2\sqrt{w}} > 1$$

One can assume that the argument of the second complementing error factor will be of the order of $[(u_sL)/(\varepsilon D_L)]^{1/2}$. For cake washing, if one assumes that $(u_sL)/(\varepsilon D_L)$ is of the order of 10^2, the value of $\frac{1+w}{2\sqrt{w}}\sqrt{\frac{u_sL}{\varepsilon D_L}}$ is of the order of 10. We also know that erfc (2.8) is less than 10^{-4}. Accordingly, the second complementary error function may be replaced by the asymptotic expression of erf(x) for large x, which is given as (Carslaw and Jaeger, 1959, p. 483):

$$\text{erfc}(x) \cong \pi^{-1/2}e^{-x^2}\left(\frac{1}{x} - \frac{1}{2\cdot3} + \frac{1.3}{2^2\cdot5}\cdots\right) \tag{iv}$$

The second term of Equation (i) can be written as, with only the leading term of Equation (iv)

$$\left[\exp\left(\frac{u_sL}{\varepsilon D_L}\right)\right]\text{erfc}\left\{\frac{1+w}{2\sqrt{w}}\sqrt{\frac{u_sL}{\varepsilon D_L}}\right\} \cong \left[\exp\left(\frac{u_sL}{\varepsilon D_L}\right)\right]\cdot\left[\exp\left\{-\frac{(1+w)^2}{4w}\left(\frac{u_sL}{\varepsilon D_L}\right)\right\}\right]$$

$$\times\left(\frac{1}{\sqrt{\pi}}\right)\left(\frac{2\sqrt{w}}{1+w}\sqrt{\frac{\varepsilon D_L}{u_sL}}\right) = \exp\left[-\frac{(1-w)^2}{4w}\frac{u_sL}{\varepsilon D_L}\right]\frac{2\sqrt{w}}{\sqrt{\pi}(1+w)}\sqrt{\frac{\varepsilon D_L}{u_sL}} \tag{v}$$

Equation (i) now becomes

$$(c_w)_0^* = 1 - \frac{1}{2}\left[\text{erfc}\left\{\frac{1-w}{2\sqrt{w}}\sqrt{\frac{u_sL}{\varepsilon D_L}}\right\} + \left[\exp\left\{-\frac{(1-w)^2}{4w}\frac{u_sL}{\varepsilon D_L}\right\}\right]\frac{2\sqrt{w}}{\sqrt{\pi}(1+w)}\sqrt{\frac{\varepsilon D_L}{u_sL}}\right] \tag{vi}$$

The magnitude of the second term of Equation (vi) decreases rapidly with the increase of $(u_sL)/(\varepsilon D_L)$ and can be ignored except at values close to $w=1$. Values of $(c_w)_0^*$ obtained from Equation (vi) for the case $u_sL/\varepsilon D_L = 25$ with and without the second term are shown below.

w	$(c_w)_0^*$	
	Equation (vi)	**Equation (vi) without the second term**
0.8	0.744	0.785
0.9	0.594	0.648
1.0	0.444	0.5
1.1	0.316	0.369
1.2	0.214	0.260

3.3.4 Re-slurrying Cake

A more direct method of removing residue solute from filter cakes can be made by re-slurrying filter cake with wash liquid and then de-watering the slurry. The procedure may be repeated if high solute removal is required. While this method is undoubtedly more process-intensive and less economical than cake washing presented in Section 3.3, it is more effective for high packing density cakes (i.e., cakes with low permeability) or in situations with significant cake cracking, which renders cake washing ineffective.

A simple analysis of re-slurry washing based on instantaneous and complete mixing of re-slurry liquid and cake is given below. Consider a general case consisting N-stages of re-slurry shown in Fig. 3.12. Let $(V_f)_{i-1}$ denote the volume of the residue associated with filter cakes liquid fed into the i-stage of the system with a solute concentration of

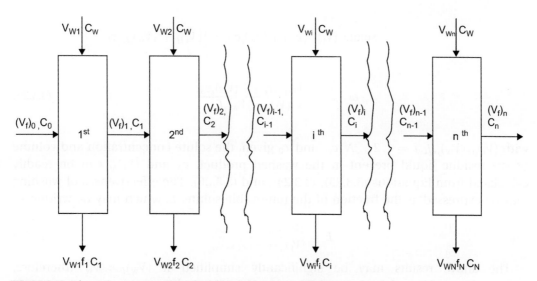

FIGURE 3.12 Schematic representation of caking washing by re-slurrying.

c_{i-1}. $(V_w)_i$ is the volume of the re-slurry liquid used in the i-th step and f_i is the fraction of the re-slurry liquid de-watered. Initially, the residue filtrate volume of the cake is $(V_f)_0$ with a concentration of c_0. Simple mass balance considerations yield the following expression

From overall mass balance of liquid

$$(V_f)_0 + \sum_{i=1}^{N}(V_w)_i = (V_f)_N + \sum_{i=1}^{N}(V_w)_i f_i$$

$$\text{or} \quad (V_f)_N = (V_f)_0 + \sum_{i=1}^{N}(V_w)_i(1 - f_i) \tag{3.3.23}$$

Overall mass balance of solute:

$$(V_f)_0 c_0 + c_w \sum_{i=1}^{N}(V_w)_i = (V_f)_N c_N + \sum_{i=1}^{N}(V_w)_i f_i c_i$$

$$\text{or} \quad C_N = \frac{(V_f)_0 c_0 - \sum_{i=1}^{N}(V_w)_i(f_i c_i c_w)}{(V_f)_0 + \sum_{i=1}^{N}(V_w)_i(1 - f_i)} \tag{3.3.24}$$

From i-th stage mass balance

$$\text{Liquid: } (V_f)_{i-1} + (V_w)_i = (V_f)_i + (V_w)_i f_i$$

$$\text{or} \quad (V_f)_i = (V_f)_{i-1} + (1 - f_i)(V_i)_i \tag{3.3.25}$$

$$\text{Solute: } (V_f)_{i-1}c_{i-1} + (V_w)_i c_w = \left[\left(V_f\right)_i + (V_w)_i f_i\right]c_i$$

$$\text{or} \quad c_i = \frac{(V_f)_{i-1}c_{i-1} + (V_w)_i c_w}{(V_f)_0 + (V_w)_i f_i} \tag{3.3.26}$$

with $(V_f)_0, (V_w)_i, f_i, i = 1, 2, \ldots N, c_w$ and c_0 given, the solute concentration and volume of the residue liquid present in the washed product, c_N and $(V_f)_N$ can be readily calculated from Equations (3.3.23), (3.3.24), and (3.3.26). The effectiveness of washing may be expressed as the fraction of the removable solute, E, which may be written as

$$E = \frac{(V_f)_0 c_0 - (V_f)_N c_N}{(V_f)_0\, c_0 - (V_f)_N c_w} \tag{3.3.27}$$

The above results may be significantly simplified if $(V_w)_i = V_w$ (therefore, $(V_f)_i = (V_f)_0 = V_f$) and $f_i = 1$. From Equation (3.3.26), one has

$$c_{i-1} - c_w - c_i + c_w = (V_w/V_f)(c_i - c_w) \Big/ \left(1 + \frac{V_w}{V_f}\right)$$

and

$$\frac{c_i - c_w}{c_{i-1} - c_w} = \left(1 + \frac{V_w}{V_f}\right)^{-1} \tag{3.3.28}$$

Therefore

$$\frac{c_N - c_w}{c_0 - c_w} = \frac{c_1 - c_N}{c_0 - c_N} \cdot \frac{c_2 - c_N}{c_1 - c_N} \cdots \frac{c_N - c_N}{c_{N-1} - c_N} = \left(1 - \frac{V_w}{V_f}\right)^{-N} \tag{3.3.29}$$

and E becomes

$$E = \frac{c_0 - c_N}{c_0 - c_w} = 1 - \frac{c_N - c_w}{c_0 - c_w} = 1 - \left(1 - \frac{V_w}{V_f}\right)^{-N} \tag{3.3.30}$$

Problems

3.1. Derive filtration diffusion equations with p_s (compressive stress) as the dependent variable. What is the definition of the filtration diffusivity? What is the difference between this definition and that given in Illustrative Example 2.3?

3.2. According to Leclerc and Rebouillat (1985), Equation (3.1.19) with $\nu = 2.8$ gives results (u_c vs. t^*) in agreement with Equation (3.1.15) (within 3%). Confirm this with numerical calculations.

3.3. Estimation of consolidation performance can be made by applying Equations (3.1.15), (3.1.16), or (3.1.19) with the knowledge of the effective filtration diffusivity, D_{av}. On an intuitive basis, D_{av} may be calculated, for the case $f' = -1$, as

(a) $D_{av} = [(D)_{p_s=0} + (D)_{p_s=p_0}]/2$

(b) $D_{av} = \dfrac{1}{p_0} \displaystyle\int_0^{p_0} D \cdot dp_s$

with D defined as in Illustrative Example 2.3 of Chapter 2.

  1. Calculate D_{av} according to the above two expressions for $CaCO_3$–H_2O mixtures. The constitutive relationships of ε_s vs. p_s and k vs. p_s are the same as those of Illustrative Examples 2.2 and 2.4 of Chapter 2.

  2. Using the D_{av} values of (1), estimate consolidation of a column of $CaCO_3$–H_2O mixtures according to Equation (3.1.19) with $\nu = 2.8$ and the following conditions:

  Initial column height   0.5 m
  Mass fraction of solids   0.4

$$p_0 = 5 \times 10^5 \text{ Pa}$$
$$\rho_s = 2665 \text{ kg m}^{-3}$$
$$\mu = 10^{-3} \text{ Pa s}$$
$$\rho = 1000 \text{ kg m}^{-3}$$

3.4. A washed cake is blown with air at 7 bar abs. for 150 s. Calculate the moisture content of the cake. The air passed below the cake is 1 bar. The values of the various relevant variables are:

Cake Length	4.78×10^{-3} m
Cake Porosity	045
Cake Particle Diameter	6.61×10^{-7} m
Surface Tension of Filtrate	0.075 N m^{-1}
Density of Filtration	1000 kg m^{-3}

3.5 An aqueous suspension containing 10% (by mass) solids is filtered at 4 bar pressure through a medium with 0.4 m^2 surface area. The filtrate volume collected at the end of filtration is 0.02 m^3 (or $0.02/0.4 = 0.05$ m^3 m^{-2}). The cake is then given a wash with water at the same pressure (4 bar) for 30 sec. Calculate

1. The wash rate (note: Since wash is carried out at the filtration pressure, wash rate is the same as the filtration rate at the end of filtration).
2. The wash ratio.
3. The fraction of filtrate residue remaining in cake after washing. Make your calculates using Equation (3.3.10) with $(F)_{w=1} = 0.7$.
   The conditions of the cake are:

Average Specific Cake Resistance	$\alpha_{av} = 1.24 \times 10^{12}$ m kg^{-1}
Medium Resistance	$R_m = 4 \times 10^{10}$ m^{-1}
Wet to Dry Cake Mass Ratio	$\overline{m} = 1.41$
Solids Mass Fraction of Suspension	$s = 0.1$
Filtrate Density	1000 kg m^{-3}
Filtrate Viscosity	10^{-3} Pa s
Solids Density	2000 kg m^{-3}

3.6. Develop an empirical expression of α of Equation (3.3.8) in terms of $N_{Pe'_m}$ cake properties, and operating variables as necessary.
[Hint: Compare Equations (3.3.8) with (3.3.18). The asymptotic expression of the complementary error functions given in Illustrative Example 3.6 may be used.]

3.7. For a three-stage cake washing by cake re-slurrying, the total wash liquid available is V_T. Determine its optimum distribution to the three stages in order to achieve maximum washing performance. Assume that $f_i = 1$, $i = 1, 2, 3$.

References

Carleton, A.J., Salway, A.G., 1993. Filtration and Separation 30, 641.

Carslaw, H.S., Jaeger, J.C., 1959. Conduction of Heat in Solids, second ed. Oxford University Press.

Choudhury, R., Dahlstrom, D.A., 1957. AIChE J. 3, 433.

Condie, D.J., Hinkel, M., Vead, C.J., 1996. Filtration and Separation 33, 825.

Delgado, J.M.P.Q., 2007. Trans. I. ChemE. Part A, Chem. Eng. Res. Design 85 (A9), 1245.

Eriksson, G., Rasmuson, A., Theliander, H., 1996. Separations Technology 6, 201.

Fetter, C.W., 1999. Contaminant Hydrology, second ed. Prentice-Hall.

Han, C.D., Bixler, H.J., 1967. AIChE J. 13, 1058.

Han, C.D., 1967. Chem. Eng. Sci. 22, 837.

Holdich, R.G., 1996. In: Rushton, A., Ward, A.S., Holdich, R.G. (Eds.), Chapter 9 Post Treatment Processes in Solid-Liquid Filtration and Separation Technology. VCH Publishers.

Lapidus, L., Amundson, N.R., 1952. J. Phys. Chem. 56, 984.

Leclerc, D., Rebouillat, S., 1985. Dewatering by compression. In: Rushton, A. (Ed.), Mathematical Models and Design Methods in Solid-Liquid Separation, Series E, Applied Science No. 88. Martinus Nijhoff, Publisher.

Moncrieff, A.G., 1965. Filtration and Separation 2, 88.

Nelson, P.A., Dahlstrom, D.A., 1957. Chem. Eng. Prog. 53, 320.

Ramarao, B.V., Tien, C., Satyadev, C.N., 2002. Determination of the constitutive relationship for filter cakes in cake filtration using the analogy between filtration and diffusion. In: Dekec, D., Chhabra, R.P. (Eds.), Transport Processes in Bubbles, Drops and Particles, second ed. Taylor and Francis.

Rhodes, F.H., 1934. Ind. Eng. Chem. 26, 1331.

Sherman, W.R., 1964. AIChE J. 10, 855.

Shirato, M., Murase, T., Kato, H., Fukaya, S., 1967. Kagaka Kogaku 31, 1125.

Shirato, M., Murase, T., Negawa, M., Senda, T., 1970. J. Chem. Eng. Japan 3, 105.

Shirato, M., Murase, T., Tokunaga, A., Yamada, O., 1974. J. Chem. Eng. Japan 7, 229.

Shirato, M., Murase, T., Atsumi, K., Nagami, T., Suzuki, H., 1978. J. Chem. Eng. Japan 11, 334.

Shirato, M., Murase, T., Atsumi, K., Aragaki, T., Noguchi, T., 1979. J. Chem. Eng. Japan 12, 51.

Shirato, M., Murase, T., Inatani, E., Tiller, F.M., Alciatore, A.F., 1987. Filtration in chemical process industry. In: Matteson, M.J., Orr, C. (Eds.), Filtration: Principles and Practices, second ed. Marcel Dekker.

Sivaram, B., Swames, P.K., 1977. J. Japan Society Soil Mech. Found. Eng. 12, 48.

Wakeman, R.J., 1976. Chem. Eng. J. 16, 73.

Wakeman, R.J., 1979a. Int. J. Miner Process 5, 379.

Wakeman, R.J., 1979b. Int. J. Miner Process 5, 395.

Wakeman, R.J., 1979c. Filtration and Separation 16, 655.

Wakeman, R.J., Tarleton, E.S., 1999. Filtration: Equipment Selection Modelling and Process Simulation. Elsevier Advanced Technology.

Wakeman, R.J., Attwood, G.J., 1990. Trans. I Chem E 68, 161.

4

Fabric Filtration of Gas–Solid Mixtures

Notation

A	maximum amplitude appearing in Equation (4.4.1) (m)
A_c	filter area (m^2)
a	acceleration (m s^{-2}) or empirical constant of Equation (4.3.1)
b	empirical constant of Equation (4.3.1)
c_{in}	particle concentration (mass) (kg m^{-3})
D_{max}	deceleration (m s^{-2})
d_g	particle diameter (m)
erf(x)	error function of argument x
F_{adh}	adhesion force (N m^{-2})
F_{disl}	dislodgement force (N m^{-2})
F_{50}	medium adhesion force (N m^{-2})
f	shaking frequency, see Equation (4.4.1) (s^{-1})
f_r	fraction of dust cake removed
$f(F_{adh})$	frequency function of adhesion force (N m^{-2})
K_1	defined by Equation (4.2.6a) (N s m^{-3})
K_2	defined by Equation (4.2.6b) (s^{-1})
k	permeability (m^2)
k_1	constant of Equation (4.3.2)
k_m	fabric permeability (m^2)
L_m	fabric thickness (m)
n_c	number of compartments
p_s	compressive stress (Pa)
p_{s_m}	value of p_s at cake/fabric interface (Pa)
R_m	fabric resistance (m^{-1})
Q	total gas volumetric flow rate (m^3 s^{-1})
Q_i	gas volumetric flow rate through the i-th compartment (m^3 s^{-1})
S_f	a quantity given as $(K_1 + K_2 w)\, A$ or $(K_1 + K_2 w)$
t	time (s)
t_ℓ	cycle time, equal to $t_r + t_c$ (s)
t_c	filter cleaning time (s)
t_f	defined by Equation (4.5.1) (s)
t_r	filtration run time (s)
v	gas (filtration) velocity (m s^{-1})
w	areal cake mass (kg m^{-2})
w_i	areal cake mass of the i-th compartment (kg m^{-2})
w_0	initial value of w (kg m^{-2})
$(w_i)_0$	initial value of w_i (kg m^{-2})
x	coordinate (m)

Principles of Filtration, DOI: 10.1016/B978-0-444-56366-8.00004-9

113

Greek Letters

α	specific cake resistance (m kg^{-1})
α_{av}	average specific cake resistance defined by Equation (4.2.2) (m kg^{-1})
ΔL_m	filter bag thickness (m)
Δp	pressure drop (Pa)
Δp_c	pressure drop across dust cake (Pa)
Δp_m	pressure drop across fabric (Pa)
Δp_M	maximum pressure drop (Pa)
Δp_{over}	over pressure drop (Pa)
ε_s	cake solidosity (–)
ε_{s_m}	maximum value of ε_s (–)
ε_{s_0}	particle volume fraction of gas–solid mixture (–)
μ	filtrate viscosity (Pa s)
ρ_s	particle density (kg m^{-3})
σ	standard deviation (–)
ϕ	defined by Equation (4.5.7) (–)
ϕ_ℓ	value of ϕ at $t = t_\ell$ (–)
ϕ'_ℓ	value of ϕ at $t = t_\ell - t_c$ (–)

Cake filtration of gas–solid mixtures, in most cases, is carried out in fabric filters. By fabric filters, we refer to a class of devices composed of woven or felted materials in the shape of a tube or sleeve (and therefore the name of baghouse filters and baghouse filtration). In fabric filtration, gas–solid mixtures to be treated are passed through filter bags either inward or outward with dust retained at the upstream side of filter bags to form dust cakes. Periodically, dust cakes are removed by applying external forces. For an individual filter bag, its operation is intermittent. However, by employing a large number of bags placed in several compartments and arranging their operation in a certain manner, continuous operation corresponding to a given total gas throughput can be achieved. A schematic diagram depicting the operation of a filter bag is shown in Fig. 4.1.

4.1   Dust Cakes of Fabric Filtration vs. Cakes Formed from Liquid/Solid Suspensions

The description and analysis of cake formation and growth resulting from the flow of fluid-particle suspensions through media permeable to fluid but not solid particles, given in Chapter 2, apply to both liquid and gas. These results, therefore, at least in principle, can be used to describe fabric filtration. However, cake filtration of liquid/solid suspensions does differ from fabric filtration of gas–solid mixtures in several aspects. The physical properties (viscosity and density) of liquid are significantly different from those of gas. For example, under normal temperature and pressure, the density and viscosity of air (which is the most common found in fabric filtration) are only a small fraction of those of water (which is the common suspending liquid in cake filtration): 1.8×10^{-5} Pa s for air

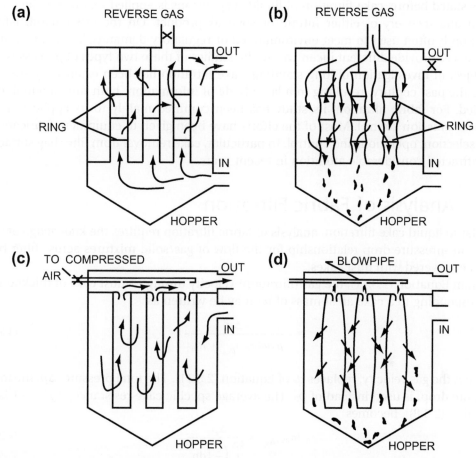

FIGURE 4.1 Operation of fabric baghouse filtration system: (a) filtration, (b) cleaning, (c) inside collecting filter, and (d) outside collecting filter.

viscosity vs. 10^{-3} Pa s for water, and 1.29 kg m^{-3} for air density vs. 10^3 kg m^{-3} for water). This kind of difference is also found in the operating conditions: Gas flow rate in fabric filtration is generally about 10^{-1} m s^{-1} while the upper limit in liquid cake filtration is approximately 10^{-3} m s^{-1}. Similar differences are also found in feed particle concentration and total throughput.

Another major difference between fabric filtration and liquid cake filtration arises from the importance of filter cleaning for fabric filtration. In order to sustain continuous operation of relatively large throughput, individual bags of a fabric filtration system must undergo periodic cleaning. Achievement of efficient dust cake removal with minimum energy requirement is, therefore, an important concern. In contrast, cake dislodgement in liquid cake filtration, generally speaking, is not of importance except in the crossflow case.

As stated before, cake filtration of liquid suspensions is applied mainly for the separation and recovery of either filtrate or solid as product. On the other hand, fabric filtration is often used to meet environmental or regulatory demands. These functional differences provide different incentives to the studies of these two types of processes. As a subject of investigation of long standing, cake filtration has been extensively studied since the past century, resulting in a large body of information, both fundamental and applied. For fabric filtration, its study has been confined mainly to its application in particle emission control. Most of the efforts have been given to equipment fabrication, test, selection, operation, and control. In particular, cake removal from filter bag surfaces has attracted considerable attention in recent years.

4.2   Analysis of Fabric Filtration

Similar to liquid cake filtration, analysis of fabric filtration requires the knowledge of the flow rate–pressure drop relationship for the flow of gas/solid mixtures across filter bag surfaces covered with dust cakes.

From Equation (2.3.11), the gas (filtration) velocity through a dust cake of thickness L (corresponding to a cake areal mass of w) may be written as

$$v = \frac{\Delta p}{\mu\left[w(\alpha_{av})_{p_{sm}} + R_m\right]} \tag{4.2.1}$$

where v, the gas velocity, replaces q_ℓ of Equation (2.3.11). The total pressure, Δp, the total pressure drop, is used instead of p_0. The average specific cake resistance, if $f' = -1$ [see Equation (2.3.8)], becomes

$$(\alpha_{av})_{\Delta p_c} = \frac{\Delta p_c}{\int_0^{\Delta p_c} \left(\frac{1}{\alpha}\right) dp_s} \tag{4.2.2}$$

where Δp_c is the pressure across the dust cake and the areal cake mass, w, is (see Equation (2.3.15b))

$$w = \int_0^L \rho_s \varepsilon_s \cdot dx \tag{4.2.3}$$

As dust cake is formed from particle retention associated with the flow of the gas–solid mixture, w may also be expressed as

$$w = w_0 + \int_0^t v \cdot \varepsilon_{s_0} \rho_s dt \tag{4.2.4}$$

where w_0 is the cake present at bag surface initially.[1]

---

[1] w_0 vanishes for new filter bags. For cleaned filter bags, the value of w_0 depends upon the efficacy of cleaning.

Equation (4.2.1) may be rearranged to give

$$\Delta p = \Delta p_c + \Delta p_m = \mu(\alpha_{av})w\,v + \mu\,R_m v$$

or

$$\Delta p = K_1 v + K_2 v\,w \tag{4.2.5}$$

where

$$K_1 = \mu\,R_m \tag{4.2.6a}$$

$$K_2 = (\mu)(\alpha_{av}) \tag{4.2.6b}$$

Equation (4.2.5) is the standard expression used in describing fabric filtration performance and design calculations (Leith and Allen, 1986; Rothwell, 1986; Seville et al, 1989). As this equation is based on the conventional filtration theory, all the assumptions used in developing the conventional cake filtration theory, including that the state of dust cake (namely, cake porosity and permeability) depends upon the compressive stress, hold here as well. However, since filter bags used in fabric filtration systems are cleaned frequently, the validity of this assumption may be in doubt. Presence of the "age effect" on cake properties may be important, thus invalidating the definition of the average specific cake resistance given by Equation (4.2.2). Nevertheless, Equation (4.2.5) has been commonly considered to be at least approximately correct and is widely used in fabric filtration studies.

However, it is important to keep in mind that the conditions leading to dust cake formation and growth may not conform to those used in developing the conventional cake filtration theory. For a number of reasons, dust cake thickness over a filter bag may not be uniform and dust cake may have cracks, which makes the uniform flow assumption invalid. More important, for bags made of felt materials, cake formation does not begin instantaneously. Applying Equation (4.2.5) is acceptable only if K_1 (or R_m) is considered as a function of time. However, the knowledge of the variation of K_1 with time is generally not available. The values of K_1 and K_2 obtained from fitting experimental data with Equation (4.2.5), therefore, may not display the physical significance as given by the conventional theory. Rather, they should be regarded only as fitting parameters.

The parameter K_1 which has the unit of Pa s m^{-1} is directly proportional to the bag resistance to gas flow, R_m, which may be expressed as

$$R_m = \frac{\Delta L_m}{k_m} \tag{4.2.7}$$

where ΔL_m is the bag thickness and k_m its permeability.

The parameter K_2 characterizes the gas flow resistance through dust cakes formed over filter bag surfaces. Combining Equations (4.2.6a) and (4.2.2), one has

$$K_2 = \frac{\mu(\Delta p_c)}{\displaystyle\int_0^{\Delta p_c}\left(\frac{1}{\alpha}\right)dp_s} = \frac{\mu(\Delta p_c)}{\displaystyle\int_0^{\Delta p_c}(k)(\varepsilon_s)(\rho_s)dp_s} \tag{4.2.8}$$

In principle, if the constitutive relationships (k vs. p_s, ε_s vs. p_s) are known, K_2 can be readily evaluated according to the above expression if Δp_c is given together with cake areal mass data. However, such information, in most cases, is not available. In practice, K_1 and K_2 are determined from results of constant-rate test data. The procedures used and their determinations are shown in the following illustrative example.

■ ■ ■ ▬▬▬▬▬▬▬▬▬▬▬▬▬▬▬▬▬▬▬▬▬▬▬▬▬▬

Illustrative Example 4.1

Based on the following test data of a clean filter bag, estimate the pressure drop after 70 min of operation with $v = 1.5 \times 10^{-2}$ m s^{-1}. The dust concentration of the gas stream to be treated is 5×10^{-3} kg m^{-3}.

Time, (min)	Δp, Pa
0	150
5	380
10	505
20	610
30	690
60	990

Solution

Equation (4.2.5) can be used to estimate the pressure drop if the values of K_1 and K_2 are known.

To determine K_1 and K_2, there are two possible procedures:

(1) Initially (i.e. $t = 0$), $w = 0$ (since for fresh bag, $w_0 = 0$, see Equation (4.2.4)):

$$K_1 = (\Delta p/v)|_{t=0} \qquad (i)$$

Also, the pressure drop across the cake, Δp_c, is

$$\Delta p_c = \Delta p - \Delta p_m = \Delta p - (\Delta p)_{t=0} \qquad (ii)$$

and

$$K_2 = \frac{(\Delta p)_c}{(v)(w)} \qquad (iii)$$

In other words, from the given data, K_2 as a function of Δp_c can be determined.

(2) If one assumes that K_2 is constant (or the dust cake is incompressive) as well as K_1, for constant-rate filtration, Δp can be seen to be a linear function of w. From the data given, K_1 and K_2 can be determined accordingly.

The results obtained from using these two procedures are

(A) Procedure (1)

The areal cake mass, w, is (from Equation (4.2.4) with $w_0 = 0$)

$$w = \int_0^t (v)(\rho_s \varepsilon_{s_0}) \mathrm{d}t = (v)(\rho_s \varepsilon_{s_0})t \qquad (iv)$$

and

$$v = 1.5 \times 10^{-2} \text{ m s}^{-1}$$

$$\rho_s \varepsilon_{s_0} = 5 \times 10^{-3} \text{ kg m}^{-3}$$

From Equations (i)–(iv), the following results are obtained:

Time(s)	Δp(Pa)	$\Delta p_m = K_1 v$(Pa)	Δp_c(Pa)	w(kg m^{-2})	K_2(S^{-1})
0	150	150	0	0	-
300	380	150	230	2.25×10^{-2}	6.815×10^5
600	505	150	355	4.50×10^{-2}	5.250×10^5
1200	610	150	460	9.00×10^{-2}	3.407×10^5
1800	690	150	540	0.135	2.667×10^5
3600	990	150	840	0.270	2.333×10^5

The above results show that K_2 decreases with the increase of w and with the increase of Δp_c. As K_2/w is proportional to α_{av} [see Equation (4.2.6b)], these results suggest that the calculated average specific resistance decreases as the compressive stress increases, which, of course, is unrealistic.

In order to explain these anomalous results, it should be noted that calculations were made with the assumptions of R_m (or K_2) being constant and cake formation coinciding with the start of filtration operation at $t = 0$. While these assumptions may be reasonable for filter bags with smooth surfaces, they may not be valid for felt bags. For such bags, there may be a significant incubation period for the formation of filter cake to cover the entire bag surface. During this initial (or incubation) period, R_m and Δp_m, the pressure drop across the fabric medium, may increase. For a hypothetical case, assume that Δp_m had increased to 350 Pa before cake formation was complete. For such a situation, the K_2 values at $t = 300$ s and $t = 3600$ s would be 8.889×10^{-4} (corresponding to a $\Delta p_c = 30$ Pa) and 1.718×10^5 (corresponding to a $\Delta p_c = 640$ Pa) instead of 6.815×10^5 and 2333×10^5 s^{-1}, conforming to the expected compressive behavior.

The tabulated results also suggest that in spite of the unrealistic physical behavior, the calculated K_2 seems to approach to a constant value as time increases. If one takes the last K_2 (one corresponding to $w = 0.27$ kg m^{-2}), 2.333×10^5, to be the correct value, one has

$$\Delta p = 150 + 2.333 \times 10^5 \ wv$$

At $t = 4200$ s (or 70 min),

$$w = (4200)(1.5 \times 10^{-2})(5 \times 10^{-3}) = 0.315 \text{ kg m}^{-2}$$

The pressure drop is found to be

$$\Delta p = (150) + (2.333 \times 10^5)(0.315)(1.5 \times 10^{-2}) = 150 + 1102 = 1252 \text{ Pa}$$

(B) Procedure (2)

If one assumes that K_1 and K_2 are constant for constant-rate filtration, the relationship between Δp and w is linear [see Equation (4.2.5)]. A plot of Δp vs. w is shown in Fig. i. It is clear that the expected linear behavior is not observed. However, for data of $w > 4.5 \times 10^{-2}$ kg m^{-2} (or $t > 600$ s), the relationship between Δp and w is nearly linear and can be expressed as

$$\Delta p = 408 + 2.156 \times 10^3 w \quad \text{for } v = 1.5 \times 10^{-2} \text{ m s}^{-1} \qquad \text{(v)}$$
$$w > 4.5 \times 10^{-2} \text{ kg m}^{-2}$$

If the above expression is used to estimate Δp at $t = 4200$ s, (or $w = 0.315$ kg m^{-2}), one has

$$\Delta p = 408 + 2.156 \times 10^3 \times 0.315 = 408 + 680 = 1088 \text{ Pa}$$

which is approximately 15% less than the value given in (A).

(C) Modification of Equation (4.2.5). The conventional cake filtration theory used to obtain Equation (4.2.5) is based on the assumption that cake formation coincides with the onset of filtration operation. Both the seemingly unrealistic behavior of the calculated results of (A) and the Δp vs. w behavior shown in fig. i suggest that this assumption may not be correct. As an alternative, one may assume that cake formation does not begin at $t = 0$, but $t = t_0$, or there is a time lag. For the period $t < t_0$, dust particle penetration into fabric medium results is an increase of medium resistance; the corresponding ΔP_m is simply

$$\Delta p = \Delta p|_{t=t_0}$$

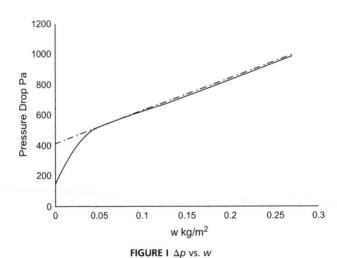

FIGURE I Δp vs. w

Similarly, with w given by Equation (4.2.4), the effective cake areal mass, $\overline{w}$, is

$$\overline{w} = w - (w)_{t_0}$$

$\Delta p|_{t_0}$ and $(w)_{t_0}$ are the values of ΔP and w at $t = t_0$.
Equation (4.2.5) can be modified to give

$$\Delta p = \Delta p|_{t_0} + K_2\, v\left[w - (w)_{t_0}\right] \qquad\qquad (vi)$$

For the present problem, one may assume $t_0 = 600$ s, $\Delta p|_{t_0} = 505$ Pa, and $(w)_{t_0} = 4.5 \times 10^{-2}$ kg m^{-2}; Δp at $t = 4200$ s according to Equation (vi) is

$$\Delta p = 505 + 2.156 \times 10^3\ (0.315 - 0.045) = 1087\ \text{pa}$$

It is impossible, at this stage, to say unequivocally which one of the three estimated Δp is more accurate. However, it is clear that using Equation (4.2.5) in interpreting experimental data and predicting filtration performance invariably require the use of certain extraneous assumptions. While Equation (4.2.5) undoubtedly has its physical significance, one should not overlook its inherent limitations.

■ ■ ■

4.3   Dust Cake Structure and Properties

Fundamental studies on the structure and properties of dust cakes formed in fabric filtration, in contrast to studies of cake formation in liquid filtration, have not received sufficient attention from researchers. However, a few recent studies have produced information about the compressive behavior of dusts cakes using an experimental setup shown in Fig. 4.2. Schmidt (1995) conducted measurements of dust cake formation by passing air–limestone particle mixtures (with medium particle diameter of 3.5 μm) through a multiply calendered polyester needle-felt medium under constant-rate condition with specified limiting (maximum) pressure drop. The pressure drop-time data were recorded. At the end of each measurement, the porosity profile of the cake was determined using a technique developed by Schmidt and Loffler (1990, 1991). The porosity profile results are shown in Fig. 4.3 and the pressure drop history data are given in Fig. 4.4.

The porosity profile results shown in Fig. 4.3 are those of three measurements with maximum pressure drops of 500, 1,000, and 2,000 Pa. In Fig. 4.3a, the results of porosity vs. distance from cake/medium interface are shown separately for each measurement, and a comparison between experimental data with simulation is given in Fig. 4.3a (Schmidt, 1997). If one assumes that $dp + dp_s = 0$, and the medium resistance is negligible, the compressive stress at the cake–medium interface equals Δp, the overall pressure drop, and these results give the cake solidosity values corresponding to $p_s = 500$, 1,000, and 2,000 Pa to be $1 - 0.75 = 0.25$, $1 - 0.713 = 0.287$, and $1 - 0.672 = 0.328$, respectively, demonstrating rather convincingly that the limestone cake is compressive.

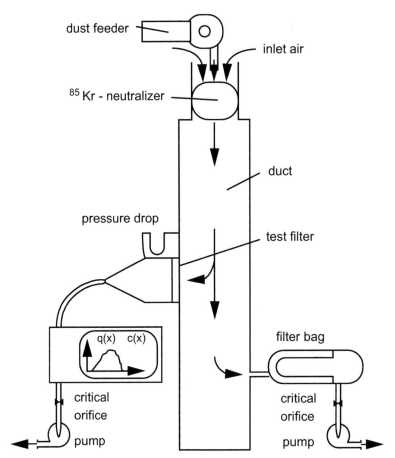

FIGURE 4.2 Experimental apparatus used in studying dust cake formation. *[Reprinted from E. Schmidt, "Experimental investigation into the compression of dust cake deposit in filter media", Filtration and Separation, Sept. 1995, 789–793, with permission of Elsevier].*

In fact, Schmidt showed that the constitutive relationship of ε_s vs. p_s for limestone cake may be given as (Schmidt, 1997)

$$p_s = \frac{a\,\varepsilon_s^b}{(\varepsilon_s)_m - \varepsilon_s} \qquad (4.3.1)$$

with p_s given in Pa, $a = 115{,}000$, $b = 5$, and $(\varepsilon_s)_{max}$ is the maximum packing density of cube packing or 0.524.

As discussed previously in Chapter 2, the cake solidosity value at the cake surface gives the value of ε_s at the zero-stress state. Therefore, the value of ε_s at cake surface should be the same irrespective of the pressure drop applied. To demonstrate that this is indeed the case, the same results of Fig. 4.3.a are now plotted as ε vs. distance from cake surface

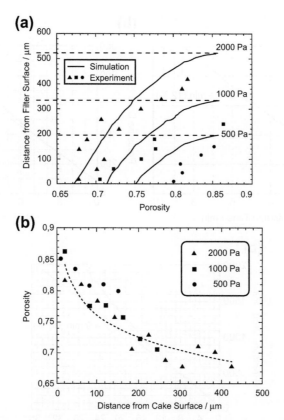

FIGURE 4.3 (a). Porosity profiles obtained by Schmidt for three different operating conditions. *[Reprinted from E. Schmidt, "Experimental investigation into the compression of dust cake deposition in filter media", Filtration and Separation, Sept. 1995, 789–793, with permission of Elsevier].* (b). Porosity profile and the estimation of cake porosity at the zero-stress state. *[Reprinted from E. Schmidt, "Experimental investigation into the compression of dust cake deposit in filter media", Filtration and Separation, Sept. 1995, 789–793, with permission of Elsevier].*

(see Fig. 4.3b). While there are considerable scatterings among the data points shown, they tend to converge to a fixed value at cake surface seems indisputable.[2]

The effect of cake compression can also be seen from the pressure drop history data. Fig. 4.4a–c show the pressure drop vs. time of filtration corresponding to the measurements mentioned previously. In all cases, the Δp vs. time relationship is linear initially and then deviates from linearity with the increase in $d(\Delta p)/dt$ as t increases. Referring to Equation (4.2.5) with w directly proportional to time, this implies an increase of α_{av} with time or increase of ΔP suggesting that the cake underwent compression as filtration proceeds.

[2]Equation (4.3.1), however, does not define a value of ε_s for $p_s = 0$. If one assumes $\varepsilon_s = 0.1$ (or $\varepsilon = 0.9$) based on Fig. 4.3., the corresponding p_s is $1.15/0.376 = 3.06$ Pa, a value sufficiently close to zero.

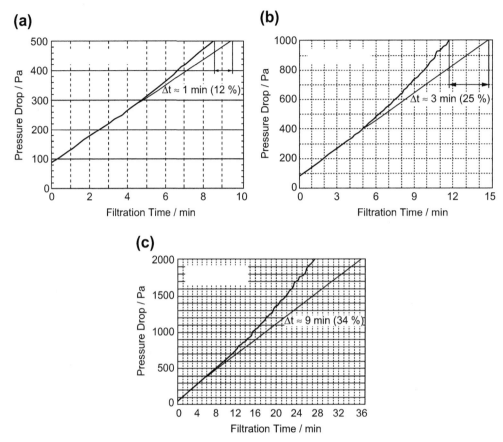

FIGURE 4.4 Pressure drop history of fabric filtration: Deviation from linear behavior at different maximum pressure drop. *[Reprinted from E. Schmidt, "Experimental investigation into the compression of dust cake deposit in filter media", Filtration and Separation, Sept. 1995, 789–793, with permission of Elsevier].*

■ ■ ■ ▬▬▬▬▬▬▬▬▬▬▬▬▬▬▬▬▬▬▬▬▬▬▬▬

Illustrative Example 4.2

Estimate the compression effect on the specific cake resistance of limestone cakes using Schmidt's results discussed in 4.3.

Solution

The results of ε_s vs. p_s of limestone cakes are (see Fig. 4.3(a) for values of ε at cake surface)

P_s(Pa)	$\varepsilon_s(-) = 1 - \varepsilon$
500	0.25
1,000	0.287
2,000	0.328

According to the Kozeny–Carmen equation of the pressure drop–flow rate relationship through porous media, [3] the permeability of a packed medium with solidosity ε_s is given as

$$k = \frac{1}{k_1} \frac{d_g^2 (1 - \varepsilon_s)^3}{\varepsilon_s^2} \tag{4.3.2}$$

where k_1 is an empirical constant and d_g is the packing grain diameter.

Since the specific cake resistance, $\alpha = (k \rho_s \varepsilon_s)^{-1}$, is proportional to $\varepsilon_s / (1 - \varepsilon_s)^3$.

The ratios of α at $\Delta p = 1{,}000$ and $2{,}000$ Pa to that at 500 Pa are

$$\frac{(\alpha)_{\Delta p = 1{,}000 \text{ Pa}}}{(\alpha)_{\Delta p = 500 \text{ Pa}}} = \frac{0.287}{0.25} \left(\frac{1 - 0.25}{1 - 0.287} \right)^3 = 1.336$$

$$\frac{(\alpha)_{\Delta p = 2{,}000 \text{ Pa}}}{(\alpha)_{\Delta p = 500 \text{ Pa}}} = \frac{0.328}{0.25} \left(\frac{1 - 0.25}{1 - 0.328} \right)^3 = 1.824$$

or an increases of α of approximately 34% and 82%, respectively.

---

[3]For a more detailed discussion, see Chapter 5.

4.4   Filter Bag Cleaning

Cleaning of filter bags in order to remove dust cake formed on their surfaces is essential to the operation of fabric filtration. Generally speaking, three different cleaning methods – shaking, backflow of clean gas, and pulse air jet – are currently in use. While a large body of information about their uses has been reported in the literature, only a few studies have yielded quantitative results. The following section gives a brief presentation of the results of such studies.

4.4.1   Cleaning by Shaking

The principle of removing dust cakes formed over fabric surface by shaking is simple and straightforward. Acceleration experienced by dust cakes subject to shaking produces a detachment force which may overcome the adhesion force between cake and medium surface (a more or less complete removal) or the internal cake cohesive force (a partial removal). It is therefore plausible to relate the extent of removal (in terms of residue cake mass) with the acceleration caused by shaking and the length of shaking. Such a correlation was established by Dennis and Wilder (1975) and is shown in Fig. 4.5 which gives the dust cake residue as a function of number of shake and the mechanical acceleration, a. The acceleration, a, caused by shaking can be found as

$$a = 4\pi f^2 A \tag{4.4.1}$$

where f is the shaking frequency and A the maximum amplitude (displacement) of the bag from its rest position.

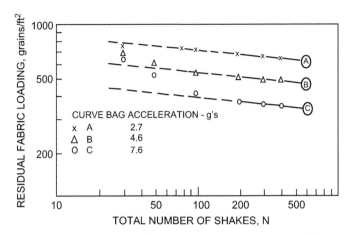

FIGURE 4.5 Residue fabric loading as a function of shaking acceleration and number of shakes *Dennis and Wilder, 1975*

4.4.2 Cleaning by Reverse Flow

Sievert and Loeffler (1987) conducted extensive measurements on dust cake release from non-woven fabric by reverse air flow. Their results (see Fig. 4.6a–c) show that the extent of dust removal depends on the reverse gas flow rate applied, the cake dust areal mass (w_E, in Sievert and Loeffler's notation), and the type of filter surface condition. Alternatively, the extent of cake release can be related to the pressure difference applied for reverse flow. Fig. 4.7 gives such results for polyester needle-felt with singed surface, which is a re-plot of the data shown in Fig. 4.6a. The most striking feature of the results of Fig. 4.7 is the achievement of nearly complete removal through the application of modest pressure drop for relatively thick dust cakes (or higher values of w, the cake areal mass). The results also indicate clearly that the efficiency of reverse flow varies greatly with the surface conditions of fabric media.

4.4.3 Cleaning by Pulse-Jet

In more recent years, the use of pulse-jet baghouse systems for emission control has become popular. For this type of system, gas (air) stream to be treated is passed through bag surface from outside to inside with cake formed on the outside surface. Cleaning is effected by short burst of high-pressure air into bags which sets up shock waves, flexing bags, and breaking up outside cakes. A schematic diagram of pulse-jet cleaning is given in Fig. 4.8.

De Ravin et al (1988) proposed a simple mechanism for pulse-jet cleaning and outlined a procedure for estimating the extent of cake removal in terms of operating variables. The basic premise used is rather simple, namely, in order to affect cake release; the dislodgement force applied, F_{disl}, must be greater than the cake/medium

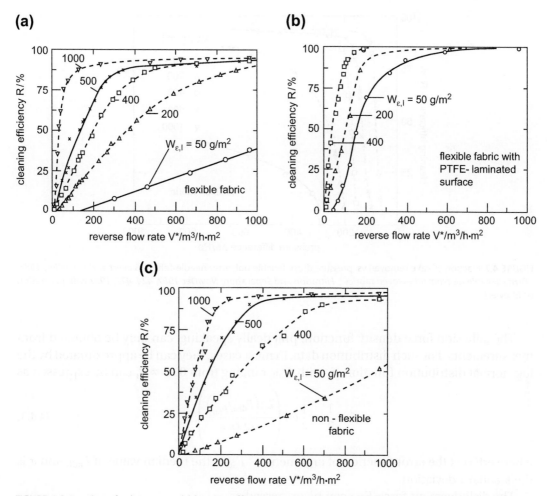

FIGURE 4.6 Fraction of cake removed (cleaning efficiency) vs. reverse flow rate for three different types of fabrics: (a) flexible polyester needle-felt; (b) polyester needle-felt with ptpe-laminated surface; (c) nonflexible polyester needle-felt. *[Reprinted from J. Sievert and F. Loeffler, "Dust cake release from non-woven fabrics", Filtration and Separation, Nov/Dec 1987, 424–427, with permission from Elsevier].*

adhesion force, F_{adh}. Because of cake/medium surface heterogeneity, the adhesion force can be expected to cover a range of values. If $f(F_{adh})$ is the probability density function of the adhesion force, the fraction of cake removal with the application of a dislodgement force, F_{disl}, f_r is

$$f_r = \frac{\int_{(F_{adh})_m}^{F_{disl}} f(F_{adh}) dF_{adh}}{\int_{(F_{adh})_m}^{(F_{adh})_M} f(F_{adh}) dF_{adh}} \quad \text{for } F_{disl} < (F_{adh})_M \tag{4.4.2}$$

where $(F_{adh})_m$ and $(F_{adh})_M$ are the minimum and maximum values of F_{adh}.

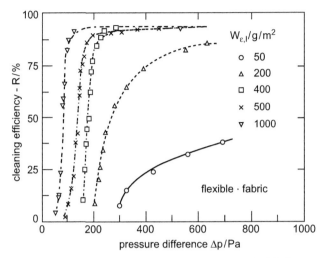

FIGURE 4.7 Fraction of cake removed vs. pressure drop, flexible polyester needle-felt. *[J. Sievert and F. Loffler, 1987, "Dust cake release from non-woven fabrics", Filtration and Separation, Nov/Dec 1987, 424–427, 1987 with permission of Elsevier].*

The adhesion force density function, practically speaking, can only be obtained from measurements. For such distribution data, in most cases, they can be approximated by the log-normal distribution function, f_r, with F_{adh} ranging from 0 to ∞. f_r can be expressed as

$$f_r = \frac{1}{2} + \frac{1}{2}\,\mathrm{erf}\left(\frac{\ell n\left(F_{disl}/F_{50}\right)}{\sqrt{2}\,\ell n\,\sigma}\right) \tag{4.4.3}$$

where erf(x) is the error function of argument x.[4] F_{50} is the median value of F_{adh} and σ is the standard deviation.

The dislodgement force F_{disl} can be expressed as

$$F_{disl} = (w)D_{max} \tag{4.4.4}$$

where w is the cake areal mass as defined before. D_{max} is the maximum fabric deceleration.

Based on the measurements obtained by using a particular device for generating air jet pulse, De Ravin et al. (1988) found the relationship between D_{max} and the air pulse over pressure (namely, the difference between the pressure drop generated by air jet and the pressure drop across filter bag prior to bag cleaning) as shown in Fig. 4.9. This relationship can be approximated as (Ju et al., 1990)

$$D_{max} = 26.46 + 29.33(\Delta P_{over}) + 275.7(\Delta P_{over})^2 \tag{4.4.5}$$

where D_{max} is given in m s^{-2} and ΔP_{over}, the pulse over pressure, in kPa.

---

[4]erf(x) = 1 − erfc(x). Equation (3.1.18b) gives the expression of erfc(x).

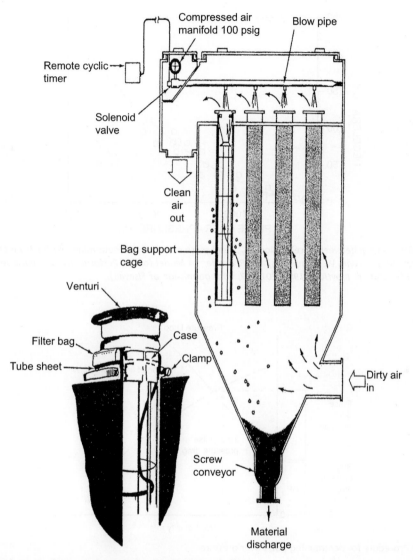

FIGURE 4.8 Schematic diagram of a typical pulse-jet baghouse filter.

The pulse over pressure can be estimated from the jet pump characteristic curve (which relates the air pressure developed vs. air flow rate) and the pressure drop–flow rate relationship across a filter bag prior to its cleaning. The procedure is illustrated in Fig. 4.10. As shown in Fig. 4.10, curve A is the jet pump characteristic curve. From point "a" (with $Q = 0$, $\Delta p = (\Delta p)_b$, where $(\Delta p)_b$ is the pressure drop across filter bag before cleaning, a line can be drawn from point "a" with a slope of S_f given as

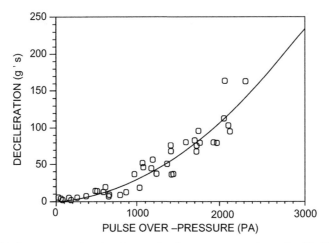

FIGURE 4.9 Relationship between deceleration, D_{max} and pulse over pressure determined by De Ravin et al (1988). *[Reported from M. De Ravin, W. Humphries, and R. Postle, "A Model for the Performance of a Pulse-Jet Filter", Filtration and Separation, May/June 1986, 201–207, with permission of Elsevier].*

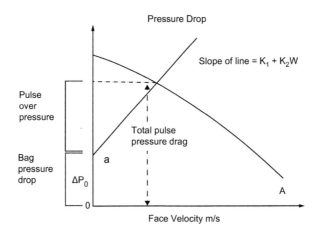

FIGURE 4.10 Procedure for determining pulse over pressure.

$$S_f = (K_1 + K_2\ w)\ A = \frac{\Delta P}{vA}\ 5$$

where A is bag surface area and $(K_1 + K_2 w)$ equals to the total resistance to air flow across a filter bag [see Equation (4.2.5)]. The pressure drop coordinate value of the intersecting point of the line from point "a" and curve A is the total pulse pressure drop and the difference between the total pressure drop and $(\Delta P)_b$ gives the value of the over pressure.

An example illustrating the prediction of cake removed by pulse-jet is given below.

---

[5]If the face velocity instead of Q is used in Fig. 4.9, S_f, then becomes $(K_1 + K_2 w)$.

■ ■ ■ ▬▬▬▬▬▬▬▬▬▬▬▬▬▬▬▬▬▬▬▬▬▬▬▬▬▬▬▬▬▬▬▬▬▬▬▬▬

Illustrative Example 4.3

(A) Estimate the fraction of cake removal under the following conditions:

(1) The cake areal mass prior to cleaning: 0.1 kg m^{-2}

(2) The cake/bag adhesion force obeys the log-normal distribution function with $F_{(50)} = 250$ N m^{-2} and $\log \sigma = 0.4$.

(3) The pressure drop across filter bag prior to cleaning: 4 KPa

(4) $K_1 = 20,000$ Ns m^{-3}, $K_2 = 30,000$ s^{-1}

(5) For estimating the pulse over pressures, the jet jumping characteristic data obtained by De Ravin et al. [see Fig. 7 for the case of reservoir pressure being 690 kPa of De Ravin et al (1988)] may be applied.

(B) Discuss the effect of w on cake removal.

Solution

(A) First, the data given by De Ravin et al. are those of Δp vs. Q with Δp in mm Hg and Q in m^3 s^{-1}. It may be more convenient to make calculations using the SI units. Also instead of Q, the flow rate, one may express the flow rate in terms of the equivalent face velocity, $v = Q/A_c$, where A_c is the surface area. The dimensions of the PVC tube used in the work of De Ravin et al is 0.1 m in diameter and 2 m length, or $A = (\pi)(0.1)(2) = 0.628$ m^2. By reading off the figure, the following results were obtained.

$Q(\text{m}^3\ \text{s}^{-1})$	Face Velocity (Q/A_c, m s^{-1})	Pressure Drop	
		mm Hg	KPa
0	0	375	50
0.019	0.03	360	48
0.059	0.094	275	36.7
0.098	0.156	210	28
0.121	0.193	165	22
0.135	0.215	120	16
0.137	0.218	100	13

From the above results of pressure drop vs. face velocity, the jet jump characteristic curve can be established as shown in Fig. i. To estimate $(\Delta p)_{over}$, the procedure described in 4.4.3 (also Fig. 4.10) may be followed. Referring to Fig. i, the coordinates of point "a" are (0, 4,000 Pa). The slope of the line emanating from point "a" is $K_1 + K_2 w = 20,000 + 30,000 (0.1) = 23,000$. The ordinate of the intersection point between the line emanating from point "a" and the jet jump characteristic curve is 9,000 Pa. $(\Delta p)_{over}$ is found to be

$$(\Delta p)_{over} = 9,000 - 4,000 = 5,000 \text{ Pa or 5 kPa}$$

From Equation (4.4.5), D_{max} is found to be

$$D_{max} = 26.46 + 29.93.(5) + 245.7.(5)^2 = 6317 \text{ m s}^{-2}$$

and the dislodgement force, F_{disl}, is

$$F_{disl} = (0.1)(6317) = 631.7 \text{ or } 632 \text{ N m}^{-2}$$

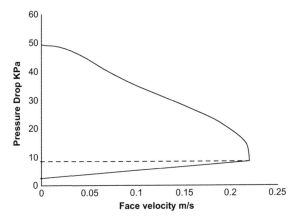

FIGURE I Determination of $(\Delta p)_{over}$

The fraction of cake removal, from Equation (4.4.3), is

$$f_r = \frac{1}{2} + \frac{1}{2} \text{erf} \left\{ \frac{\log \frac{632}{250}}{\sqrt{2}(0.4)} \right\} = \frac{1}{2} + \frac{1}{2}(0.712) = 0.843$$

and fraction of cake residue is $1 - 0.843 = 0.157$.

(B) To examine the effect of w on f_r, assume that $w = 0.07$ instead of 0.1, the slope of the line emanating from point a now becomes

$$20,000 + (0.07)(30,000) = 22,100$$

which is less than 23,000. Therefore, the ordinate of the intersecting point should be less than before. However, since the difference is rather insignificant (less than 5%) and may therefore be ignored, $(\Delta P)_{over}$ may be assumed to be unchanged. With the same D_{max}, the dislodgement force, F_{disl}, now becomes

$$F_{disl} = (6317)(0.07) = 492 \text{ N m}^{-2}$$

and the friction of cake removal, f_r, is

$$f_r = \frac{1}{2} + \frac{1}{2} \text{erf} \left\{ \frac{\log \frac{492}{250}}{\sqrt{2}(0.4)} \right\} = 0.7333$$

and fraction of cake residue is $1 - 0.7333 = 0.2607$ which is significantly different from the value of 0.157 with $w = 0.1$.

■ ■ ■

4.5   Fabric Filtration Design Calculations

Crawford (1976) developed a method for calculating multi-compartment fabric filtration in which filter cleaning is implemented by withdrawing compartments from service sequentially for fixed duration before returning them to service. The assumptions used are:

1. The operation is cyclic and steady-state.
2. The total amount of gas to be treated (total throughput) is fixed.
3. The pressure drop across filter bags cannot exceed a maximum value, Δp_m.
4. Equation (4.2.5) can be used to calculate the instantaneous filtration rate. K_1 and K_2 are known and remain constant.
5. Filter cleaning is complete.

By steady-state, it is assumed that the state of the system remains the same from cycle to cycle. A steady-state behavior of a system of four compartments is shown schematically in Fig. 4.11. At the beginning of a cycle, i.e. $t = 0$, compartment no. 1 is free of deposited cake while the cake areal mass of compartments nos. 2, 3, and 4 are $(w_2)_0$, $(w_3)_0$, and $(w_4)_0$,

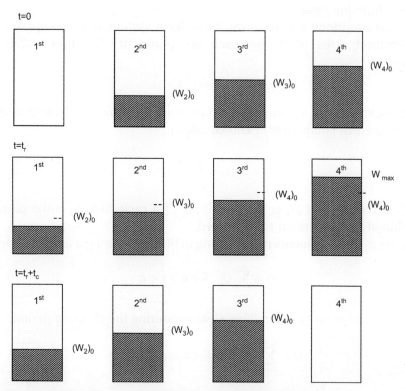

FIGURE 4.11 Steady-state fabric filtration cake built-up in a 4-compartment system.

respectively, as shown in Fig. 4.11(a). During a filtration run, $0 < t < t_r$, where t_r is the filtration run time, all four compartments participate in filtration until $t = t_r$ (see Fig. 4.11(b)), when the cake areal mass of compartment no. 4 reaches the maximum value, w_{max} (corresponding to a pressure drop of Δp_m). Compartment no. 4 is then withdrawn from service for a period of t_c. Therefore during $t_r < t < t_r + t_c$, only three compartments (namely nos. 1, 2, 3) participate in filtration. At $t = t_r + t_c = t_\ell$, cake areal mass of the three active compartments reaches the values of $(w_2)_0$, $(w_3)_0$ and $(w_4)_0$ for compartments nos. 1, 2 and 3, respectively. With the return of compartment no. 4 to service, the condition of the system at $t = t_r + t_c = t_\ell$ is identical to that at $t = 0$, which can be seen by comparing Fig. 4.11(a) with Fig. 4.11(c). t_ℓ is the cycle time of the system. Specifically, the pressure drop varies with time during one cycle, but repeats itself from cycle to cycle, namely, Δp vs. t for $(i-1)t_\ell < t < it_\ell$ is the same as Δp vs. t for $it_\ell < t < (i+1)t_\ell$, where i is a positive integer.

It should be noted that for a particular compartment in a n_c-compartment system, the time interval between two successive cleaning, t_f is

$$t_f = n_c(t_r + t_c) - t_c \tag{4.5.1}$$

In other words, a cleaned compartment which enters into series at $t = 0$ participates in filtration for $0 < t < t_f$. The built-up of dust cake is also a cyclic process such that for a given compartment, w vs. t for $0 < t < t_f$ is the same as w vs. t for $t_f + t_c < t < (t_f + t_c) + t_f$. t_f is termed the filtration time.

With the assumptions stated above, consider a system of n_c compartments with surface (filtration) area of A_c for each compartment. Equation (4.2.5) may be rewritten as

$$\Delta p = \frac{Q_i}{A_c}(K_1 + K_2 w_i) \tag{4.5.2}$$

where Q_i is the volumetric air flow rate through the i-th compartment and w_i is the cake areal mass of the i-th compartment and equals to

$$w_i = (w_i)_0 + \int_0^t \left(\frac{Q_i}{A_c}\right) \cdot c_{in} dt \tag{4.5.3}$$

where $(w_i)_0$ is the value of w_i at the beginning of a cycle and c_{in} is the particle mass concentration of the gas stream to be filtered.

As there are n_c compartments participating in filtration during a filtration run, one has

$$Q = \sum_{i=1}^{n_c} Q_i \quad \text{for } 0 < t < t_r \tag{4.5.4a}$$

and, with one compartment undergoing cleaning during the cleaning period,

$$Q = \sum_{i=1}^{n_c-1} Q_i \quad \text{for } t_r < t < t_r + t_c = t_\ell \tag{4.5.4b}$$

where Q is the total gas throughput.

Differentiating Equation (4.5.3), one has

$$\frac{dw_i}{dt} = c_{in}\frac{Q_i}{A_c}$$

Substituting Equation (4.5.2) into the above expression yields

$$\frac{dw_i}{dt} = \frac{c_{in}(\Delta P)}{K_1 + K_2 w_i} \tag{4.5.5}$$

Integrating Equation (4.5.5) yields

$$K_1\left[(w_i) - (w_i)_0\right] + (K_2/2)\left[(w_i)^2 - (w_i)_0^2\right] = c_{in}\int_0^t (\Delta p)\cdot dt \tag{4.5.6}$$

The physically realistic root of the above equation, w_i, is

$$w_i = (K_1/K_2)\left\{-1 + \sqrt{1 + 2(K_2/K_1)(w_i)_0 + (K_2/K_1)^2(w_i)_0^2 + (2K_2c_{in}/K_1^2)\int_0^t (\Delta p)dt}\right\}$$

If one writes

$$\phi = (2K_2c_{in}/K_1^2)\int_0^t (\Delta p)\cdot dt \tag{4.5.7}$$

w_i now becomes

$$w_i = (K_1/K_2)\left\{-1 + \sqrt{\left[1 + (K_2/K_1)(w_i)_0\right]^2 + \phi}\right\} \tag{4.5.8}$$

The above expression may be rewritten as

$$(K_1 + K_2 w_i)^2 = K_1^2\left\{\left[1 + (K_2/K_1)(w_i)_0\right]^2 + \phi\right\} \tag{4.5.9a}$$

and

$$\left[K_1 + K_2(w_i|_{t_\ell})\right]^2 = K_1^2\left\{\left[1 + (K_2/K_1)(w_i)_0\right]^2 + \phi_\ell\right\} \tag{4.5.9b}$$

where ϕ_ℓ is the value of ϕ at $t = t_\ell = t_r + t_c$.

The steady-state cyclic condition implies (see Fig. 4.11)

$$(w_i)_{t=t_\ell} = (w_{i+1})_0 \quad i = 1,2,3,\ldots(n_c - 1)$$
$$(w_{n_c})_{t=t_\ell} = (w_1)_0 \tag{4.5.9c}$$

Combining (4.5.9b) and (4.5.9c) yields

$$\left[K_1 + K_2(w_{i+1})_0\right]^2 = \left[K_1 + K_2(w_i)_0\right]^2 + K_1^2\phi_\ell \tag{4.5.10}$$

which may be rewritten for $i = 1, 2, \ldots (n_c - 1)$ to give

$$\left[K_1 + K_2(w_2)_0\right]^2 = K_1^2 + K_1^2\phi_\ell$$

$$\left[K_1 + K_2(w_3)_0\right]^2 = \left[K_1 + K_2(w_2)_0\right]^2 + K_1^2\phi_\ell = K_1^2 + 2K_1^2\phi_\ell$$

$$\vdots$$

$$\left[K_1 + K_2(w_{n_c})_0\right]^2 = K_1^2 + (n_c - 1)K_1^2\phi_\ell$$

A general representation of the above equations may be written as

$$\left[K_1 + K_2(w_i)_0\right]^2 = K_1^2 + (i - 1)K_1^2\phi_\ell = K_1^2[1 + (i - 1)\phi_\ell]$$

and

$$(w_i)_0 = -(K_1/K_2) + (K_1 K_2)\sqrt{1 + (i - 1)\phi_\ell} \tag{4.5.12}$$

Substituting Equations (4.5.12) into (4.5.8), one has

$$w_i = (K_1/K_2)\left[\sqrt{1 + \phi + (i - 1)\phi_\ell} - 1\right] \tag{4.5.13}$$

Combining Equations (4.5.13) and (4.5.2) yields

$$Q_i = \frac{(A_c)(\Delta P)}{K_1\sqrt{1 + \phi + (i - 1)\phi_\ell}} \tag{4.5.14}$$

Substituting Equations (4.5.14) into (4.5.4a) [or (4.5.4b)], one has

$$Q = \sum_{i=1}^{n_c} \frac{(A_c)(\Delta P)}{K_1\sqrt{1 + \phi + (i - 1)\phi_\ell}} \qquad 0 < t < t_r = t_\ell - t_c \tag{4.5.15a}$$

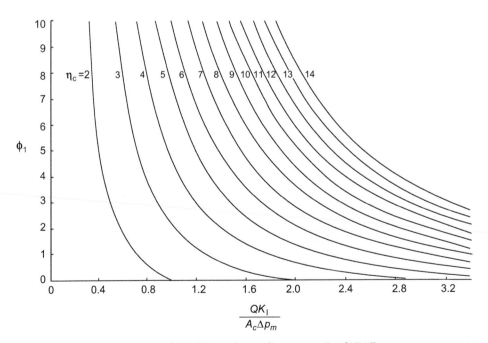

FIGURE 4.12 Relationship between ϕ_1 and $(QK_1)/(A_c\Delta p_m)$ according to equation (4.5.16).

$$= \sum_{i=1}^{n_c-1} \frac{(A_c)(\Delta P)}{K_1\sqrt{1+\phi+(i-1)\phi_\ell}} \qquad t_\ell - t_c < t_r < t_1 \qquad (4.5.15b)$$

From Equation (4.5.15b), at $t = t_\ell$, $\Delta P = (\Delta P)_m$, one has

$$\frac{QK_1}{A_c(\Delta P)_m} = \sum_{i=1}^{n_c-1} \frac{1}{\sqrt{1+i\phi_\ell}} \qquad (4.5.16)$$

Equation (4.5.16) gives the relationship between $(QK_1)/(A_c)(\Delta P)_m$, n_c, and ϕ_ℓ (they are all dimensionless). The results are shown in Fig. 4.12.

For $\phi_\ell > 10$, an approximate expression of ϕ_ℓ as a function of n_c and $QK_1/A_c(\Delta P)_m$ is found to be

$$\phi_\ell \cong \frac{\left(\sum_{i=1}^{n_c-1} 1/\sqrt{i}\right)^2}{\left[(QK_1)/(A_c(\Delta p)_m)\right]^2} \qquad (4.5.17)$$

The values of the series $\sum_{i=1}^{n_c-1} 1/\sqrt{i}$ and its square as a function of n_c are

n_c	$\left(\sum_{1}^{n_c-1} 1/\sqrt{i}\right)$	$\left(\sum_{1}^{n_c-1} 1/\sqrt{i}\right)^2$
2	1	1
3	1.707	2.914
4	2.285	5.219
5	2.784	7.753
6	3.281	10.44
7	3.640	13.25
8	4.017	16.17
9	4.371	19.11
10	4.704	22.13
11	4.744	22.51
12	5.323	28.33
13	5.612	31.49
14	5.888	34.67

To obtain a relationship between ϕ and t, differentiating Equation (4.5.7), one has

$$(\Delta p) = \left[K_1^2/(2K_2c_{in})\right](d\phi/dt) \qquad (4.5.18)$$

Substituting Equation (4.5.18) into (4.5.15a), the following expression is obtained:

$$Q = \sum_{i=1}^{n_c} \frac{A_c}{K_1\sqrt{1+\phi+(i-1)\phi_\ell}} \frac{K_1^2}{2K_2c_{in}} \frac{d\phi}{dt}$$

Integrating the above expression from $t = 0$ to t (i.e., from $\phi = 0$, to ϕ), after rearrangement, one has

$$t = \frac{A_c K_1}{2 K_2 c_{in} Q} \sum_{i=1}^{n_c} \int_0^{\phi} \frac{d\phi}{\sqrt{1 + \phi + (i-1)\phi_\ell}} \qquad 0 < \phi < \phi_\ell' \qquad (4.5.19)$$

and ϕ_1' is the value of ϕ at $t = t_\ell - t_c = t_r$. Similarly from Equation (4.5.15b), one has (for $t > t_1 - t_c$)

$$t = (t_\lambda - t_c) + \frac{A_c K_1}{2 K_2 c_{in} Q} \sum_{i=1}^{n_c} \int_{\phi_\ell'}^{\phi_\ell} \frac{d\phi}{\sqrt{1 + \phi + (i-1)\phi_\lambda}} \qquad \phi_\lambda' < \phi < \phi_\lambda \qquad (4.5.20)$$

If the integrals of Equations (4.5.19) and (4.5.20) are written explicitly, one has

$$t = \frac{A_c K_1}{K_2 c_{in} Q} \sum_{i=1}^{n} \left[\sqrt{1 + \phi + (i-1)\phi_\ell} - \sqrt{1 + (i-1)\phi_\ell} \right] \quad \phi < \phi < \phi_\ell' \qquad (4.5.21)$$

$$t = (t_\lambda - t_c) + \frac{A_c K_1}{K_2 c_{in} Q} \sum_{i=1}^{n_{c-1}} \left[\sqrt{1 + \phi + (i-1)\phi_\lambda} - \sqrt{1 + (i-1)\phi_\lambda} \right] \quad \phi_\lambda' < \phi < \phi_\lambda \qquad (4.5.22)$$

Note that

$$\phi_\ell' = (2 K_2 c_{in} / K_1^2) \int_0^{t_\ell - t_c} (\Delta p) dt \qquad (4.5.23)$$

From Equation (4.5.21) at $\phi = \phi_\ell'$ (or $t = t_\ell - t_c$), one has

$$t_\ell - t_c = \frac{A_c K_1}{K_2 c_{in} Q} \sum_{i=1}^{n_c} \left[\sqrt{1 + \phi_\ell' + (i-1)\phi_\ell} - \sqrt{1 + (i-1)\phi_\ell} \right]$$

From Equation (4.5.22), at $\phi = \phi_1$ (or $t = t_\ell$), one has

$$t_c = \frac{A_c K_1}{K_2 c_{in} Q} \sum_{i=1}^{n_{c-1}} \left[\sqrt{1 + i\phi_\lambda} - \sqrt{1 + \phi_\lambda' + (i-1)\phi_\lambda} \right] \qquad (4.5.25)$$

Adding Equations (4.5.25) and (4.5.24) yields

$$t_\ell = \frac{A_c K_1}{K_2 c_{in} Q} \left[\sqrt{1 + \phi_\ell' + (n_c - 1)\phi_\ell} - 1 \right] \qquad (4.5.26)$$

The above expression can be used to calculate the cycle time, $t_\ell = t_r + t_c$, if the value of ϕ_ℓ' is known (ϕ_ℓ can be found from Equation (4.5.16), (4.5.17), or Fig. 4.12). Equation (4.5.25) gives the relationship between t_c and ϕ_ℓ'. However, using it to determine ϕ_ℓ' for a given t_c can only be done by iteration. To establish an approximate expression of ϕ_ℓ' as a function t_c, the series present in Equation (4.5.25) may be approximated as

$$\sum_{i=1}^{n_{c-1}} \left[\sqrt{1 + i\phi_\lambda} - \sqrt{1 + \phi_\lambda' + (i-1)\phi_\lambda} \right] = \sum_{i-1}^{n_{c-1}} \left[\sqrt{1 + i\phi_\lambda} - \sqrt{1 + i\phi_\lambda - (\phi_\lambda - \phi_\lambda')} \right]$$

$$\cong \sum_{i=1}^{n_{c-1}} (1/2)(1 + i\phi_\ell)^{-1/2} (\phi_\ell - \phi_\ell')$$

$$= (1/2)(\phi_\ell - \phi_\ell') \frac{Q K_1}{A_c (\Delta P)_m}$$

and Equation (4.5.25) becomes

$$t_c = (1/2)(\phi_\ell - \phi'_\ell)\frac{QK_1}{A_c(\Delta P)_m}\cdot\frac{A_cK_1}{K_2c_{in}Q}$$

or

$$\phi'_\ell = \phi_\ell - \frac{2K_2c_{in}(\Delta P)_mt_c}{K_1^2} \tag{4.5.27}$$

For a given t_c with ϕ'_λ known, the cycle time t_ℓ, and cleaning time can be readily found from Equations (4.5.25) and (4.5.26).

For design calculations, assuming that Q, A_c, n_c, $(\Delta P)_m$, t_c, K_1, and K_2 are given, one is interested in obtaining the cycle time, t_ℓ, and the history of the areal cake mass built-up, and pressure drop history. The following procedure may be applied:

1. Based on the given values of Q, K_1 $(\Delta P)_m$, and A_c, ϕ_ℓ can be determined according to Equation (4.5.16) or Fig. 4.12.
2. ϕ'_ℓ can be estimated according to Equation (4.5.27).
3. The cycle time, t_ℓ, can be found from Equation (4.5.26).
4. The pressure drop history can be found from Equations (4.5.15a), (4.5.15b), (4.5.21), and (4.5.22). Specifically, from the last two equations, ϕ as a function of t, $0 < t < t_\ell$, can be determined. With ϕ vs. t known, the pressure drop history, Δp, vs. t can be determined by using Equations (4.5.15a) and (4.5.15b).
5. With ϕ vs. t known, the areal cake mass built-up of a given compartment can be readily found from Equations (4.5.8) or (4.5.13).

■ ■ ■ ▬▬▬▬▬▬▬▬▬▬▬▬▬▬▬▬▬▬▬▬▬▬▬▬▬▬▬▬

Illustrative Example 4.4[6]

A fabric filtration system has five compartments with 1,000 m² filter area in each compartment. Determine:

1. The cycle time, t_ℓ.
2. Δp vs. time during one cycle.
3. Cake build-up for each compartment.
4. Q_i vs. t, $i = 1, 2, \ldots, 5$

corresponding to the following conditions:

$$Q = 100 \text{ m}^3 \text{ s}^{-1}$$

$$c_{in} = 0.01 \text{ kg m}^{-3}$$

$$(\Delta P)_m = 2500 \text{ Pa}$$

$$K_1 = 40,000 \text{ Ns m}^{-3}$$

$$K_2 = 60,000 \text{ s}^{-1}$$

$$t_c = 180 \text{ s}$$

Solution

1. Equation (4.5.26) is used to calculate t_ℓ. The value of ϕ'_ℓ can be found from Equation (4.5.27) with ϕ_ℓ known. To obtain the value of ϕ_ℓ, one may use either Equation (4.5.17) or Fig. 4.12. First,

$$\frac{QK_1}{A_c(\Delta P)_m} = \frac{(100)(40,000)}{(1000)(2500)} = 1.6$$

From Fig. 4.12, with $n_c = 5$, $\phi_\ell = 2.45$.
From Equation (4.5.27), ϕ'_ℓ is found to be

$$\phi'_\ell = 2.45 - \frac{2(6 \times 10^4)(0.01)(2500)(180)}{(4 \times 10^4)^2} = 2.45 - 0.34 = 2.11$$

and t_ℓ is found from Equation (4.5.26), or

$$t_\ell = \frac{(1,000)(4 \times 10^4)}{(6 \times 10^4)(0.01)100}\left[\sqrt{1 + 2.11 + (5-1)245} - 1\right]$$

$$= 666.67\left[\sqrt{12.61} - 1\right] = 1729 \text{ s}$$

2. To obtain the pressure drop history, Δp is given by Equation (4.5.18):

$$(\Delta p) = \left[K_1^2/(2K_2 c_{in})\right]\frac{d\phi}{dt} \qquad \text{(i)}$$

From Equations (4.5.21) and (4.5.22), $\dfrac{d\phi}{dt}$ is found to be

$$\frac{d\phi}{dt} = \frac{(2K_2 c_{in} Q)/(A_c K_1)}{\sum_{i=1}^{n_c}[1 + \phi + (i-1)\phi_\ell]^{-1/2}} \qquad 0 < \phi < \phi'_\ell \qquad \text{(iia)}$$

$$\frac{d\phi}{dt} = \frac{(2K_2 c_{in} Q)/(A_c K_1)}{\sum_{i=1}^{n_c-1}[1 + \phi + (i-1)\phi_\ell]^{-1/2}} \qquad \phi'_\ell < \phi < \phi_\ell \qquad \text{(iib)}$$

Therefore,

$$\Delta p = \frac{K_1 Q/A_c}{\sum_{i=1}^{n_c}[1 + \phi + (i-1)\phi_\ell]^{-1/2}} \qquad 0 < \phi < \phi'_\ell \qquad \text{(iiia)}$$

$$\Delta p = \frac{K_1 Q/A_c}{\sum_{i=1}^{n_c-1}[1 + \phi + (i-1)\phi_\ell]^{-1/2}} \qquad \phi'_\ell < \phi < \phi \qquad \text{(iiib)}$$

The relationship between ϕ and t is given by Equations (4.5.21) and (4.5.22):

$$t = \frac{A_c K_1}{K_2 c_{in} Q}\sum_{i=1}^{n_c}\left[\sqrt{1 + \phi + (i-1)\phi_\ell} - \sqrt{1 + (i-1)\phi_\ell}\right] \qquad 0 < \phi < \phi'_\ell \qquad \text{(iva)}$$

or

$$t = (t_\ell - t_c) + \frac{A_c K_1}{K_2 c_{in} Q} \sum_{i=1}^{n_{c-1}} \left[\sqrt{1 + \phi + (i-1)\phi_\ell} - \sqrt{1 + \phi'_\ell + (i-1)\phi_\ell} \right] \phi'_\ell < \phi < \phi_\ell \qquad \text{(ivb)}$$

In other words, from Equations (iva) and (ivb), ϕ can be determined as a function of t. In particular, the pressure drop history over a given cycle (say for $0 < t < t_\ell$, $t_\ell < t < 2\,t_\ell$, etc) remains the same. Since Equations (iiia) and (iiib) give ΔP as a function of ϕ, combining the results yields the pressure drop history.

The numerical results of ϕ vs. t and Δp vs. ϕ are as follows:

ϕ	t	Δp
0	0	1538
0.2	169	1610
0.4	332	1674
0.6	489	1732
1.0	808	1835
1.5	1140	1948
2.0	1474	2049
2.11^-	1545^-	2071
2.11^+	1575^+	2420
2.2	1594	2441
2.45	1729	2500

The discontinuity of ΔP at $\phi = 2.11$ or $t = 1545$ corresponds to the withdrawal of one compartment. The results of Δp vs. t are shown in Fig. i.

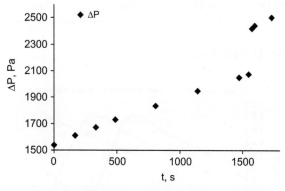

FIGURE I Pressure Drop vs. Time

3. To obtain cake build-up of a five-compartment system, cake growth continuously for $t = 0$ to $t = 5t_\ell - t_c$. On the other hand, as shown before, the $\phi - t$ relationship repeats itself over the cycle $(i - 1)\, t_\ell < t < t_\ell$. Accordingly, over the former $0 < t < 5t_\ell$, one has

ϕ	T	ϕ	t	ϕ	t	ϕ	t	ϕ	t
0	0	2.45	1729	4.9	3458	7.35	5187	9.8	6916
0.2	169	2.65	1898	5.1	3627	7.55	5356	10.0	7088
0.4	332	2.85	2061	5.3	3990	7.75	5519	10.2	7251
1.0	808	3.45	2537	5.9	4266	8.35	5995	10.8	7727
2.0	1474	4.45	3203	6.9	4932	9.35	6661	11.8	8393
2.11	1545	4.56	3274	7.01	5003	9.46	6732	11.91	8764
2.45	1729	4.90	3458	7.25	5187	9.8	6916	12.25	8648

The cake and mass history is given by Equation (4.5.8) or

$$w_i = (K_1/K_2)\left\{ -1 + \sqrt{\left[1 + (K_1/K_2)(w_i)_0^2 + \phi\right]} \right\} \tag{v}$$

and $(w_i)_0$ can be found from Equation (4.5.12) or

$$(w_i)_0 = -(K_1/K_2) + (K_1 K_2)\sqrt{1 + (i-1)\phi_1} \tag{vi}$$

The initial values of w_i, $(w_i)_0$ are
$(w_1)_0 = 0$, $(w_2)_0 = 0.572$, $(w_3)_0 = 0.953$, $(w_4)_0 = 1.26$, and $(w_5)_0 = 1.525$
The calculated results of w_i vs. t are given below.

ϕ	$i = 1,$	2	3	4	5
$1t_\ell$	0.572	0.953	1.26	1.525	1.964,0
$2t_\ell$	0.953	1.26	1.525	2.764,0	0.572
$3t_\ell$	1.26	1.525	1.764,0	0.572	0.953
$4t_\ell$	1.527	1.764,0	0.572	0.453	1.26
$5t_\ell$	1.764,0	0.572	0.953	3.26	1.525

The results are also shown in Fig. ii.

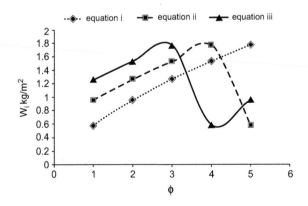

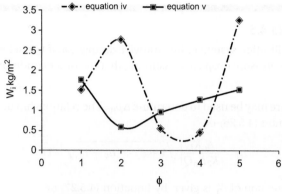

FIGURE I AND II w_i vs. ϕ, $i = 1, 2, 3, 4$

3. To calculate the volumetric gas flow rate through each compartment, consider the first cycle, $0 < t < t_\ell$; Q_i is given by Equation (4.5.14):

$$Q_i = \frac{(A_c)(\Delta P)}{K_1 \sqrt{1 + \phi + (i-1)\phi_1}}$$

$$A_c/K_1 = 1000/4000 = 2.5 \times 10^{-2}$$

From the previous $t - \phi - \Delta P$ results, Q_i can be obtained as follows:

				Q_i				
t	ϕ	Δp	$i = 1$	2	3	4	5	$\sum Q_i$
0	0	1538	33.45	21.8	16.83	14.02	12.5	103.78
332	0.4	1674	35.4	22.34	17.65	15.05	13.34	103.78
808	1.0	1835	39.4	22.63	18.39	15.89	14.35	103.66
1545^-	2.11	2081	29.4	22.66	19.12	16.85	15.23	103.26
1545^+		2420	34.3	26.48	22.25	19.69	–	102.82
1729	2.45	2500	33.65	26.5	27.57	19.98	–	102.7

The results apply to $0 < t < t_\ell$. For $t_\ell < t < t_{2\ell}$, the flow rate history of compartment no. 5 is the same as that of compartment no. 1 of the previous cycle. The flow histories of compartment no. 1, no. 2, no. 3, and no. 4 coincide with those of compartment no. 2, no. 3, no. 4, and no. 5 as shown in Fig. 4.11. This behavior repeats itself from cycle to cycle.

A check of the accuracy of the computation is made by comparing the sum of Q_i's with the given value of $Q = 100$ m² s⁻¹. As shown in the last column of the above tabulated results, $\sum Q_i$ is found to vary approximately between 102.7 and 103.78, a difference of approximately 3% from the corrected value. In this regard, one should realize that the value of ϕ_1 was read off Fig. 4.12 and ϕ'_λ was estimated from the approximate expression of Equation (4.5.27). However, even with these approximations, sufficiently accurate results were obtained.

---

[6]Taken from Crawford, M., "Air Pollution Control Theory", p. 499; McGraw-Hill, 1976.

■ ■ ■ ▬▬▬▬▬▬▬▬▬▬▬▬▬▬▬▬▬▬

Illustrative Example 4.5

For the same fabric filtration system of Illustrative Example 4.4, if t_ℓ is set to be 2,000 s, find the maximum allowable pressure drop ΔP_m assuming that the other conditions remain the same.

Solution

An iteration procedure may be used to determine Δp_m. The relationship between t_ℓ and ϕ_ℓ (and ϕ'_ℓ) is given by Equation (4.5.26) or

$$t_\ell = \frac{A_c K_1}{K_2 c_{in} Q}\left[\sqrt{1 + \phi'_\ell + (n_c - 1)\phi_\ell} - 1\right] \tag{i}$$

An approximate expression of ϕ'_1 is given by Equation (4.5.27) or

$$\phi'_\ell = \phi_\ell - \frac{2 K_2 c_{in}(\Delta p)_m t_c}{K_1^2} \tag{ii}$$

An iteration calculation can be used to determine Δp_m. First assume

$$\phi'_\ell = \phi_\ell$$

From (i), one has

$$\left[\frac{K_2 c_{in} Q t_\ell}{A_c K_1} + 1\right]^2 = 1 + n_c \phi_\ell$$

and

$$\phi_\ell = (1/n_c)\left[\left(\frac{K_2 c_{in} Q t_\ell}{A_c K_1} + 1\right)^2 - 1\right]$$

$$= (1/5)\left[\left(\frac{(60,000)(10^{-2})(100)(2000)}{(1000)(4000)} + 1\right)^2 - 1\right] = 3$$

From Fig. 4.12, with $\phi_\ell = 3$, $n_c = 5$

$$\frac{Q K_1}{A_c(\Delta p_m)} = 14.8$$

$$\Delta p_m = \frac{(100)(40,000)}{(100)(14.8)} = 2703 \text{ say } 2700 \text{ Pa}$$

Next make a new estimation of ϕ'_1. From (ii),

$$\phi'_\ell = \phi_\ell - \frac{2(60,000)(10^{-2})(2900)(180)}{(40000)^2}$$

$$= \phi_\ell - 0.365$$

Therefore, from Equation (i),

$$t_\ell = \frac{A_c K_1}{K_2 c_{in} Q} \left[\sqrt{(1 - 0.365) + n_c \phi_\ell} - 1 \right]$$

Solving for ϕ_ℓ, one has

$$\phi_\ell = \left(\frac{1}{n_c} \right) \left[\left\{ \frac{K_2 c_m Q t_\ell}{A_c K_1} + 1 \right\}^2 - 0.635 \right] = 3.073$$

From Fig. 4.11 with $\phi_1 = 3.073$, $n_c = 5$

$$\frac{Q K_1}{A_c (\Delta p_m)} = 17.5$$

and

$$\Delta p_m = 2758 \text{ say } 2760 \text{ Pa}$$

■ ■ ■

4.6   Simplified Calculation of Multi-Compartment Fabric Filtration

The method described in the previous section considers the built-up of dust cake in each compartment individually subject to the constraint that the total gas throughput (namely, the sum of the throughputs of individual bags) is kept constant. Cooper and Alley (2002) suggested a simplification of Crawford's method by assuming that some of the important variables such as the length of the cycle time, t_ℓ, or the maximum pressure drop can be estimated using an average filtration velocity. This simplified procedure of calculation is presented below:

Consider a system with n_c compartments in service with a cycle time of $t_\ell = t_r + t_c$. During the period $0 < t < t_r$ when all compartments are in series, the average filtration velocity v_{n_c} is

$$v_{n_c} = \frac{Q}{(n_c)(A_c)} \tag{4.6.1}$$

During the period $t_r < t < t_r + t_c = t_\ell$, with one compartment undergoing cleaning and out of service, the average filtration velocity, $v_{n_c - 1}$ is

$$v_{n_c - 1} = \frac{Q}{(n_c - 1)(A_c)} \tag{4.6.2}$$

For a compartment starting from its clean state to its being ready for cleaning, the total amount of dust cake formed (or cake areal mass), $(w)_{max}$ is

$$w_{max} = c_{in} \left[v_{n_c} n_c t_r + v_{n_c - 1} (n_c - 1) t_c \right] \tag{4.6.3}$$

and the corresponding maximum allowable pressure drop, Δp_m, is

$$\Delta p_m = \{K_1 + K_2 w_{max}\} v_{n_c - 1} \tag{4.6.4}$$

To improve the accuracy of this simplified procedure, Cooper and Ally suggested that the value of Δp_m obtained from (4.6.4) be multiplied by a correction factor f_{n_c} which is a function of n_c given as follows:

n_c	f_{n_c}	n_c	f_{n_c}
3	0.87	10	0.67
4	0.80	12	0.65
5	0.76	15	0.64
7	0.71	20	0.62

■ ■ ■ ▬▬▬▬▬▬▬▬▬▬▬▬▬▬▬▬▬▬▬▬▬▬▬▬▬▬▬▬▬▬

Illustrative Example 4.6

Same as Illustrative Example 4.4, calculate the cycle time, t_ℓ, using the simplified procedure discussed in 4.6.

Solution

The average filtration velocities, v_5 and v_4, are

$$v_5 = \frac{100}{(5)(1000)} = 0.02 \text{ m s}^{-1}$$

$$v_4 = \frac{100}{(4)(1000)} = 0.025 \text{ m s}^{-1}$$

The maximum cake areal mass, $(w)_m$, is

$$w_{max} = 0.01[5(t_r)(0.02) + 4t_c(0.025)]$$

$$= 0.01[0.1 t_r + 18]$$

From Equation (4.6.4), one has

$$\Delta p_m = (K_1 + K_2 w_{max}) v_4 \cdot f_4$$

or

$$2500 = [40,000 + (60,000)(0.01)(0.1 t_r + 18)](0.025)(0.76)$$

$$t_r = \left[\frac{2500}{(0.025)(0.76)} - 40,000 - (600)(18)\right] \Big/ 60$$

$$= 1346 \text{ s}$$

and $t_\ell = 1346 + 180 = 1526$ s instead of 1729 s found in Illustrative Example 4.4.

■ ■ ■

Problems

4.1. A filter has 30,000 m^2 surface area and is used to treat a stream of 250 m^3 s^{-1} air (standard conditions) with a dust concentration of 0.025 kg m^{-3}. If $K_1 = 80,000$ N s m^{-3} and $K_2 = 50,000$ s^{-1}, determine the pressure drop immediately after cleaning and estimate the pressure drop as a function of time thereafter until the pressure drop becomes twice of the initial value.

4.2. A filter system is to be designed to treat 25 m^3 s^{-1} of standard air with $c_{in} = 8 \times 10^{-3}$ kg m^{-3}. For the dust and the type of filter proposed, $K_1 = 35,000$ N s m^{-3} and $K_2 = 60,000$ s^{-1}. The designed value of the filtration velocity is set to be 1.2×10^{-2} m s^{-1}.
  (1) Determine the number of filter bags required. The bag has a diameter of 0.2 m and height 8 m.
  (2) What is the pressure drop immediately after cleaning?
  (3) The maximum allowable pressure drop is 3000 N m^{-2}; how many service hours the system provides before it requires cleaning?

4.3. An experimental filter was used to conduct measurements for the determination of K_1 and K_2. The data obtained were as follows: $Q = 1.0$ m^3 s^{-1}, $\Delta P = 2500$ Pa just before the filter cleaning and $\Delta P = 700$ N m^{-2} immediately after cleaning; 100 kg of dust was removed for each cleaning. The filter area was 75 m^2. What values of K_1 and K_2 were found?

4.4. A baghouse system has four compartments and handles air at the rate of 50 m^3 s^{-1} with $c_{in} = 6 \times 10^{-3}$ kg m^{-3}. Each compartment has 500 m^2 of filtration surface area, $K_1 = 12,000$ N s m^{-3}, and $K_2 = 15,000$ s^{-1}. If $t_c = 150$ s and $-\Delta p_m = 500$ N m^{-2}, determine t_ℓ.

4.5. Adopt the conventional cake filtration theory to fabric filtration and assume that dust cake are compressible, from Equations (2.4.3) and (2.4.2b) and using the notation of this chapter, one has

$$v = \frac{\Delta p}{\mu \left[(\alpha_{av})_{\Delta P_c} w + R_m \right]}$$

The above expression may be rewritten to give

$$\Delta P = K_1 v + K_2 w \cdot v$$

with

$$K_1 = \mu R_m$$
$$K_2 = \mu (\alpha_{av})_{\Delta p_c}$$

Consider the filtration of air–limestone mixture using a multiple-calendered polyester needle-felt fabric as filter medium. Filtration rate is kept constant (4 $\times$ 10^{-2} m s^{-1}). K_1 is estimated to be 2.25×10^3 Ns m$^{-3.}$
  (1) Estimate the value of K_2 as Δp increases from an initial value of 90 Pa to 1000 Pa. Equation (4.3.1) may be used for estimating ε_s of limestone cake and

the Kozeny–Carmen equation for estimating cake permeability. ρ_s may be assumed to be 2,400 kg m^{-3}.

(2) Based on your calculation, is it reasonable to assume K_2 being constant? (Note that this assumption is commonly used in fabric filtration as discussed in this chapter.)

4.6. Determine the extent of filter cleaning and the residue cake areal mass by pulse air jet using the method of De Ravin et al (see Section 4.4.3) corresponding to the following conditions:

Jet Jump Characteristic Data

$\dfrac{\text{Volumetric Air Flow Rate}}{\text{Filter Bag Surface Area}}$ (m s^{-1})	Δp(kPa)
0	18
0.35	17.7
0.53	17
0.74	16
0.98	14.5
0.105	14
0.115	13
0.12	11.9
0.125	9
0.126	6

The cake/fabric adhesion force follows the log-normal distribution equation with $F_{50} = 200$ log, $F_{50} = 2.301$, and log $\sigma = 0.4$.
Pressure Drop Across Bag before Cleaning: 5 KPa
Cake Areal Mass before Cleaning, $w = 0.05, 0.1, 0.15, 0.20$ kg m^{-2},

$$K_1 = 20{,}000 \text{ N s m}^{-3}$$

$$K_2 = 30{,}000 \text{ s}^{-1}$$

Based on your results, what is your conclusion on the effect of w on filter cleaning?

4.7. Consider a fabric filter system with five compartments. The filtration area for each compartment is 4000 ft^2. The system is designed to service 40,000 ft^3/min of air with a dust concentration of 10 grains ft^{-3}. The filtration time (namely, the time between cleaning for a given compartment) is 60 minutes and the cleaning of a compartment requires 4 min to complete. $K_1 = 1.00$ in H$_2$O – min/ft and $K_2 = 0.003$ in H$_2$O – min-ft/grain. Calculate the maximum pressure drop which must be supplied using the simplified procedure discussed in 4.6.
Hint: You may wish to convert the data given into SI units. Note:

$$1 \text{ grain} = 6.48 \times 10^{-5} \text{ kg}$$

$$1 \text{ in H}_2\text{O pressure} = 249 \text{ Pa}$$

References

Cooper, C. David, Alley, F.C., 2002. Air Pollution Control: A Design Approach, third ed. Waveland Press. p. 190.

Crawford, Martin, 1976. Air Pollution Control Theory. McGraw Hill Ins.

Dennis, B., Wilder, J.W., Fabric Filter Cleaning Studies, EPA-650/2-75-009 (NTIS no. P8-240-372/3G1), Jan. 1975.

De Ravin, M., Humphries, W., Postle, R., 1988. Filtrat Separ, 25, 201.

Ju, J., Chiu, M.-S., Tien, C., 1990. J. Air. Waste. Manag. Assoc, 50, 600.

Leith, David, Allen, R.W.K., 1986. Dust Filtration by Fabric Filters. In: Wakeman, R.J. (Ed.), Progress in Filtration and Separation. Elsevier.

Rothwell, E., (March–April) 1986. Filtrat Separ, 113.

Schmidt, E., (Sept.) 1995. Filtrat Separ, 789.

Schmidt, E., (May) 1997. Filtrat Separ, 365.

Schmidt, E., Loffler, F., 1990. Powder Tech, 60, 173.

Schmidt, E., Loffler, F., 1991. Part. Part. Syst. Charact, 8, 105.

Seville, J.P., Cheung, W., Clift, R., (May/June), 1989. Filtrat Separ, 187.

Sievert, J., Loffler, F., (Nov/Dec) 1987. Filtrat Separ, 424.

Deep Bed Filtration

5

Deep Bed Filtration: Description and Analysis

Notation

A	coefficient of Equation (5.4.8) ($m^3\ s^{-1}$) or coefficient of Equation (5.4.25) ($m\ s^{-1}$)
$A(\theta)$	ratio of c^+ to σ defined by Equation (5.7.18) (−)
A_p	projected area of a collector (m^2)
a	coefficient of filtration rate expression (see footnote on p. 159)
a_c	radius of solid sphere inside Happel's cell (m)
B, C, D	coefficients of Equation (5.4.25)
b	radius of fluid envelop of Happel's cell (m)
C_D	drag coefficient defined by Equation (5.6.5) (−)
c	particle concentration (vol/vol)
c_{eff}, c_{in}	effluent and influent particle concentration (vol/vol)
c_i	effluent particle concentration of the i-th unit bed element
c^+	defined as c/c_{in}
$(c_i)_j$	effluent concentration of the j-th collector of the i-th UBE
D	diameter of filter bed (m)
d_f	fiber diameter (m)
d_g	filter-grain diameter
$<d_g>$	mean value of d_g (m)
$<d_f>$	mean fiber diameter (m)
E	total collection efficiency defined by Equation (5.3.2)
E^2	differential operator given by Equation (5.4.4)
e	unit bed element collection efficiency (−)
$F(\alpha, \sigma)$	function defined by Equation (5.2.3a) (−)
F_D	drag force acting on filter grain (N)
$G(\beta, \sigma)$	function defined by Equation (5.2.9) (−)
J_i	particle deposition flux of the i-th collector
K	defined as $k_1/16$ (−) [see Equation (5.5.5)]
K_1, K_2, K_3, K_4	coefficients of the stream function of Equation (5.4.8)
Ku	Kuwabara's hydrodynamic factor given by Equation (5.4.27e)
k	permeability of filter medium (m^2)
k_0	initial value of k (m^2)
k_1	coefficient of Equation (5.5.1) (−)
k_2	coefficient of Equation (5.5.3.) (−)
L	filter medium height (m)
L_0	filter medium height in fixed-bed state (m)
ℓ	length of periodicity (m)
$\bar{\ell}$	gas molecule mean free path (m)
ℓ_f	fiber length (m)
$(\ell_f)_{av}$	average fiber length

Principles of Filtration, DOI: 10.1016/B978-0-444-56366-8.00005-0
Copyright © 2012 Elsevier B.V. All rights reserved

M	molecular weight
N	total number of granules in a sample $(-)$ or the total number of unit bed elements given by L/ℓ $(-)$
N_c	number of collectors in a unit bed element $(-)$
N_{Re}	defined as $<d_g>\varphi_r u_s \rho / \mu$ $(-)$
N_{Re_1}	defined by Equation (5.6.10b) $(-)$
N_{Kn}	defined as $2\bar{\ell}/d_f$
n	exponent of Equation (5.6.2)
P	penetration defined as $1-E$ $(-)$
p	pressure (Pa), or a quantity defined by Equation (5.4.11b) $(-)$
Q	volumetric flow rate into the a given UBE
$(q_i)_j$	volumetric flow rate through the j-th collector of the i-th unit bed element $(\mathrm{m^3\ s^{-1}})$
R	gas law constant 8.315 $\mathrm{J\ K^{-1}\ mol^{-1}}$
R_1, R_2	radial positions of the inlet and exit of the a radial filter (m)
r	radial coordinate (m)
r^*	defined as r/a_c $(-)$
$\mathrm{Re}_{t\infty}$	defined by Equation (5.6.6) $(-)$
s_p	particle surface area $(\mathrm{m^2})$
T	absolute temperature (K)
t	time (s)
u	fluid approach velocity $(\mathrm{m\ s^{-1}})$
u_r, u_θ	r- and θ-components of fluid velocity $(\mathrm{m\ s^{-1}})$
u_s	superficial velocity $(\mathrm{m\ s^{-1}})$
V	upflow velocity in back washing $(\mathrm{m\ s^{-1}})$
V_i	velocity term of Equation (5.6.2) $(\mathrm{m\ s^{-1}})$
$V_{t\infty}$	terminal velocity $(\mathrm{m\ s^{-1}})$
v_p	particle volume $(\mathrm{m^3})$
w	defined by Equation (5.4.11a)
z	axial coordinate

Greek Letters

$\alpha\ \beta$	parameter vectors of Equations (5.2.3a) and (5.2.9), respectively $(-)$
$-\Delta p$	pressure drop (Pa)
$(-\Delta p)_0$	initial value of $(-\Delta p)$
ε	porosity $(-)$
ε_0	initial value of ε $(-)$
ε_d	deposit porosity $(-)$
η_f	fiber collector efficiency $(-)$
ε_s	solidosity $(-)$
$(\eta_f)_s$	single fiber efficiency $(-)$
η	collector efficiency $(-)$
η_s	single collector efficiency defined by Equation (5.4.18b)
$(\eta_s)_j$	efficiency of the j-th collector of the i-th unit bed element $(-)$
θ	corrected time defined by Equation (5.1.3) (s) or angular coordinate $(-)$
λ	filter coefficient $(\mathrm{m^{-1}})$
λ_0	initial filter coefficient $(\mathrm{m^{-1}})$
μ	fluid viscosity (Pa, S)
ρ	fluid density $(\mathrm{kg\ m^{-3}})$

ρ_p	particle density (kg m^{-3})
σ	particle specific deposit (−)
σ_{in}	value of σ at filter inlet (−)
ϕ	quantity defined by Equation (5.6.10a) (−)
ϕ_s	particle sphericity factor defined by Equation (5.5.2)
ψ	stream function (m^3 s^{-1})

As stated in 1.2, deep bed filtration refers to the process in which removal of particles from a particle suspension (either liquid or gas) is effected by passing the suspension through a medium composed of granular or fibrous entities with particle deposition taking place at the surface of the entities (granule or fiber) throughout the entire medium. The accumulation of deposited particles within the medium causes a continuous structural change of the medium, which, in turn, affects the rate of filtration as well as the flow resistance of the medium. Accordingly, both the effluent particle concentration and the pressure drop necessary to maintain a fixed fluid throughput may vary with time (or in the case of constant pressure operations, the throughput diminishes with time). Since in most applications of deep bed filtration, there exists a standard of effluent quality to be met as well as the maximum allowable pressure drop, periodic filter cleaning becomes necessary once these limits are reached. Accordingly, deep bed filtration performance is described by the histories of the effluent concentration and pressure drop as well as the frequency and conditions used for filter cleaning. The calculations and estimations of filter performance will be the main focus of discussions of this and the following chapters.

5.1   Macroscopic Conservation Equation

As an introductory text, the discussions given below will be restricted to the simple one-dimensional and rectilinear case. The equations which may be used to describe the dynamics of deep bed filtration consist of the relationships based on the conservation principle, the assumed filtration rate expression, and the mechanics of fluid flow through porous media. The one-dimensional particle conservation equation may be written as

$$u_s \frac{\partial c}{\partial z} + \frac{\partial(\sigma + \varepsilon c)}{\partial t} = 0 \tag{5.1.1}$$

The independent variables are the axial distance (z) and time (t). The dependent variables are the particle concentration of the suspension (c) and the specific deposit of particles in the medium (σ). Both c and σ may be expressed on a vol/vol basis. The assumptions involved, in addition to that of rectilinear flow, are negligible axial dispersion effect and uniform constant fluid superficial velocity (u_s). The medium porosity (ε) may be expressed as

$$\varepsilon = \varepsilon_0 - \frac{\sigma}{1 - \varepsilon_d} \tag{5.1.2}$$

where ε_0 is the initial value of ε and ε_d is the porosity of the particle deposit. Strictly speaking, ε_d is a function of both time and position as the extent and manner of particle deposition vary both spatially and temporally. In practice, ε_d is often given an arbitrary value as a fitting parameter. A schematic representation is shown in Fig. 5.1.

Equation (5.5.1) is the conservation equation commonly used in fixed-bed processes. Common to these processes, a corrected time, θ, defined as

$$\theta = t - \int_0^z \frac{dz}{u_s/\varepsilon} \tag{5.1.3}$$

may be used instead of t. Equation (5.1.1) now becomes

$$u_s \frac{\partial c}{\partial z} + \left(1 - \frac{c}{1 - \varepsilon_d}\right) \frac{\partial \sigma}{\partial \theta} = 0 \tag{5.1.4a}$$

or

$$u_s \frac{\partial c}{\partial z} + \frac{\partial \sigma}{\partial \theta} = 0 \tag{5.1.4b}$$

since $c \ll 1$ in most deep bed filtration applications.

To complete the description, a filtration rate expression is required and will be given in the next section. In addition, proper initial and boundary conditions must be specified.

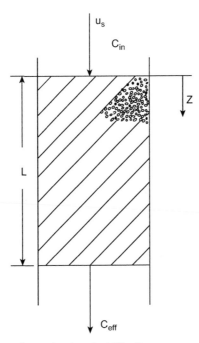

FIGURE 5.1 Schematic representation of granular deep bed filtration.

For a clean, homogeneously packed filter medium and subject to an effluent with constant particle concentration, c_{in}, these conditions are

$$c = 0, \quad \sigma = 0 \quad \text{for } z \geq 0, \ \theta \leq 0 \tag{5.1.5}$$

$$c = c_{in} \quad \text{at } z = 0, \tag{5.1.6}$$

■ ■ ■ ▬▬▬▬▬▬▬▬▬▬▬▬▬▬▬▬▬▬▬▬▬▬▬▬▬▬▬▬▬▬▬▬▬

Illustrative Example 5.1

Derive the macroscopic particle conservation equation for radial flow filters. A schematic representation of radical flow filtration is given in Fig. i. The filter medium extends from $r = R_1$ to $r = R_2$. Suspension flows outward and the volumetric flow rate (m³ s⁻¹) is constant. The assumptions used to derive Equation (5.1.1) may be applied.

Solution

The overall mass balance is

$$2\pi r \, u_r = 2\pi (u_r)_{R_1} R_1 = Q, \quad \text{or} \quad u_r = Q/(2\pi r) \tag{i}$$

Consider a volume element, $r < r < r + \Delta r$ and unit height, the particle flux entering (at $r = r$) and leaving (at $r = r + \Delta r$) the element over a time period of Δt are

Particle Flux Entry $\qquad (2\pi r \, u_r c)_r \, \Delta t$

Particle Flux $\qquad (2\pi r \, u_r c)_{r + \Delta r} \, \Delta t$

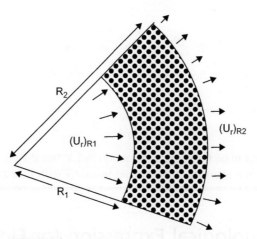

FIGURE I Schematic Representation of a Radial Filter.

The amount of particle accumulated in this volume element during the period Δt is

$$(2\pi r)(\Delta r)[\Delta(\varepsilon c) + \Delta \sigma]$$

Accordingly,

$$\left[(2\pi r \, u_r c)_r - (2\pi r \, u_r c)_{r+\Delta r}\right] \Delta t = (2\pi r)(\Delta r)[\Delta(\varepsilon c) + \Delta\sigma]$$

which gives

$$\left(\frac{1}{r}\right)\frac{\partial}{\partial r}(r \, u_r c) + \frac{\partial(\varepsilon c)}{\partial t} + \frac{\partial\sigma}{\partial t} = 0$$

or

$$u_r \frac{\partial c}{\partial r} + \frac{\partial(\varepsilon c)}{\partial t} + \frac{\partial\sigma}{\partial t} = 0 \qquad\qquad\text{(ii)}$$

If a set of new variables, $\tilde{r}$ and θ, is introduced with

$$\tilde{r} = r \qquad\qquad\text{(iii.a)}$$

$$\theta = t - \int_0^r \frac{\varepsilon \, dr}{u_r} \qquad\qquad\text{(iii.b)}$$

One has

$$\frac{\partial}{\partial r} = \frac{\partial}{\partial \tilde{r}}\frac{\partial \tilde{r}}{\partial r} + \frac{\partial}{\partial \theta}\frac{\partial \theta}{\partial r} = \frac{\partial}{\partial \tilde{r}} - \frac{\varepsilon}{u_r}\frac{\partial}{\partial \theta}$$

$$\frac{\partial}{\partial t} = \frac{\partial}{\partial \tilde{r}}\frac{\partial \tilde{r}}{\partial t} + \frac{\partial}{\partial \theta}\frac{\partial \theta}{\partial t} = \frac{\partial}{\partial \theta}$$

Equation (ii) becomes

$$u_r \frac{\partial c}{\partial \tilde{r}} - (u_r)\frac{\varepsilon}{(u_r)}\frac{\partial c}{\partial \theta} + \frac{\partial(\varepsilon c)}{\partial \theta} + \frac{\partial\sigma}{\partial \theta} = 0$$

Since the change of ε, for deep bed filtration, is slight, the above expression becomes, after dropping the overbar "~",

$$u_r \frac{\partial c}{\partial r} + \frac{\partial\sigma}{\partial \theta} = 0 \qquad\qquad\text{(iv)}$$

Equation (iv) may appear to be the same as Equation (5.1.1) but there is one major difference, namely, unlike u_s, u_r is not a constant, and varies inversely with r, or $u_r = Q/2\pi r$.

■ ■ ■

5.2   Phenomenological Expression for Filtration Rate

The specific deposit σ, gives the amount (volume) of deposited particles per unit medium volume. Therefore, the quantity $\partial\sigma/\partial\theta$ is the rate of filtration.

A simple expression of filtration rate was first suggested by Iwasaki (1937), based on his experimental observations. Experimental data for slow sand filters show that the particle concentration profile throughout a filter (c vs. z) can often be described by the logarithmic law, this is,

$$\frac{\partial c}{\partial z} = -\lambda c \tag{5.2.1}$$

and λ is known as the filter coefficient.

Thus, one can find the filtration rate expression by combining Equations (5.1.4b) and (5.2.1), or

$$\frac{\partial \sigma}{\partial \theta} = \lambda u_s c \tag{5.2.2}^{[1]}$$

Generally speaking, the logarithmic behavior observed by Iwasaki holds true only during the initial period of filtration. The fact that λ does not remain constant but varies with time during the course of filtration suggests that one may express λ as

$$\lambda = \lambda_0 F(\alpha, \sigma) \tag{5.2.3a}$$

with

$$F(\alpha, 0) = 1 \tag{5.2.3b}$$

and the filtration rate expression becomes

$$\frac{\partial \sigma}{\partial \theta} = u_s \lambda_0 F(\alpha, \sigma) c \tag{5.2.4}$$

where λ_0 is the initial value of λ (or the initial or clean filter coefficient) and $F(\alpha, \sigma)$, the correcting factor to account for the deviation from the logarithmic law of the concentration profile. α is the relevant parameter vector.

Two interpretations of the filter coefficient may be offered. Referring to Equation (5.2.1), the rate of filtration is first order with respect to the particle concentration of the suspension to be treated. λ, therefore, may be viewed as the first-order reaction rate constant. Alternatively, λ may be considered as the probability of a particle's being captured (deposited) during a time interval of $1/u_s$ in its flowing through the medium (or traveling a unit distance). In the case of λ being not constant (or varies with σ), λ may be considered as a pseudo-first-order rate constant.

The selection of a particular form of expression for F depends, of course, upon the filter medium as well as the suspension to be filtered. If a filter's performance is enhanced with the increase in deposition, then the suspension particle concentration (c vs. z) obtained at different times can be expected to display a systematic downward displacement as time passes. An upward displacement of the concentration profiles corresponding to increasing time implies that the filter's performance deteriorates with deposition. A downward displacement followed by an upward displacement means the filter first improves its performance with deposition and then deteriorates. These different behaviors are shown in Fig. 5.2. A partial listing of $F(\alpha, \sigma)$ proposed by various investigators is given in Table 5.1.

---

[1]This expression is based on the assumption that there is no re-entrainment of deposited particles. If re-entrainment is present, additional terms must be added. For example, $\partial \sigma / \partial \theta$ may be expected as $\lambda u_s c - a\sigma$ where a is the re-entrainment rate constant.

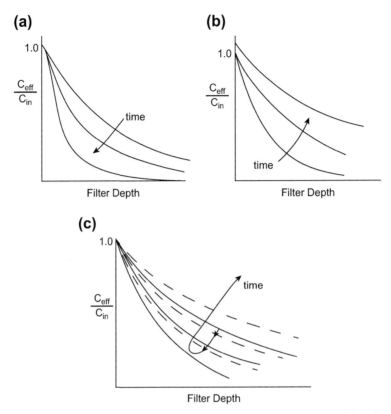

FIGURE 5.2 Effect of deposition on filter performance (a) performance enhanced with time, (b) performance declined with time, (c) mixed behavior.

The various behaviors displayed in Fig. 5.2 may be explained as follows: As a result of deposition, the geometry and size of filter grains are modified. The flow field near filter grains may be altered, and the magnitude and nature of the particle–filter grain (including deposited particles) interactions may also be changed. The combined effect of these changes may enhance deposition, retard deposition, or have no effect. If the net effect remains favorable (enhancing deposition), the behavior shown in Fig. 5.2(a) is observed. In contrast, Fig. 5.2(b) depicts the situation if the net effect is unfavorable (retarding deposition). The mixed behavior of Fig. 5.2(c) corresponds to the case of transition from favorable to unfavorable (curve 1) or from unfavorable to favorable (curve 2).

Equations (5.2.2) and (5.2.4), together with appropriate initial and boundary conditions [such as those of Equations (5.1.5) and (5.1.6)], give a complete description of the particle concentration distributions in the suspension and filter medium as functions of position and time. Specifically these results enable the prediction of the effluent concentration history (c at $z = L$ vs. time) and the extent of deposition throughout the medium (σ vs. z at various times). This, of course, assumes that the values of λ_0 and the functional form of F are known.

Table 5.1   List of Expressions for $F(\alpha, \sigma)$

$$\frac{\lambda}{\lambda_0} = F(\alpha, \sigma)$$

	Expression	Adjustable Parameters	Investigators
(1)	$F = 1 + b\sigma;\ b > 0$	b	Iwasaki (1937) Stein (1940)
(2)	$F = 1 - b\sigma;\ b > 0$	b	Ornatski et al. (1955) Mehter et al. (1970)
(3)	$F = 1 - \dfrac{\sigma}{\varepsilon_0}$	-	Shekhtman (1961) Heertjes and Lerk (1967)
(4)	$F = 1 - \dfrac{\sigma}{\sigma_{\text{ult}}}$	-	Maroudas and Eisenklam (1965)
(5)	$F = \left(\dfrac{1}{1 + b\sigma}\right)^n;\ b > 0, n > 0$	b, n	Mehter et al. (1970)
(6)	$F = \left[\dfrac{\phi(\sigma)/\phi_0}{\varepsilon_0 - \dfrac{\sigma}{1 - \varepsilon_d}}\right]^n$	n	Deb (1969)
(7)	$F = \left(\dfrac{b\sigma}{\varepsilon_0}\right)^{n_1} \left(1 - \dfrac{\sigma}{\varepsilon_0}\right)^{n_2};\ b > 0$	b, n_1, n_2	Mackrle et al. (1965)
(8)	$F = 1 + b\sigma - \dfrac{a\,\sigma^2}{\varepsilon_0 - \sigma};\ b > 0,\ a > 0$	a, b	Ives (1960)
(9)	$F = \left(1 + \dfrac{b\sigma}{\varepsilon_0}\right)^{n_1} \left(1 - \dfrac{\sigma}{\varepsilon_0}\right)^{n_2} \left(1 - \dfrac{\sigma}{\sigma_{\text{ult}}}\right)^{n_3};\ b > 0$	b, n_1, n_2, n_3	Ives (1969)

The accumulation of deposited particles within a filter medium causes a change of the structure of the medium, which, in turn, results in a decrease of the medium permeability. Since the extent of deposition is not uniform, this implies that the medium permeability during filtration, in general, is not constant, but varies along the z-direction (i.e., the direction of suspension flow). If the fluid throughput is kept constant, the pressure drop required to maintain a constant throughput varies with time and can be calculated as follows.

Assuming that the flow of fluid through porous media (either granular or fibrous) may be described by Darcy's law, or

$$u_s = -\frac{k}{\mu}\frac{\partial p}{\partial z} \tag{5.2.5}[2]$$

The pressure drop $(-\Delta p)$ across a filter medium of height L, is

$$(-\Delta p) = \mu\,u_s \int_0^L \frac{dz}{k} \tag{5.2.6}$$

[2]Equation (5.2.5) can be seen to be a special case of Equation (2.2.12) with $q_s = 0$ and $q_\ell = u_s$. A more detailed discussion of the flow rate–pressure drop relationship for fluid flow in porous media is given in 5.5.

Initially, the pressure drop is $(-\Delta p)_0$ and the filter permeability, k_0, is constant throughout the entire medium, or

$$(-\Delta p)_0 = \frac{\mu\, u_s}{k_0} L \qquad (5.2.7)$$

Dividing Equations (5.2.6) by (5.2.7) yields

$$\frac{(-\Delta p)}{(-\Delta p)_0} = \frac{1}{L}\int_0^L \left(\frac{k_0}{k}\right) dz \qquad (5.2.8a)$$

The differential form of the above expression is

$$\frac{(\partial p/\partial z)}{(\partial p/\partial p)_0} = \frac{k_0}{k} \qquad (5.2.8b)$$

where $(\partial p/\partial z)$ is the local pressure gradient necessary to maintain a given constant suspension throughput.

The change of local permeability (or pressure gradient) results from the accumulation of deposited particles. Similar to Equation (5.2.3a), one may write

$$\frac{k_0}{k} = \frac{(\partial p/\partial z)}{(\partial p/\partial z)_0} = G(\beta,\ \sigma) \qquad (5.2.9)$$

Table 5.2   List of Expressions for $G(\beta, \sigma)$

$$\frac{(\partial P/\partial z)}{(\partial P/\partial z)_0} = G(\beta, \sigma)$$

	Expression	Adjustable Parameters	Investigators
(1)	$G = 1 + d\sigma;\ d > 0$	d	Mehter et al. (1970)
(2)	$G = 1 + d\dfrac{\sigma}{\varepsilon_0};\ d > 0$	d	Mints (1966)
(3)	$G = \left(\dfrac{1}{1 - d\sigma}\right)^{m_1};\ d > 0,\ m_1 > 0$	d, m_1	Mehter et al. (1970)
(4)	$G = \left(1 - \dfrac{2\sigma}{\tilde{\beta}}\right)^{-1/2};\ \tilde{\beta} > 0$	-	Maroudas and Eisenklam (1965)
(5)	$G = \left\{1 + d[1 - 10^{-m_1\sigma/(1-\varepsilon_d)}]\right\}\left\{\dfrac{\varepsilon_0}{\varepsilon_0 - \dfrac{\sigma}{1 - \varepsilon_d}}\right\}^3$	a, m_1	Deb (1969)
(6)	$G = \left(1 + \dfrac{d\sigma}{\varepsilon_0}\right)^{m_1}\left(1 - \dfrac{\sigma}{\varepsilon_0}\right)^{m_2};\ d > 0,\ m_1 > 0,\ m_2 > 0$	d, m_1, m_2	Ives (1969)
(7)	$G = 1 + f\left\{\left(\lambda_0 + d\varepsilon_0\right)\sigma + \left(\dfrac{e + d}{2}\right)^2 + d\varepsilon_0^2 \ln\left(\dfrac{\varepsilon_0 - \sigma}{\varepsilon_0}\right)\right\};\ f, d, e > 0$	f, d, e	Ives (1961)
(8)	$G = \left(\dfrac{\varepsilon_0}{\varepsilon_0 - \sigma}\right)^3 \left(\dfrac{1 - \varepsilon_0 + \sigma}{1 - \varepsilon_0}\right)^2 \left\{\sqrt{\dfrac{\sigma}{3(1 - \varepsilon_0)} + \dfrac{1}{4}} + \dfrac{\sigma}{3(1 - \varepsilon_0)} + \dfrac{1}{2}\right\}$		Camp (1964)

and Equation (5.14a) becomes

$$\frac{(-\Delta p)}{(-\Delta p)_0} = \int_0^1 G(\beta, \sigma)\mathrm{d}\left(\frac{z}{L}\right)$$

(5.2.10)

Table 5.2 gives a list of some of expressions of $G(\beta, \sigma)$ proposed previously.

The dynamic behavior of deep bed filtration is, therefore, given by Equations (5.1.4b), (5.2.4), (5.1.5), (5.1.6), and (5.2.8a) or (5.2.10). The parameters and functions present in this system of equations are λ_0, k_0, $F(\alpha, \sigma)$ and $G(\beta, \sigma)$. Their determinations, correlations, and estimations will be discussed in the following sections.

■ ■ ■ ▬▬▬▬▬▬▬▬▬▬▬▬▬▬▬▬▬▬▬▬▬▬▬▬▬▬▬▬▬▬▬

Illustrative Example 5.2

For deep bed filtration, assuming that $F=1$ or $\lambda = $ constant (see Equation (5.2.3a)) and $G = 1 + b\sigma$.

(a) Obtain an expression of $(-\Delta p)/(-\Delta p)_0$ from Equation (5.2.10) in terms of operating and system variables (namely, filter medium depth, media porosity, suspension flow rate, etc.).

(b) Calculate $(-\Delta p)/(-\Delta p)_0$ vs. time for the following conditions:

$$b = 1200$$

$$\lambda = 1.02 \times 10^{-2}\ \mathrm{m}^{-1}$$

$$L = 1\ \mathrm{m}$$

$$u_s = 1.5 \times 10^{-3}\ \mathrm{m\,s}^{-1}$$

$$c_{in} = 10^{-4}\ \mathrm{vol/vol}$$

$$\varepsilon = 0.49$$

Solution

(a) From Equation (5.2.10), $(-\Delta p)/(-\Delta p)_0$ may be rewritten as

$$\frac{(-\Delta p)}{(-\Delta p)_0} = \frac{1}{L}\int_0^L G(\beta, \sigma)\mathrm{d}z = \frac{1}{L}\int_0^L (1 + b\sigma)\mathrm{d}z$$

$$= 1 + \frac{b}{L}\int_0^L \sigma\, \mathrm{d}z$$

(i)

The expression of the solidosity distribution, σ vs. z, can be found from the solution of the macroscopic conservation equation and the filtration rate equation (i.e., Equations (5.1.4b) and (5.2.4) with appropriate initial and boundary conditions, or

$$u_s\frac{\partial c}{\partial z} + \frac{\partial \sigma}{\partial \theta} = 0$$

(ii)

$$\frac{\partial \sigma}{\partial \theta} = u_s \lambda c \qquad \text{(iii)}$$

with $\lambda = \lambda_0 = $ constant and $c = c_{in}$ at $z = 0$:

$$\frac{c}{c_{in}} = \exp(-\lambda_0 z) \qquad \text{(iv)}$$

with c given by Equation (iv). The specific deposition profile is found to be

$$\sigma = [\lambda \, u_s c_{in} \exp(-\lambda_0 z)]\theta \quad \text{for } \theta > 0$$

or

$$\sigma = [\lambda_0 \, u_s c_{in} \exp(-\lambda_0 z)] \left[t - \int_0^z \frac{dz}{u_s/\varepsilon} \right] \quad \text{for } t > t_0 = \int_0^L \frac{dz}{u_s/\varepsilon} \qquad \text{(v)}$$

To obtain the σ-profile for $t < t_0$, note that it requires a finite time for the feed to reach a certain depth. Corresponding to t, the filter depth to be traveled, z_A is given as

$$\int_0^{z_A} \frac{dz}{u_s/\varepsilon} = t \qquad \text{(vi)}$$

and

$$\sigma = 0, \qquad z_A < z < L$$

$$\sigma = [\lambda_0 \, u_s c_{in} \exp(-\lambda_0 z)] \left[t - \int_0^z \frac{dz}{u_s/\varepsilon} \right] \quad 0 < z < z_A \qquad \text{(vii)}$$

To obtain the expression of $(-\Delta p)/(-\Delta p)_0$, Equation (i) may be applied. For $t > t_0$, with the σ-profile given by Equation (v), the integral $\int_0^L \sigma \, dz$ is found to be

$$\int_0^L \sigma \, dz = u_s c_{in} \lambda_0 \left\{ \int_0^L [\exp(-\lambda_0 z)] \left(t - \frac{\varepsilon z}{u_s} \right) dz \right\}$$

$$= u_s c_{in} \lambda \left\{ \left[-\frac{1}{\lambda_0} \exp(\lambda_0 z) \right] \left(t - \frac{\varepsilon z}{u_s} \right) \Big|_0^L - \int_0^L \left[\frac{-1}{\lambda_0} \exp(-\lambda_0 z) \right] \left[-\frac{\varepsilon}{u_s} \right] dz \right\}$$

$$= u_s c_{in} \left[t - \left(t - \frac{\varepsilon L}{u_s} \right) \exp(-\lambda_0 L) \right] - u_s c_{in} (\varepsilon/u_s) \int_0^L \exp(-\lambda_0 z) dz$$

$$= u_s c_{in} \left[t - \left(t - \frac{\varepsilon L}{u_s} \right) \exp(-\lambda_0 L) \right] + (c_{in} \varepsilon/u_s)[\exp(-\lambda_0 L) - 1]$$

and the pressure drop expression $(-\Delta p)/(-\Delta p)_0$ may be expressed as

$$\frac{(-\Delta p)}{(-\Delta p)_0} = 1 + \frac{b}{L}\int_0^L \sigma\, dx$$

$$= 1 + (bc_m)(u_s t/L)\left[1 - \left(1 - \frac{\varepsilon}{(u_s t/L)}\right)\exp(-\lambda L)\right] + \frac{(bc_{in})\varepsilon}{\lambda_0 L}[\exp(-\lambda_0 L) - 1] \quad \text{(viii)}$$

$$\text{for } t > t_0 = \int_0^L \frac{dz}{(u_s/\varepsilon)}$$

To obtain $(-\Delta p)/(-\Delta p)_0$ for $t < t_0$, the σ-profile of Equation (vii) may be used to evaluate $\int_0^L \sigma\, dz$ or

$$\int_0^L \sigma\, dz = \int_0^{z_A} u_s\lambda c_{in}[\exp(-\lambda_0 z)]\left(t - \frac{\varepsilon z}{u_s}\right)dz$$

$$= u_s c_{in}\left[t - \left(t - \frac{\varepsilon z_A}{u_s}\right)\exp(-\lambda_0 z_A)\right] + \frac{c_{in}\varepsilon}{\lambda_0}[\exp(-\lambda_0 z_A) - 1]$$

$$= u_s c_{in} t + \frac{c_{in}\varepsilon}{\lambda_0}[\exp(-\lambda_0 z_A) - 1]$$

and the pressure drop ratio expression is

$$(-\Delta p)/(-\Delta p)_0 = 1 + (bc_{in})\left(\frac{u_s t}{L}\right) + \frac{(bc_{in})\varepsilon}{\lambda L}[\exp(-\lambda_0 z_A) - 1] \quad \text{(ix)}$$

$$z_A = (t/t_0)L \quad \text{for } t < t_0$$

(b) With the given conditions, one has

$$t_0 = \frac{\varepsilon L}{u_s} = \frac{(0.46)(1)}{1.5 \times 10^{-3}} = 327 \text{ s}$$

$$bc_{in} = (1200)(10^{-4}) = 1.2 \times 10^{-3}$$

$$\lambda_0 L = 1.02 \times 10^{-3}$$

The results of $(-\Delta p)/(-\Delta p)_0$ are found from Equations (viii) and (v) are

Time(s)	Pressure Drop Ratio	Time(s)	Pressure Drop Ratio	Time(s)	Pressure Drop Ratio
50	1.0002	300	1.0030	6000	1.1070
100	1.0004	500	1.0060	10000	1.1796
150	1.0009	1000	1.0152	20000	1.3622
200	1.0005	2000	1.0334	30000	1.5488
250	1.0023	4000	1.0700	40000	1.7276

The results are also shown in the following figures.

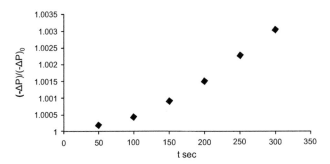

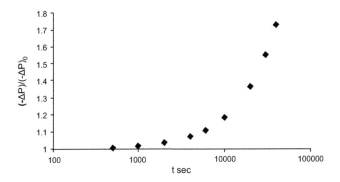

5.3   Physical Significance of the Filter Coefficient

In formulating the filtrate rate expression of Equation (5.2.2), λ is used as a lumped parameter. Its interpretation as being either a reaction rate constant or a particle deposition probability does not yield any insight which enables its prediction from fundamental considerations. In fact, according to Equation (5.2.2), a filter medium can only be regarded as a "black box" with particle collecting capabilities.

To provide a theoretical conceptual framework of λ, Payatakes (1973) and Tien and Payatakes (1979) suggested that a homogeneous, random packed granular medium may be considered as a number of unit bed elements (UBE) connected in series. Each UBE has a unit cross section and a thickness, ℓ, known as the length of periodicity defined as follows: Consider a cubic volume element of $N\ell \times N\ell \times N\ell$, where N is a large integer. N^3 granular (spheres) may be accommodated within the cube as $N \rightarrow \infty \cdot \ell$, therefore, is expressed as

$$(1 - \varepsilon)(N\ell)^3 = \sum_{i=1}^{N} \left(\frac{\pi}{6}\right)(d_{g_i})^3 \quad \text{as } N \rightarrow \infty$$

where d_{g_i} is the size (diameter) of the i-th granule. The above expression may be re-arranged to give

$$\ell = \left[\frac{\pi}{6(1-\varepsilon)} \right]^{1/3} <d_g> \tag{5.3.1a}$$

$$\text{with} \quad <d_g> = \lim_{N \to \infty} \frac{\left[\sum_{i=1}^{N} (d_{g_i})^3 \right]^{1/3}}{N} \tag{5.3.1b}$$

where $<d_g>$ is the average filter-grain diameter. (In other words, the granules which form the medium are not uniform in size.) Note that by this definition, ℓ is of the magnitude of $<d_g>$ but can be either greater or less than $<d_g>$ depending upon the porosity value.

For a granular medium so described, for each UBE, there present within it a number of collectors of specified size and geometry. The particle collection efficiency of a UBE (or unit collector efficiency), e, depends upon the suspension flow field around (or through) the collectors, the nature and mechanism of the transport of particles from the suspensions to the collector surface, and the adhesion of particles on collector surfaces. A schematic representation based on the UBE concept is shown in Fig. 5.3.

With the schematic representation given by Fig. 5.3, the overall collection efficiency of a deep bed filter is

$$E = \frac{c_{in} - c_{eff}}{c_{in}} \tag{5.3.2}$$

where c_{in} and c_{eff} denote the influent and effluent particle concentrations of a filter bed, respectively. One should also note that in most aerosol science literatures, filtration performance is often given by the so-called penetration, P, defined as the effluent and influent concentration ratio, c_{eff}/c_{in}, or

$$E = 1 - P \tag{5.3.3}$$

The overall collection efficiency of a filter can be expressed in terms of the unit collector efficiencies of the unit bed element constituting the filter. Referring to Fig. 5.3, let c_i denote the particle concentration of the suspension exiting the i-th UBE, the collection efficiency of the i-th UBE, e_i, is

$$e_i = \frac{c_{i-1} - c_i}{c_{i-1}} \tag{5.3.4}$$

The overall collection efficiency, E, can be written as

$$E = 1 - \prod_{i=1}^{N} (1 - e_i) \tag{5.3.5}$$

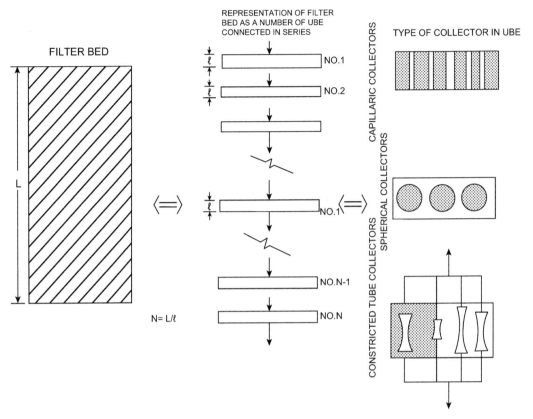

FIGURE 5.3 Schematic representation of filter medium.

where N is the total number of UBE's connected in series. For a filter of height L, the total number of UBE's counted in series, N, is

$$N = L/\ell \tag{5.3.6}$$

or, more precisely, N should be taken as the integer closest to the value of L/ℓ.

From the phenomelogical filtration rate expression given before, the suspended particle concentration gradient, $\partial c/\partial z$, is proportional to c with λ being the proportionality constant. Integrating Equation (5.2.1) from $z = (i-1)\ell$ to $z = i\ell$, and recognizing that ℓ is a small quantity so that over a distance of ℓ, λ may be considered a constant, or

$$\ell n \, \frac{c_{i-1}}{c_i} = \lambda_i \ell \tag{5.3.7}$$

A comparison of Equations (5.3.7) with (5.3.4) and dropping the subscript yields

$$\lambda \ell = -\ln(1 - e)$$

or

$$\lambda = \frac{1}{\ell} \ell n \frac{1}{1-e} \tag{5.3.8a}$$

For $e \ll 1$, the above expression becomes

$$\lambda \cong e/\ell \tag{5.3.8b}$$

To relate the filter coefficient, λ, with the collector efficiencies of the collectors present in a UBE, for the general case that collectors within a UBE are of different sizes, the collection efficiency of a unit bed element represents the combined particle collection by all collectors in that unit bed element. Based on particle mass balances one has, for the i-th unit bed element,

$$u_s(c_{i-1} - c_i) = \sum_{j=1}^{N_c}(q_i)_j \left[c_{i-1} - (c_i)_j \right] \tag{5.3.9}$$

where c_{i-1} and c_i are the particle concentrations of the suspension entering and leaving the i-th unit bed element and $(q_i)_j$ is the suspension flow rate through (or over) the j-th collector of the i-th unit bed element and there are N_c collectors in each UBE. All these collectors have the same suspension influent, c_{i-1} and, generally speaking, different effluent concentration, $(c_i)_j$, $j = 1, \ldots N_c$. The mixing-cup mean effluent concentration gives the value of c_i or

$$c_i = \frac{\sum_{j=1}^{N_c}(q_i)_j \ (c_i)_j}{\sum_{j=1}^{N_c}(q_i)_j} \tag{5.3.10}$$

Since

$$e_i = (c_{i-1} - c_i)/c_{i-1} \tag{5.3.11a}$$

$$(\eta_i)_j = \left[c_{i-1} - (c_i)_j \right] \Big/ c_{i-1} \tag{5.3.11b}$$

and $(\eta_i)_j$ is the collector efficiency of the j-th collector present in the i-th unit bed element.

Combining Equations (5.3.9), (5.3.11a) and (5.3.11b), one has

$$e_i = (1/u_s) \sum_{j=1}^{N_c}(q_i)_j(\eta_i)_j \tag{5.3.12}$$

Substituting Equations (5.3.12) into (5.3.8b) yields

$$\lambda = \left[\sum_{j=1}^{N_c}(q_i)_j(\eta_i)_j \right] \Big/ (u_s \ell) \tag{5.3.13}$$

Equation (5.3.13) expresses the filter coefficient in terms of the collector efficiencies of the N_c collectors present in a UBE. Considering the flow of a suspension past a collector surface, a suspended particle may be removed from the suspension and deposited onto the collector surface if the particle, in its movement, makes contact with the surface and the particle-collector surface interactions are favorable for particle adhesion. Thus, by tracking particle trajectories, the collector efficiency of a given collector may be predicted. Once the collector efficiencies of the collectors are known, the filter coefficient can then be determined through Equation (5.3.13). This approach, in principle, can be applied to collectors free of deposited particles as well as those with specific degree of deposition. In fact, based on these efficiencies, F as defined by Equation (5.2.3a) can be determined.

The principle as outlined above is conceptually simple, although its implementation may be is computationally demanding or even impractical. Through studies carried out during the past several decades, a degree of success has been achieved, mainly in the estimation of λ_0, the clean filter coefficients. This will be discussed in Chapter 6. Equation (5.3.13) may be significantly simplified if specific models are used for filter media characterization as shown in the following sections.

■ ■ ■ ▬▬▬▬▬▬▬▬▬▬▬▬▬▬▬▬▬▬▬▬▬▬▬▬▬▬▬▬

Illustrative Example 5.3

Obtain an expression of ℓ for fibrous media.

Solution

The length of periodicity of fibrous media can be obtained by following the same procedure given in 5.3. Assuming that fibers constituting fibrous media are cylinders of diameter d_f and length ℓ_f. A simple volume element of $N\ell \times N\ell \times \sum_{i=1}^{N}(\ell_f)_i/N$ where $(\ell_f)_i$ is the length of the i-th cylindrical collector present in a UBE containing N (a large integer) fibers. By definition

$$(N\ell)(N\ell)\left[\sum_{i=1}^{N}(\ell_f)_i/N\right](1-\varepsilon) = \sum_{i=1}^{N}\left(\frac{\pi}{4}\right)\left(d_{f_i}\right)^2\ell_{f_i}$$

and

$$\ell = \left[\frac{\pi}{4(1-\varepsilon)}\right]^{1/2}\left[\lim_{N\to\infty}\frac{\sum_{i=1}^{N}(d_{f_i})^2\ell_{f_i}/N}{\sum_{i=1}^{N}\ell_{f_i}/N}\right]^{1/2}$$

$$= \left[\frac{\pi}{4(1-\varepsilon)}\right]^{1/2}<d_f>$$

with

$$<d_f> = \left[\frac{\sum_{i=1}^{N}(d_{f_i})^2\ell_{f_i}}{(\ell_f)_{av}N}\right]^{1/2}$$

and

$$(\ell_f)_{av} = \sum_{i=1}^{N} \ell_{f_i}/N$$

It is interesting to compare the value of ℓ given by the above expression with that of Equation (5.3.2a) assuming $<d_f> = <d_g>$ and all fibers have the same length for two media with equal porosity. The rate of the two ℓ's is simply

$$\frac{(\ell)_{fibrous\ media}}{(\ell)_{granular\ media}} = \frac{(\pi/4)^{1/2}}{(\pi/6)^{1/3}} = \frac{0.8862}{0.806} \cong 1.10$$

or a difference of approximately 10%.

■ ■ ■

5.4   Representation of Filter Media with Cell Models

5.4.1   Happel's Model for Granular Media

For studies of filtration in granular media, Happel's model is often applied. According to Happel's model (Happel, 1958), a granular medium is represented as an assembly of identical cells consisting of a solid sphere of radius a_c surrounded by a fluid envelope of b (see Fig. 5.4). a_c is taken to be

$$a_c = \frac{1}{2} <d_g> \tag{5.4.1}$$

and $\quad a_c/b = (1 - \varepsilon)^{1/3}$ (5.4.2)

The fluid flow is assumed to be incompressible and axisymmetric and is sufficiently slow so that the inertial effect may be ignored. The stream function of the flow ψ satisfies the following equation:

$$E^4\psi = 0 \tag{5.4.3}$$

where

$$E^2 = \frac{\partial^2}{\partial r^2} + \frac{\sin\theta}{r^2} \frac{\partial}{\partial\theta} \left(\frac{1}{\sin\theta} \frac{\partial}{\partial\theta}\right) \tag{5.4.4}$$

The radial and tangential velocity components, u_r and u_θ, are

$$u_r = \frac{-1}{r^2 \sin\theta} \frac{\partial\psi}{\partial\theta} \tag{5.4.5}$$

$$u_\theta = \frac{1}{r^2 \sin\theta} \frac{\partial\psi}{\partial r} \tag{5.4.6}$$

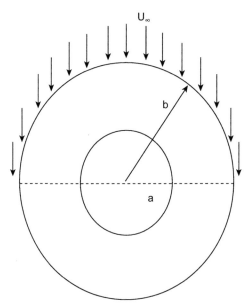

FIGURE 5.4 Happel's model for granular media.

with the following boundary conditions

$$u_r = u_\theta = 0, \; r = a_c \tag{5.4.7a}$$

$$u_r = -u_s \cos\theta, \; r = b \tag{5.4.7b}$$

$$\frac{1}{r}\frac{\partial u_r}{\partial \theta} + r\frac{\partial}{\partial r}\left(\frac{u_\theta}{r}\right) = 0, \; r = b \tag{5.4.7c}$$

with the linear velocity entering the fluid envelope taken to be the superficial velocity u_s.

The general solution of ψ satisfying Equation (5.4.3) may be written as (Happel and Brenner, 1965)

$$\psi = A\left(\frac{K_1}{r^*} + K_2 r^* + K_3 r^{*2} + K_4 r^{*4}\right)\sin^2\theta \tag{5.4.8}$$

$$r^* = r/a_c \tag{5.4.9}$$

Applying the boundary conditions of Equations (5.4.7a)–(5.4.7b) to Equations (5.4.3), (5.4.5a) and (5.4.5b), the coefficients A, $K_1 - K_4$ are found to be

$$A = (u_s/2)a_c^2 \tag{5.4.10a}$$

$$K_1 = 1/w \tag{5.4.10b}$$

$$K_2 = -(3 + 2p^5)/w \qquad (5.4.10c)$$

$$K_3 = (2 + 3p^5)/w \qquad (5.4.10d)$$

$$K_4 = -p^5/w \qquad (5.4.10e)$$

and

$$w = 2 - 3p + 3p^5 - 2p^6 \qquad (5.4.11a)$$

$$p = a_c/b = (1 - \varepsilon)^{1/3} \qquad (5.4.11b)$$

Using Happel's model for filter medium description implies that the N_c collectors present in a UBE of a filter bed are spherical and identical. The spherical collector has a radius $a_c(= d_g/2)$ surrounded by a fluid envelope of radius b. The total number of the spherical collectors present in a UBE, N_c, is

$$N_c = \frac{\ell(1 - \varepsilon)}{\frac{\pi}{6} <d_g>^3} = \left[\frac{\pi}{6(1 - \varepsilon)}\right]^{1/3} \frac{(1 - \varepsilon)}{\left(\frac{\pi}{6}\right) <d_g>^3} <d_g>$$

$$\qquad (5.4.12)$$

$$= \left[\frac{6(1 - \varepsilon)}{\pi}\right]^{2/3} <d_g>^{-2}$$

The fluid flow rate over a collector, q_i, is

$$q_i = \pi b^2 u_s \qquad (5.4.13a)$$

and all collectors have the same collector efficiency, or

$$(\eta_i)_j = \eta_i \qquad (5.4.13b)$$

Substituting Equations (5.4.13a) and (5.4.13b) into Equation (5.3.9), one has

$$e_i = \frac{N_c}{u_s} (\pi)(b^2) u_s \eta_i$$

$$\qquad (5.4.14)$$

$$= \left[\frac{6(1 - \varepsilon)}{\pi}\right]^{2/3} <d_g>^{-2} \pi \left(\frac{b}{a_c}\right)^2 \eta_i = \left(\frac{9}{16}\pi\right)^{1/3} \eta_i$$

Combining Equations (5.3.8b), (5.4.14), and (5.3.1a), for granular media, the relationship between the filter coefficient and the collector efficiency based on Happel's model is found to be

$$\lambda = \frac{3(1 - \varepsilon)^{1/3}}{2<d_g>} \eta \qquad (5.4.15)$$

and the filtration rate is

$$\frac{\partial \sigma}{\partial \theta} = u_s \frac{3(1 - \varepsilon)^{1/3}}{2<d_g>} \eta c \qquad (5.4.16)$$

The collector efficiency, η, by definition, gives the fraction of the total amount of particles passing over a collector, which has been deposited, or

$$\eta_i = \frac{J_i}{u_s(\pi b^2)c_{i-1}} \tag{5.4.17}$$

where J_i is the total deposition flux over a collector, or

$$J_i = u_s(\pi b^2)c_{i-1}\,\eta_i$$

Historically, in aerosol filtration studies, J is commonly expressed as

$$J_i = (u_s)(\pi a_c^2)c_{i-1}\,\eta_{s_i} \tag{5.4.18a}$$

or

$$\eta_{s_i} = \frac{J_i}{u_s(\pi a_c^2)c_{i-1}} \tag{5.4.18b}$$

η_{s_i} is known as the single collector efficiency. In reality, η_{s_i}, in spite of being termed efficiency, is actually a dimensionless flux.[3] The relationship between η_ℓ and η_{s_i} is

$$\frac{\eta_{s_i}}{\eta_i} = \left(\frac{b}{a_c}\right)^2 = (1-\varepsilon)^{-2/3} \tag{5.4.19}$$

The relationship between λ and η_s is simply

$$\lambda = \frac{3(1-\varepsilon)}{2<d_g>}\,\eta_s \tag{5.4.20}$$

Based on the drag force acting on the solid sphere, the pressure drop per unit medium length, $(-\Delta p/\Delta L)$ is found to be (Happel, 1958)

$$-\frac{\Delta p}{\Delta L} = \frac{3+2p^5}{3-(9/2)p+(9/2)p^5-3p^6}\,\frac{(18)\mu(1-\varepsilon)}{<d_g>^2}\,u_s \tag{5.4.21}$$

The above expression gives the pressure drop-flow rate relationship of fluid flow through a granular medium. Comparing the above expression with Darcy's law given by Equation (5.2.5), the medium permeability, k, is found to be

$$k = \frac{<d_g>^2}{18(1-\varepsilon)}\,\frac{3-(9/2)p+(9/2)p^5-3p^6}{3+2p^5} \tag{5.4.22}$$

5.4.2   Kuwabara's Model for Fibrous Media

For deep bed filtration in fibrous media, Kuwabara's model is often used for media characterization. Similar to Happel's model, Kuwabara's model is also of the cell model type. A fibrous medium is represented by a collection of solid fiber (of

---

[3]Note that with this definition, η_{s_i} may be greater than unity. Physically speaking, any properly defined efficiency should not be greater than 100%.

random a_c) surrounded by a fluid envelope of radius b with the cells placed normal to the direction of the main flow. Analogous to Equations (5.4.1) and (5.4.2), we now have

$$a_c = <d_f> \tag{5.4.23}$$

$$(a_c/b) = (1 - \varepsilon)^{1/2} \tag{5.4.24}$$

The flow is assumed to be incompressible and axisymmetric. With negligible inertial effect, the stream function expressed in the cylindrical polar coordinate, ψ, is given as

$$\psi = \left[Ar + \frac{B}{r} + Cr \, \ell n \frac{r}{a_c} + Dr^3 \right] \sin \theta \tag{5.4.25}$$

and the velocity components are

$$u_r = \frac{1}{r} \frac{\partial \psi}{\partial \theta}, \qquad u_\theta = -\frac{\partial \psi}{\partial r} \tag{5.4.26}$$

The coefficient A, B, C, and D are given as (Kuwabara, 1959; Brown, 1993)

$$A = \left(\frac{-1 + \varepsilon_s}{2} \right) \frac{u_s}{Ku} \tag{5.4.27a}$$

$$B = \frac{a_c^2}{2} \left(1 - \frac{\varepsilon_s}{2} \right) \frac{u_s}{Ku} \tag{5.4.27b}$$

$$C = \frac{u_s}{Ku} \tag{5.4.27c}$$

$$D = \left(\frac{-\varepsilon_2}{4a_c^2} \right) \frac{u_s}{Ku} \tag{5.4.27d}$$

$$Ku = \left[- (1/2)\ell n \, \varepsilon_s - (3/4) + \varepsilon_s - (\varepsilon_s^2/4) \right] \tag{5.4.27e}[4]$$

The above results were obtained using boundary conditions similar to those of Happel's model with the exception that, instead of zero shear stress at the fluid envelope [i.e., Equation (5.4.4)], the vorticity, $\nabla \times u$ is assumed to vanish at $r = b$.

To determine the number of collectors present in a UBE, from the result of Illustrative Example 5.3, the length of the periodicity ℓ, of fibrous media, is

$$\ell = \frac{\pi^{1/2} a_c}{(1 - \varepsilon)^{1/2}} = \left(\frac{\pi}{1 - \varepsilon} \right)^{1/2} \frac{<d_f>}{2} \tag{5.4.28}$$

---

[4]Assuming that the fluid velocity entering into the fluid envelope is u_s.

Assuming that all fibers are identical (namely, same diameter d_f and same length, ℓ_f) the number of fiber collectors present in a unit bed element ($\ell \times 1 \times 1$) is

$$N_c = \frac{\ell(1-\varepsilon)}{(\pi/4)<d_f>^2 \ell_f} \tag{5.4.29}$$

The relationship between the unit bed element collection efficiency, e, and the fiber collector efficiency, η_f, can be obtained from Equation (5.3.9) and the condition that all the collectors are cylindrical and identical in size, or

$$(u_s)(e) = (N_c)(u_s)\ell_f\left[<d_f>/(1-\varepsilon)^{1/2}\right]\eta_f$$

$$= (u_s)\frac{\ell(1-\varepsilon)}{(\pi)(<d_f>/2)^2}\frac{<d_f>}{(1-\varepsilon)^{1/2}}\eta_f \tag{5.4.30}$$

$$= (u_s)\frac{(1-\varepsilon)}{(\pi)(<d_f>/2)^2}\frac{<d_f>}{(1-\varepsilon)^{1/2}}\left(\frac{\pi}{1-\varepsilon}\right)^{1/2}\left(\frac{<d_f>}{2}\right)\eta_f$$

$$\text{or} \quad e = \frac{2\eta_f}{\pi^{1/2}} \tag{5.4.31}$$

and

$$\lambda = \frac{c}{\ell} = \frac{4\eta_f}{\pi^{1/2}}\left(\frac{1-\varepsilon}{\pi}\right)^{1/2}\frac{1}{<d_f>} = \frac{4(1-\varepsilon)^{1/2}\eta_f}{\pi<d_f>} \tag{5.4.32a}$$

and the particle concentration gradient, $\partial c/\partial z$, can be expressed as

$$\frac{\partial c}{\partial z} = -\frac{4(1-\varepsilon)^{1/2}\eta_f \, c}{\pi<d_f>} \tag{5.4.32b}$$

The single fiber efficiency, $(\eta_f)_s$, is related to the fiber efficiency, η_f, as

$$(\eta_f)_s = \eta_f/(1-\varepsilon)^{1/2} \tag{5.4.33a}$$

Equations (5.4.32a) and (5.4.32b) now become

$$\lambda = \frac{4(1-\varepsilon)(\eta_f)_s}{\pi<d_f>} \tag{5.4.33b}$$

$$\text{and} \quad \frac{\partial c}{\partial z} = -\frac{4(1-\varepsilon)(\eta_f)_s}{\pi<d_f>}c \tag{5.4.34a}$$

$$\frac{\partial \sigma}{\partial \theta} = \frac{4(1-\varepsilon)(\eta_f)_s}{\pi<d_f>}u_s c \tag{5.4.34b}$$

Furthermore, if the fluid velocity entering the fluid envelope is often assumed to be the interstitial velocity instead of the superficial velocity, an additional term, ε, then appears in the denominator of Equation (5.4.32b) and $\partial c/\partial z$ becomes

$$\frac{\partial c}{\partial z} = -\frac{4(1-\varepsilon)(\eta_f)_s}{\pi\varepsilon<d_f>}c \tag{5.4.34c}$$

Practically speaking, for fiber media, ε is approximately 0.9 in most cases. The difference due to the presence of ε is therefore insignificant.

The pressure drop across a fibrous medium of height L, (Δp), can be shown to be (Brown, 1993)

$$\Delta p = \frac{16(1-\varepsilon)L\,\mu\,u_s}{<d_f>^2\left[-(1/2)\ell n(1-\varepsilon) - 0.75 + (1-\varepsilon) - (1-\varepsilon)^2/4\right]} \tag{5.4.35}$$

From the above expressions, the permeability of homogeneous fibrous media according to the Kuwabara model is

$$k = \frac{<d_f>^2}{16(1-\varepsilon)}\left[-(1/2)\ell n(1-\varepsilon) - 0.75 + (1-\varepsilon) - (1-\varepsilon)^2/4\right] \tag{5.4.36}$$

■ ■ ■ ▬▬▬▬▬▬▬▬▬▬▬▬▬▬▬▬▬▬▬▬▬▬▬▬▬▬▬▬▬▬▬▬▬

Illustrative Example 5.4

A granular medium is composed of particles of average grain diameter of 0.59 mm and a porosity of 0.49. The filter coefficient of the medium is found to be 1.02×10^{-3} m^{-1}. What are the corresponding values of the spherical collector efficiency and the single collector efficiency?

From Equations (5.4.15) and (5.4.20), one has

$$\eta = \frac{2<d_g>}{3(1-\varepsilon)^{1/3}}\lambda$$

$$\eta_s = \frac{2<d_g>}{3(1-\varepsilon)}\lambda$$

with $<d_g> = 0.59$ mm $= 5.9 \times 10^{-4}$ m and $\varepsilon = 0.49$

η and η_s are found to be

$$\eta = \frac{(1.18 \times 10^{-3})}{3 \times (0.51)^{1/3}}(1.02)10^{-3} = 3.071 \times 10^{-7}$$

$$\eta_s = \frac{(1.18 \times 10^{-3})}{3 \times (0.51)}(1.02)10^{-3} = 7.944 \times 10^{-7}$$

▬▬▬▬▬▬▬▬▬▬▬▬▬▬▬▬▬▬▬▬▬▬▬▬▬▬▬▬▬▬▬▬ ■ ■ ■

5.5   Flow Rate–Pressure Drop Relationships for Flow through Porous Media

In addition to Equations (5.4.21) and (5.4.35), a large number of equations expressing the flow rate–pressure drop relationship for flow through porous media have been proposed by various investigators in the past. Among them, the widely used Kozeny–Carman equation is given as

$$-\frac{\Delta p}{L} = k_1 \frac{(1-\varepsilon)^2}{\varepsilon^3} \frac{\mu\, u_{\mathrm{s}}}{\left[\phi_{\mathrm{s}}\, d_{\mathrm{g}}\right]^2} \tag{5.5.1}$$

where ϕ_{s} is known as the sphericity factor. For a granule of nominal diameter, d_{g}, (or $<d_{\mathrm{g}}>$), ϕ_{s} is defined as

$$\phi_{\mathrm{s}} = \frac{6/d_{\mathrm{g}}}{s_{\mathrm{p}}/v_{\mathrm{p}}} \tag{5.5.2}$$

where $s_{\mathrm{p}}/v_{\mathrm{p}}$ is the surface area to volume ratio of the granule. It is simple to show that for a perfect sphere, $\phi_{\mathrm{s}} = 1.0$. For commonly used substances in filtration, ϕ_{s} ranges from 0.28 (mica flakes) to 0.95 (Ottawa sand).

The constant k_1 is determined by fitting experimental data with Equation (5.5.1). Kozeny (1927), Carman (1937), and later Ergun (1952) found k_1 to be 72, 180, and 150, respectively.

The Kozeny–Carman Equation was obtained by assuming that the pore space of a porous medium may be assumed as a bundle of capillaries and fluid flow through these capillaries is laminar (given by the Hagen–Poiseuille equation). For large $N_{\mathrm{Re}}/(1-\varepsilon) = <d_{\mathrm{g}}> \phi_{\mathrm{s}} u_{\mathrm{s}}\rho/[(1-\varepsilon)\mu]$, the effect of kinetic energy loss must be included in estimating $-\Delta p$. The pressure drop resulting from kinetic energy loss was found to be (Burke and Plummer, 1928)

$$\frac{-\Delta P}{L} = k_2 \frac{\rho u_{\mathrm{s}}^2}{\phi_{\mathrm{s}}<d_{\mathrm{g}}>} \frac{1-\varepsilon}{\varepsilon^3} \tag{5.5.3}$$

with k_2 being 1.75.

A more complete expression combining Equations (5.5.2) and (5.5.3) was proposed by Ergun (1952). Ergun's equation is given as

$$\left(\frac{-\Delta P}{L}\right) \frac{\phi_s<d_{\mathrm{g}}>}{\rho\, u_{\mathrm{s}}^2} \frac{\varepsilon^3}{1-\varepsilon} = k_1 \frac{(1-\varepsilon)}{N_{\mathrm{Re}}} + k_2 \tag{5.5.4}$$

where $N_{\mathrm{Re}} = <d_{\mathrm{g}}>\phi_{\mathrm{s}} u_{\mathrm{s}}\rho/\mu$ and $k_1 = 150$, $k_2 = 1.75$. A comparison of Ergun's equations with Kozeny–Carman and Burke–Plummer Equations is shown in Fig. 5.5.

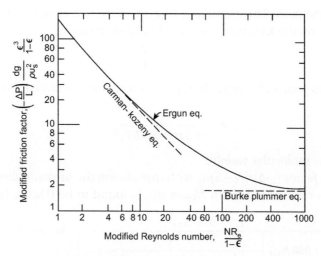

FIGURE 5.5 Pressure-Drop-Flow rate relationship given by Equations (5.5.1), (5.5.3), and (5.5.4).

Equation (5.5.4) is commonly used for liquid flow. For gas flow, McDonald et al. (1979) and Mori and Iinoya (1982) found that better results may be obtained by using $k_1 = 180$ and $k_2 = 1.8$.

For gas flow through fibrous media, Equation (5.5.1) is equally applicable if different values of k_1 are used. By rewriting Equation (5.5.1), one has

$$\frac{-\Delta P}{L} \frac{(r_f^2)}{\mu\, u_s} = \frac{k_1 \varepsilon_s^2}{4(1 - \varepsilon_s)^3} = \frac{4(k_1/16)\varepsilon_s^2}{(1 - \varepsilon_s)^3} = \frac{4K\varepsilon_s^2}{(1 - \varepsilon_s)^3} \tag{5.5.5}$$

where $K = k_1/16$ and r_f is the fiber radius.

Langmuir (1942) found experimentally that K is a function of ε_s instead of being a constant. For $0.5 < \varepsilon_s < 0.8$, K is nearly constant (ranging from 3.18 to 3.67), corresponding to a k_1 value of approximately 55 instead of the value of 64 given by Kozeny. Another expression of $(-\Delta P/L)(r_f^2/\mu\, u_s)$ as a function of ε_s, based on extensive experimental data covering ε_s ranging from 0.006 to 0.3, was given by Davis (1952) as

$$\left(\frac{-\Delta p}{L}\right)\left(\frac{r_f^2}{\mu\, u_s}\right) = 16\; \varepsilon_s^{1.5}(1 + 56\; \varepsilon_s^3) \tag{5.5.6}$$

The term, ε_s^3, however, may be neglected for low values of ε_s.

The assumption of treating gas phase as a continuum commonly used in analyzing gas flow in fibrous media becomes invalid if the mean free path of gas molecules is comparable to or greater than fiber diameter. Such a situation may occur if fibers are ultrathin or the pressure is low. The pressure drop decreases with the increase of the gas molecule mean free path. The correction (the so-called aerodynamic slip correction) to be made depends on the Knudsen number, N_{Kn}, defined as

$$N_{Kn} = \frac{2\bar{\ell}}{d_f} \tag{5.5.7}$$

where $\bar{\ell}$ is the gas molecule mean free path. $\bar{\ell}$ can be calculated from the following equation

$$\bar{\ell} = \frac{\mu}{0.5\,p(8M/\pi RT)^{1/2}} \tag{5.5.8}$$

where M is the gas molecular weight.

The correction factor $(-\Delta p)/(-\Delta p)_0$ with $(\Delta p)_0$ being the pressure drop value given by Equations (5.4.5) or (5.4.6) (i.e., with $N_{Kn} \cong 0$), is found to be (Davis, 1973)

$$\frac{(-\Delta p)}{(-\Delta p)_0} = \frac{1 + 1.996\,N_{Kn}}{1 + 1.996\,N_{Kn}\left[\dfrac{1}{1 + \left\{2(1-\varepsilon_s)^2/(1-\varepsilon_s^2 + 2\,\ell n\,\varepsilon_s)\right\}}\right]} \quad \text{for } 0 < N_{Kn} < 0.25 \tag{5.5.9}$$

■ ■ ■ ▬▬▬▬▬▬▬▬▬▬▬▬▬▬▬▬▬▬▬▬▬▬▬▬▬▬▬▬▬▬

Illustrative Example 5.5

Calculate the mean free path of air molecules at $p = 1$ atm and $T = 298$ K.
From Equation (5.5.8)

$$\bar{\ell} = \frac{\mu}{0.5\,p(8M/\pi RT)^{1/2}}$$

Solution

At 1 atm, 298 K,

$$\mu = 1.8 \times 10^{-5}\ \text{Pa. s}$$

$$p = 1\ \text{atm} = 1.013 \times 10^5\ \text{Pa}$$

$$M = 0.029\ \text{kg mol}^{-1}$$

$$R = 8.314\ \text{J K}^{-1}\,\text{mol}^{-1}$$

$$\bar{\ell} = \frac{1.8 \times 10^{-5}}{0.5(1.013 \times 10^5)\left[\dfrac{(8)(0.029)}{(\pi)(8.314)(298)}\right]^{1/2}}$$

$$= \frac{1.8 \times 10^{-10}}{0.5(1.013)(5.4596 \times 10^{-3})}$$

$$= 6.521 \times 10^{-8}\ \text{m or } 0.06521\ \mu\text{m}$$

■ ■ ■

5.6   Filter Cleaning by Back Washing and Bed Expansion

Increase of particle deposition within filter medium leads to a continuing decrease of medium permeability and increasing pressure drop necessary to maintain a constant liquid throughput. Periodic filter cleaning to remove deposited particles is therefore required.

For a filter bed with sufficient deposition, cleaning is commonly accomplished by passing cleaning water upward through the bed, causing the bed to expand and become fluidized. The extent of cleaning depends upon the upward cleaning water velocity applied, the duration of cleaning, and a host of other variables. The problem is truly complex and its understanding is far from being complete.

Traditional practice of backwashing used in water treatment is to apply upflow velocity of approximately 15–20 gallons per minute (0.025–0.034 m s^{-1}) with 20–30% bed expansion. The cleaning water used per wash varies typically from 100–150 gallons/ft^2 (or 4.1–5.1 m^3 m^{-2}) with a water consumption of not more than 5% of the water produced (Letterman, 1991). The information of the upflow velocity vs. bed expansion is important to design calculations.

A large number of investigations have been reported in the literature about the relationship of liquid fluidization and bed expansion. They are too numerous to be mentioned here. Only a brief presentation of the subject is given in the following:

Consider a homogeneous bed packed with granular materials with height L_0 and porosity ε_0. If the bed is subject to an upward liquid flow and expands its height to L with a porosity of ε, for homogeneous fluidization, the relationship between L and L_0 is given as

$$L_0(1 - \varepsilon_0) = L(1 - \varepsilon)$$

$$\text{or } L/L_0 = (1 - \varepsilon_0)/(1 - \varepsilon) \tag{5.6.1}$$

The seminal work of Richardson and Zaki (1954) on liquid fluidization and sedimentation relates the upward liquid velocity V with bed porosity ε by the following expression[5]:

$$\frac{V}{V_i} = \varepsilon^n \tag{5.6.2}$$

where V_i is the velocity at $\varepsilon = 1.0$ and can be taken as the terminal velocity of a single particle settling through a liquid medium of infinite extent, V_{t_∞}. The exponent n is an

---

[5]An earlier work of Lewis and Bowerman (1952) actually suggested the same idea but their treatment is less complete.

empirical constant and, based on experimental data, was found to be (Richardson and Zaki, 1954; Richardson, 1971)

$$n = \begin{cases} 4.65 + 20(d_g/D) & \text{for } Re_{t\infty} < 0.2 \\ \left(4.4 + 18\dfrac{d_g}{D}\right)Re_{t\infty}^{-0.03} & 0.8 < Re_{t\infty} < 1.0 \\ \left(4.4 + 18\dfrac{d_g}{D}\right)Re_{t\infty}^{-0.1} & 1.0 < Re_{t\infty} < 200 \\ 4.4\ Re_{t\infty}^{-0.1} & 200 < Re_{t\infty} < 500 \\ 2.4 & 500 < Re_{t\infty} \end{cases} \tag{5.6.3}$$

where D is the relevant characterization dimension, i.e., the diameter of the filter bed.

The terminal velocity, $V_{t\infty}$, in general, can be expressed as

$$V_{t\infty} = \sqrt{\frac{4(\rho_p - \rho)g_p\, d_g}{3\,C_D\,\rho}} \tag{5.6.4}$$

where C_D is the drag coefficient defined as

$$C_D = \frac{F_D/A_p}{(\rho\,u^2/2)} \tag{5.6.5}$$

where F_D is the total drag force acting on the particle due to fluid flow at velocity u. A_p is the projected particle area. For fluid flow over a sphere, $A_p = (\pi/4)\,d_p^2$. In the case of a sphere settling in a fluid media of infinite extent, $u = V_{t\infty}$.

The drag coefficient, C_D, is a function of the particle Reynolds number, $Re_{t\infty}$,

$$Re_{t\infty} = \frac{V_{t\infty}\, d_g\, \rho}{\mu} \tag{5.6.6}$$

The relationship between C_D and $(Re)_{t\infty}$ is shown in Fig. 5.6 (Lapple and Shepherd, 1940). An empirical fit of the relationship gives the following expressions (Bird et al., 2007):

$$C_D = \begin{cases} 24/(Re)_{t\infty} & \text{for} \quad Re_{t\infty} < 0.1 \\ \left[\sqrt{\dfrac{24}{Re_{t\infty}}} + 0.5407\right]^2 & \text{for} \quad Re_{t\infty} < 6 \times 10^3 \\ 0.44 & \text{for} \quad 500 < Re_{t\infty} < 1 \times 10^5 \end{cases} \tag{5.6.7}$$

Substituting the expressure of C_D for $(Re)_{t\infty} < 1$ (see Equation (5.6.7) into Equation (5.6.4)), one has

$$V_{r\infty} = \frac{g\, d_g^2(\rho_p - \rho)}{18\,\mu} \tag{5.6.8}$$

which is the terminal velocity according to Stoke's law.

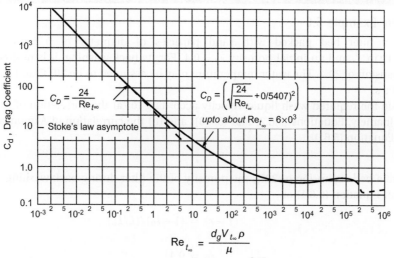

FIGURE 5.6 Drag coefficient C_d, vs particle Reynolds number $\mathrm{Re}_{t_\infty} = \dfrac{d_g V_{t_\infty} \rho}{\mu}$.

In general, Equations (5.6.1), (5.6.2), (5.6.4), and (5.6.7) can be used to estimate the extent of bed expansion, $(L - L_0)/L_0$. Calculation of V_{t_∞} from Equations (5.6.7) and (5.6.4) can be made iteratively. For example, one may first assume $C_D = 24/(\mathrm{Re})_{t_\infty}$ and then determine V_{t_∞}. The value of V_{t_∞} can then be used to determine $(\mathrm{Re})_{t_\infty}$ and to see whether or not the use of $C_D = 24/N_{\mathrm{Re}}$ is justified; if not, a different expression of C_D from Equation (5.6.7) will be applied. The procedure may be repeated until the correct value of V_{t_∞} is found.

Since the publication of the work of Richardson and Zaki more than half a century ago, a number of studies aimed at improving the accuracy of Equations (5.6.2) have appeared in the literature. Here, we will only give the most recent work of Akgiray and Soyer (2006).

The Akgiray–Soyer correlation is given as

$$\phi = 3.137\, N_{\mathrm{Re}_1} + 0.673\, (N_{\mathrm{Re}_1})^{1.766} \tag{5.6.9}$$

where

$$\phi = \frac{\varepsilon^3}{(1-\varepsilon)^2}\, \frac{(\phi_s d_g)^3 \rho(\rho_p - \rho)g}{216\,\mu^2} \tag{5.6.10a}$$

$$N_{\mathrm{Re}_1} = \frac{(\phi_s d_g)\rho V}{6\mu(1-\varepsilon)} \tag{5.6.10b}$$

To obtain the required wash water velocity for a given bed expansion from Equation (5.6.9), iteration calculation is required since both ε and V appear in all terms of the correlation. A simple iterative procedure for this purpose is given in the Illustrative Example 5.6.

■ ■ ■ ━━━━━━━━━━━━━━

Illustrative Example 5.6

Consider a filter bed packed with sand particles of diameter of 0.75 mm, $\phi_s = 1.0$, $\rho_p = 2650$ kg m^{-3}, and an initial porosity $\varepsilon_0 = 0.4$. For its back washing, it is desired to have a bed expansion of 20%. The water temperature may be assumed to be 20° C and its viscosity $\mu = 10^{-3}$ Pa s.

Calculate the required wash liquid velocity according to (a) the correlation of Richardson and Zaki [namely, Equations (5.6.2) and (5.6.3)] and (b) the correlation of Akgiray and Soyer [namely, Equation (5.6.9)].

Solution

To achieve a bed expansion of 20%, the porosity of the filter during back washing, ε, can be found from Equation (6.6.1) or

$$(1 - \varepsilon) = (1 - \varepsilon_0)/(L/L_0)$$

$$\varepsilon_0 = 0.4$$

$$L/L_0 = 1.2$$

$$(1 - \varepsilon) = (1 - 0.4)/1.2 = 0.6/1.2 = 0.5$$

$$\text{and } \varepsilon = 0.5$$

(a) Calculate the required wash velocity according to Equations (5.6.2) and (5.6.3). From Equation (5.6.2), one has

$$V = (V_{t_\infty})(\varepsilon)^n$$

To obtain V_{t_∞} and to estimate the correct value of n to be used, the value of Re_{t_∞} must be known. To start the calculation, assume that Re_{t_∞} is sufficiently small. From Equation (5.6.2) n is found to be

$$n = 4.65$$

since $d_g \ll D$ and the term 20 d_g/D may be ignored. Using the Stokes law expression of V_{t_∞} [i.e., Equation (5.6.8)], V_{t_∞} is found to be

$$V_{t_\infty} = \frac{g(d_g)^2(\rho_p - \rho)}{18\,\mu} = \frac{(9.8)(7.5 \times 10^{-4})^2(1650)}{18(10^{-3})} = 0.505 \text{ m s}^{-1}$$

We can now check whether the values of n and V_{t_∞} used are correct or not. With $V_{t_\infty} = 0.505$ m s^{-1},

$$\text{Re}_{t_\infty} = \frac{(7.5 \times 10^{-4})(0.505)(1,000)}{10^{-3}} = 379$$

From Equations (5.6.7),

$$C_D = \left[\sqrt{\frac{24}{\text{Re}_{t_\infty}}} + 0.5407\right]^2 = 0.3648$$

and from Equation (5.6.4)

$$V_{t_\infty} = \sqrt{\frac{(4)(1650)(9.8)(7.5 \times 10^{-4})}{(3)(0.3648)(1000)}} = 0.211 \text{ m s}^{-1}$$

Based on $V_{t_\infty} = 0.211$ m s^{-1}, Re_{t_∞} is found to be

$$\text{Re}_{t_\infty} = \frac{(7.5 \times 10^{-4})(0.211)(1,000)}{10^{-3}} = 158$$

The drag coefficient C_D is

$$C_D = \left[\sqrt{\frac{24}{158}} + 0.5407\right]^2 = 0.480$$

and V_{t_∞} is found to be

$$V_{t_\infty} = \sqrt{\frac{(4)(1650)(9.8)(7.5 \times 10^{-4})}{(3)(0.48)(1000)}} = 0.184 \text{ m s}^{-1}$$

A further computation gives

$$\text{Re}_{t_\infty} = \frac{(7.5 \times 10^{-4})(0.184)(1,000)}{10^{-3}} = 138$$

$$C_D = \left[\sqrt{\frac{24}{138}} + 0.5407\right]^2 = 0.511$$

$$V_{t_\infty} = \sqrt{\frac{(4)(1650)(9.8)(7.5 \times 10^{-4})}{(3)(0.511)(1000)}} = 0.179 \text{ m s}^{-1}$$

which is sufficiently close to the assumed value of 0.184 (~5% reduction). Therefore, we may assume

$$V_{t_\infty} = 0.17 \text{ m s}^{-1}$$

With $Re_{t_\infty} = 128$ (based on $V_{t_\infty} = 0.17$ m s^{-1}), the exponent n, according to Equation (5.6.3), is

$$n = \left(4.4 + 18 \frac{d_g}{D}\right) Re_{t_\infty}^{-0.1} \cong (4.4)(128)^{-0.1} = 2.71$$

The required wash velocity V is

$$V = (0.17)(0.5)^{2.71} = 0.026 \text{ m s}^{-1} \text{ or } 3.82 \text{ gpm ft}^{-2}$$

(b) Required wash liquid velocity according to Equation (5.6.9). Equation (5.6.9) may be rewritten as

$$\phi = A \frac{\varepsilon^3}{(1-\varepsilon)^2} = \left[3.137 \, N_{Re_1} + 0.673 \, N_{Re_1}^{1.766}\right] \tag{ia}$$

$$= 3.137 \left[1 + \frac{0.673}{3.137} N_{Re_1}^{0.766}\right] N_{Re_1} \tag{ib}$$

$$= 0.673 \left[1 + \frac{3.137}{0.673} N_{Re_1}^{-0.766}\right] N_{Re_1}^{1.766} \tag{ic}$$

and $$A = \frac{d_g^3 \rho(\rho_p - \rho)g}{216 \, \mu^2} \tag{ii}$$

$$N_{Re_1} = \frac{d_g \rho V}{6\mu(1-\varepsilon)} \tag{iii}$$

For iteration calculation of V for a given ε, from (ib), one may write

$$N_{Re_1} = \frac{A}{3.137} \frac{\varepsilon^3}{(1-\varepsilon)^2} \frac{1}{1 + 0.215 \, N_{Re_1}^{0.766}}$$

Accordingly, iteration calculation of V can be made according to

$$V^{(i)} = \frac{6\mu}{d_g \rho} \frac{A}{3.137} \frac{\varepsilon^3}{(1-\varepsilon)} \frac{1}{1 + 0.215 \left[\frac{d_g \rho V^{(i-1)}}{6\mu(1-\varepsilon)}\right]^{0.766}} \tag{iv}$$

where the superscript (i) denoted the value of the i-th iteration. The various quantities appearing in Equation (iv) are

$$A = \frac{(7.5 \times 10^{-4})^3 10^3 (1.65 \times 10^3) \, 9.8}{(216)(10^{-3})^2} = 31.58$$

$$\frac{6\mu}{d_g \rho} = \frac{(6)(10^{-3})}{(7.5 \times 10^{-4}) \, 10^3} = 8 \times 10^{-3}$$

$$N_{Re_1} = \frac{(7.5 \times 10^{-4})(10^3)V}{6(10^{-3})(0.5)} = 250\,V$$

To begin the iteration calculation, assume $V^{(0)} = 0$. From Equation (iv) with $V^{(0)} = 0$, one has

$$V^{(1)} = (8 \times 10^{-3}) \frac{31.58}{3.137} (0.25) = 0.02\ \text{m s}^{-1}$$

Similarly, $V^{(2)}$ is found to be

$$V^{(2)} = (8 \times 10^{-3}) \frac{31.58}{3.137} (0.25) \frac{1}{1 + 0.215\,[(250)(0.02)]^{0.766}} = 0.012\ \text{m s}^{-1}$$

$V^{(3)}$ is found to be

$$V^{(3)} = 0.013\ \text{m s}^{-1}$$

Therefore, we can take $V = 0.013$ m s^{-1}. Note that the value is only half of that of (a).

■ ■ ■

5.7   Solution of the Macroscopic Conservation Equations of Deep Bed Filtration

As discussed in 5.1 and 5.2, the macroscopic equation of deep bed filtration may be written as

$$u_s \frac{\partial c}{\partial z} + \frac{\partial \sigma}{\partial \theta} = 0 \tag{5.1.4b}$$

$$\frac{\partial \sigma}{\partial \theta} = u_s \lambda_0 F(\alpha - \sigma)c \tag{5.2.4}$$

with the following initial and boundary conditions

$$c = 0, \quad \sigma = 0, \quad \text{for} \quad z > 0, \quad \theta = 0 \tag{5.1.5}$$

$$c = c_{in} \qquad\qquad z = 0 \tag{5.1.6}$$

This system of equations can be solved numerically by a number of methods [for example, see Courant and Hilbert, (1962) or Aris and Amundson (1973)]. Perhaps the procedure suggested by Herzig et al. (1970) provides a more direct method of solution. A description of the method by Heizig et al. is given below.

Herzig et al. showed that Equations (5.1.4b) and (5.2.4) are equivalent to a pair of uncoupled ordinary differential equations. Accordingly, instead of solving the two partial differential equations simultaneously, one needs to only solve the two ordinary differential equations sequentially as shown below.

By applying the conservation principle to a filter bed of depth Z over a time interval from 0 to t_1, for a unit cross-sectional area of bed, one has

Number of particles that have entered during time internal 0 to t_1 at $z = 0$	$=$	Number of particles that have left during time interval 0 to t_1 through the cross section at $z = Z$

$$+ \text{ Number of particles retained in filter region } 0 \text{ to } Z \text{ at time } t_1 \tag{5.7.1}$$

Based on Equation (5.7.1), one has

$$\int_0^{t_1} u_s c_{in} dt = \int_0^{t_1} u_s c(Z, t)\, dt + \int_0^Z (\sigma + \varepsilon c)\, dz \tag{5.7.2}$$

The time, t_1, and axial distance, Z, are chosen arbitrarily. From Equation (5.2.4), one has

$$c(Z, t) = \frac{1}{u_s \lambda_0 F[\alpha, \sigma(Z, t)]} \left. \frac{\partial \sigma}{\partial t} \right|_Z \tag{5.7.3}$$

Since σ is a function of both time and the axial distance, $d\sigma$ can be written as

$$d\sigma = \left(\frac{\partial \sigma}{\partial t} \right)_z dt + \left(\frac{\partial \sigma}{\partial z} \right)_t dz \tag{5.7.4}$$

At $z = Z$, the total differential, $d\sigma$, becomes

$$d\sigma = \left(\frac{\partial \sigma}{\partial t} \right)_Z dt \tag{5.7.5}$$

Equations (5.7.3) and (5.7.5) can now be used to eliminate $c(Z, t)$ present in the first integral on the right side of Equation (5.7.2), or

$$\int_0^{t_1} u_s c_{in} dt = \int_0^{\sigma(Z, t_1)} \frac{d\sigma(Z, t)}{\lambda_0 F[\alpha, \sigma(Z, t)]} + \int_0^Z (\sigma + \varepsilon c)\, dz \tag{5.7.6}$$

In the above expression, $F[\alpha, \sigma(Z, t)]$ is evaluated at the position $z = Z$. Since F is a function of σ, its dependence on z and t is implicit. Thus, when Leibnitz's rule is applied to differentiate Equation (5.7.6) with respect to Z (note that Z is chosen arbitrarily), one has

$$0 = \frac{1}{\lambda_0 F(\alpha, \sigma)} \left(\frac{\partial \sigma}{\partial z} \right)_t + \sigma + \varepsilon c \tag{5.7.7}$$

in which Z and t_1, being arbitrary, have been replaced by z and t, respectively.

Equation (5.7.7) can be re-arranged to give

$$-\left(\frac{\partial \sigma}{\partial z} \right)_t = \lambda_0\, F(\alpha, \sigma)[\sigma + \varepsilon c] \tag{5.7.8}$$

If the independent variables z and θ are used instead of z and t, Equation (5.7.8) becomes

$$\left(\frac{\partial \sigma}{\partial z}\right) = \frac{\varepsilon}{u_s}\left(\frac{\partial \sigma}{\partial \theta}\right)_z - \lambda_0 F(\alpha, \sigma)[\sigma + \varepsilon c] \qquad (5.7.9)$$

Furthermore, if the filtration rate expression of Equation (5.2.4) is substituted into the above expression, one has

$$\left(\frac{\partial \sigma}{\partial z}\right)_\theta = -\lambda_0\, F(\alpha, \sigma)\sigma \qquad (5.7.10)$$

Moreover, applying Equation (5.2.4) to $z = 0$ and denoting σ_{in} to be the value of σ at $z = 0$, or $\sigma_{in} = \sigma(0, \theta)$ yields

$$\left(\frac{\partial \sigma_{in}}{\partial \theta}\right) = u_s \lambda_0\, F(\alpha, \sigma_{in}) c_{in} \qquad (5.7.11)$$

The initial condition of Equation (5.7.11) is

$$\sigma_{in} = 0, \quad \text{at } \theta = 0 \qquad (5.7.12)$$

and the boundary condition of Equation (5.7.10) is

$$\sigma = \sigma_{in}, \quad \text{at } z = 0 \qquad (5.7.13)$$

Equations (5.7.10) through (5.7.13) are equivalent to Equations (5.1.4), (5.2.4), (5.1.5), and (5.1.6). The specific deposit at the filter inlet, σ_{in}, as a function of time can be found from Equations (5.7.10) and (5.7.13). Once σ_{in} (as a function of θ) is known, the specific deposit profile, σ vs. z at any time, can be found from Equations (5.7.10) and (5.7.13). The suspension particle concentration can now be calculated from Equation (5.1.4) with σ known as a function of z and θ. However, a more direct calculation method can be made as follows: First, if one defines c^+ to be

$$c^+ = c/c_{in} \qquad (5.7.14)$$

then Equation (5.2.4) becomes

$$\left(\frac{\partial \sigma}{\partial \theta}\right) = u_s \lambda_0 F(\alpha, \sigma) c_{in}\, c^+ \qquad (5.7.15)$$

Substituting the above two expressions into Equation (5.1.4) results in

$$u_s\, \lambda_0 F(\alpha,\, \sigma)\, c_{in} c^+ + u_s c_{in}\, \frac{\partial c^+}{\partial z} = 0$$

which upon rearrangement gives

$$\frac{1}{c^+}\frac{\partial c^+}{\partial z} = -\lambda_0 F(\alpha, \sigma) \qquad (5.7.16)$$

On the other hand, Equation (5.7.10) can be rewritten as

$$\frac{1}{\sigma}\frac{\partial \sigma}{\partial z} = -\lambda_0 F(\alpha, \sigma) \qquad (5.7.17)$$

Comparing the above two expressions leads to the following results:

$$\frac{c^+}{\sigma} = A(\theta) \qquad (5.7.18)$$

where $A(\theta)$ is a function of θ only and may be evaluated at the filter inlet, or at $z = 0$.

$$A(\theta) = \frac{c^+}{\sigma} = \frac{c^+}{\sigma}\bigg|_{z=0} = \frac{1}{\sigma_{in}} \qquad (5.7.19)$$

Combining equations (5.7.18) and (5.7.19) yields

$$c^+ = \frac{\sigma}{\sigma_{in}}$$

or

$$\frac{c}{c_{in}} = \frac{\sigma}{\sigma_{in}} \qquad (5.7.20)$$

In other words, the suspension particle concentration in the fluid phase, c, can be obtained once σ and σ_i are known. A relationship of this type was also found in fixed-bed adsorption. Ives (1960) obtained an expression similar to that corresponding to a specific form of $F(\alpha, \sigma)$.

The above-outlined method offers a simple and convenient way of solving the granular filtration's governing equations. Generally speaking, Equations (5.7.10)–(5.7.13) must be solved numerically. However, under certain conditions analytical solution becomes possible as shown in the following illustrative example.

■ ■ ■ ▬▬▬▬▬▬▬▬▬▬▬▬▬▬▬▬▬▬▬▬▬▬▬▬▬▬▬▬▬▬▬▬

Illustrative Example 5.7

Ornatski *et al.* (1955) suggested that if deposition results principally in filter clogging, $F(\alpha, \sigma)$ can be expressed as

$$F(\alpha, \sigma) = 1 - k\sigma$$

where k is an arbitrary positive constant.

Solution

Equation (5.7.11) can be re-arranged to give

$$\frac{d\sigma_{in}}{1 - k\sigma_{in}} = (u_s \lambda_0 c_{in}) d\theta$$

which upon integration with the initial condition $\sigma_{in} = 0$ at $\theta = 0$, yields

$$1 - k\sigma_{in} = \exp[-u_s \lambda_0 c_{in} k\theta]$$

Also from Equation (5.7.10) and the assumed expression for F, one has

$$\frac{d\sigma}{\sigma(1 - k\sigma)} = -\lambda_0 dz$$

Integrating the above equation with the boundary condition, $\sigma = \sigma_{in}$ at $z = 0$, one has

$$\frac{\sigma(1 - k\sigma_{in})}{\sigma_{in}(1 - k\sigma)} = \exp[-\lambda_0 z]$$

Since $c/c_{in} = \sigma/\sigma_{in}$ [namely, Equation (5.7.20)], the above expression can be rewritten as

$$\frac{c}{c_{in}} \frac{1 - k\sigma_{in}}{1 - k\sigma_{in}(c/c_{in})} = \exp[-\lambda_0 z]$$

Solving for (c/c_{in}), one has

$$\frac{c}{c_{in}} = \frac{\exp[-\lambda_0 z]}{1 - k\sigma_{in} + k\sigma_{in}\exp[-\lambda_0 z]} = \frac{1}{\exp[\lambda_0 z](1 - k\sigma_{in}) + k\sigma_{in}}$$

Substituting into the above equation the expression of σ_{in} obtained previously, the solution of c/c_{in} is found to be

$$\frac{c}{c_{in}} = \frac{\exp[u_s \lambda_0 c_{in} k\theta]}{\exp[\lambda_0 z] + \exp[u_s \lambda_0 c_{in} k\theta] - 1}$$

■ ■ ■

Problems

5.1. In 5.3, we obtain an expression of ℓ by assuming ℓ being the characteristic dimension of a cube which accommodates, on the average, one filter grain. If a sphere instead of a cube is used, what is the expression of ℓ? Also, obtain the expression of N_c, the relationship between e and η and the relationship between λ and η. Compare the expressions you obtained with those given in 5.3.

5.2. A sample of eight fibers taken from a fibrous medium gives the following results

Fiber no.	$d_f(\mu m)$	$\ell_f(\mu m)$
1	7	15
2	8	20
3	6.5	14
4	10	13
5	8.2	9
6	9	10
7	11	11
8	6	8

Using the results given in Illustrative Example 5.3, calculate $<d_f>$ and $(\ell_f)_{av}$.

5.3. What are the major differences between one-dimensional linear deep bed filtration and one-dimensional radial filtration? Which one is more effective in removing particles for the same flow rate and same filter depth? Make your comparison on the basis that $\lambda = $ constant.

5.4. A simple model of porous media is to consider the pore space of a medium is equivalent to a bundle of uniform capillaries of radius, a_c, and uniform length. The flow through the capillaries is assumed to be laminar and fully developed. Obtain expressions of a_c and N_c in terms of the medium permeability. Also, what is the relationship between η and e?

5.5. Consider a filter bed of height 2 m composed of grains of diameter 0.3 mm and an initial porosity of 0.45. The suspension flow is kept at $u_s = 1.5 \times 10^{-3}$ m s^{-1}. What is the initial pressure drop associated with the flow calculated (a) according to the Kozeny–Carman equations and (b) the results based on Happel's model?

5.6. A fibrous filter of porosity 0.95 and fiber diameter of 4 μm are used for air purification. Airflow velocity is kept at 0.5 m s^{-1} at 298 K and 1 atm. Estimate the pressure drop as a function of filter depth.
If the fiber diameter is 2 μm and the inlet pressure is at 0.1 atm, what is the pressure drop?

5.7. What is the maximum value of η? What is the maximum value of η_s if the filter medium is characterized by Happel's model?

5.8. A granular filter bed of depth 1 m and initial porosity of 0.45 is used to treat a suspension with particle concentration of 100 parts per million (volume). The suspension flow rate is kept at 1.2×10^{-4} m s^{-1}. The filter coefficient λ is given as

$$\lambda = 1 \times 10^{-2}(1 + 300\sigma)$$

Predict the effluent concentration history.

5.9. Same as Problem 5.8, what is the pressure drop after 24 hours' operation? Make your calculations on the following basis.
  (i) The deposited particles are distributed uniformly.
  (ii) 80% of the deposited particles are present uniformly in the top half and the remainder 20% in the bottom half.
  (iii) 50% of the deposited particles are present on the top 1/5 of the filter, 25% in the second 1/5, 13% in the 3rd 1/5, 7% in the 4th 1/5, and 5% in the bottom 1/5.
The permeability correction function G is given as

$$F = 1 + 500\sigma$$

Which one of the three answers obtained, in your judgment, is more accurate? Why?

References

Akgiray, O., Soyer, E., 2006. J. Water Supply and Tech.-AQUA 55, 7–8.

Aris, R., Amundson, N.R., 1973. Mathematical Methods in Chemical Engineering, vol. 2. Prentice-Hall.

Bird, R.B., Stewart, W.E., Lightfoot, E.N., 2007. Transport Phenomena, revised second ed. John Wiley & Sons, pp. 186.

Brown, R.C., 1993. Air Filtration: An Integrated Approach to the Theory and Applications of Fibrous Filters. Pergamon Press, pp. 42.

Burke, S.P., Plummer, W.B., 1928. Ind. Eng. Chem. 20, 1196.

Camp, T.R., 1964. Asce, J. Sanitary Eng. Div. 90, 3.

Carman, P.C., 1937. Trans. Inst. Chem. Eng. (London) 15, 150.

Courant, R., Hilbert, D., 1962. Methods of Mathematical Physics, vol. 2. Wiley-Interscience.

Davis, C.N., 1952. Proc. Inst. Mech. Eng. 1B (5), 185.

Davis, C.N., 1973. Air Filtration. Academic Press.

Deb, A.K., 1969. Proc. ASCE J. Sanitary Eng. Div. 95, 399.

Ergun, S., 1952. Chem. Eng. Prog. 48, 89.

Happel, J., Brenner, H., 1965. Low Reynolds Number Hydrodynamics. Prentice-Hall.

Happel, J., 1958. AIChE J. 4, 197.

Heertjes, R.M., Lerk, C.F., 1967. Trans. Int. Chem. Eng. 45, T138.

Herzig, J.P., Leclerc, D.M., LeGoff, P., 1970. Ind. Eng. Chem. 62 (5), 8.

Ives, K.J., 1960. Proc. Inst. Civil Engrs, (London) 16, 189.

Ives, K.J., 1961. ASCE J. Sanitary Eng. Div. 87, 23.

Ives, K.J., 1969. Theory of Filtration, Special Subject No. 7 Int'l Water Supply Congress and Exhibition, Vienna.

Iwasaki, T., 1937. J. Am. Water Works Assoc. 29, 1591.

Kozeny, J., 1927. Sitzungsberichte der Akademia der Wissenchften in Wien, Mathematisch-Naturwissenchaftliche, Klasse, 136IIa, 271.

Kuwabara, S., 1959. J. Phys. Soc., Japan 14, 527.

Langmuir, I., 1942. Report on Smokes and Filters, Section I, U.S. Office of Scientific Research and Development, No. 865 Part IV.

Lapple, C.E., Shepherd, C.B., 1940. Ind. Eng. Chem. 32, 606.

Letterman, R.D., 1991. Filtration Strategies to Meet the Surface Water Treatment Rule. American Water Works Assoc.

Lewis, E.M., Bowerman, E.W., 1952. Chem. Eng. Prog. 48, 603.

Mackrle, V., Draka, O., Svec, J., 1965. Hydrodynamics of the Disposal of Low Level Liquid Radioactive Wastes in Soil, Int'l Atomic Energy Agency Contract Report No. 98, Vienna.

Maroudas, A., Eisenklam, P., 1965. Chem. Eng. Sci. 20, 815.

McDonald, J.F., El-Sayed, M.S., Mow, K., Dullien, F.A.L., 1979. Ind. Eng. Chem. Fundam. 18, 199.

Mehter, A.A., Turian, R.M., Tien, C., 1970. Filtration in Deep Beds of Granular Carbon, Research Report No. 70–3, FWPEA Grant No. 17020, OZO, Syracuse University.

Mints, D.M., 1966. Modern Theory of Filtration, Special Report No. 10, International Water Supply Congress, Barcelona.

Mori, Y., Iinoya, K., 1982. Proc. Int. Sym. Powder Technol., Soc. of Powder Technology, Japan., pp. 557.

Ornatski, N.V., Sergeev, E.V., Shekhtman, Y.M., 1955. Investigations of the Process of Clogging of Sands. University of Moscow.

Payatakes, A.C., 1973. A New Model for Granular Porous Media: Applications to Filtration through Packed Beds, PhD Dissertation, Syracuse University.

Richardson, J.F., Zaki, W.W., 1954. Trans. Int. Chem. Eng. 32, 35.

Richardson, J.F., 1971. Incipient Fluidization and Particulate Systems. In: Fluidization, Dawson, J.F., Harrison, D. (Eds.). Academic Press.

Stein, P.C., 1940. A Study of the Theory of Rapid Filtration of Water through Sand, D.Sc. Thesis, MIT.

Shekhtman, Y.M., 1961. Filtration of Suspension of Low Concentrations. Institute of Mechanics, U.S.S.R. Academy of Sciences.

Tien, C., Payatakes, A.C., 1979. AIChE J. 25, 737.

6

Particle Deposition Mechanisms, Predictions, Determinations and Correlations of Filter Coefficient/ Collector Efficiency

Notation

A	coefficient of Equation (6.3.8) ($-$)
A_s	defined by Equation (6.1.19b)
A'_s	a quantity given by Equation (6.4.3a)
a_c	collector radius (m)
a_p	particle radius (m)
B	a quantity defined by Equation (6.1.39)
b	radius of Happel's cell (m) or exponent of Equation (6.3.9) ($-$)
C_1	defined by Equation (6.3.6a)
C_2	defined by Equation (6.3.6b)
c	suspension particle concentration (vol/vol)
c_1, c_2, c_3, c_4	quantities given by Equation (6.4.3b)–(6.4.3e)
c_∞	particle concentration of approaching fluid (vol/vol)
c_{in}	influent particle concentration (vol/vol)
c_{eff}	effluent particle concentration (vol/vol)
c_s	Cunningham's correction factor ($-$)
D	diffusivity ($m^2\ s^{-1}$)
D_{BM}	Brownian diffusivity ($m^2\ s^{-1}$)
d_c	collector diameter, equal to d_f for fibrous media and d_g for granular media (m)
d_f	fiber diameter (m)
d_g	filter grain diameter (m)
d_p	particle diameter (m)
E	potential energy (J)
E_t	total energy (J)
e	coefficient of restitution ($-$)
e_0	value of e with no flex
F	defined as λ/λ_0 (or η/η_0)
F_e	electrostatic force vector (N)
F_x	force acting along the tangential direction (N)
F_y	force acting along the normal diameter (N)
$f_{EC}, f_{Ei}, f_{EM}, f_{EP}, f_{Ex}, f_{ES}, f_{icp}$	quantities listed in Table 6.1 ($-$)
$F_{EC}, F_{EI}, F_{EM}, F_{ES}, F_{er}, F_{icp}$	different types of electrostatic forces as detailed in Table 6.15 (N)
F_{DL}	double-layer force given by Equation (6.5.5) (N)
F_{LO}	London–van der Waals force given by Equation (6.5.2) (N)

Principles of Filtration, DOI: 10.1016/B978-0-444-56366-8.00006-2

f	expression given by Equation (6.3.2)
f_1	a quantity given by Equation (6.4.9b)
g	gravitational acceleration (m s^{-2})
H	Hamaker constant (J)
h	protrusion height (m)
h_c	critical value of h (m)
I	mass flux over a packing grain (vol/m^2)
I_p	particle flux over a collector surface (vol/m^2)
$K_{EC}, K_{EI}, K_{EM}, K_{EP}, K_{ES}, K_{icp}$	quantities listed in Table 6.1
K_1, K_2, K_3, K_4	coefficients given by Equation (5.4.10b)–(5.4.10e)
Ku	Kuwabara hydrodynamic factor given by Equation (5.4.27e) (−)
KE	kinetic energy (J)
k	Boltzmann constant or coefficient of function F (see Equation (6.2.7))
k_1, k_2, k_3, k_4	quantities listed in Table 6.3
k_d	dimensionless deposit permeability (normalized by a_c^2)(−)
k_i	defined as $(1 - \bar{v}_i^2)Y_i$
L	filter medium length (m)
$\bar{\ell}$	gas molecule mean free path (m)
M	mass of particle deposit per unit volume of filter material (kg m^{-3})
M_D	moment caused by hydrodynamic drag (see Equation (6.1.35b) (N.m)
m	mass of deposited particles per unit wire screen surface area (kg m^{-2})
m_j	number concentration of the j-th ion species, see Equation (6.5.4)
m_0	value of m which gives η_{f_s} a value equal to 2 $(\eta_{s_f})_0$
m_p	particle mass (kg)
N_{DL}	dimensionless parameter defined by Equation (6.5.8c)
N_{E1}, N_{E2}, N_{DL}	dimensionless parameters defined by Equations (6.5.8a)–(6.5.8c)
N_G	dimensionless gravitational parameter defined by Equation (6.1.14)
N_{Kn}	Knudsen number defined by Equation (5.5.7)
N_{LO}	London force parameter defined by Equation (6.4.2)
N_{Pe}	Peclet number defined by Equation (6.1.26)
N_R	Inception parameter defined as a_p/a_c (−)
N_{Re_s}	Reynolds number defined by Equation (6.3.12b)
N_{Sh}	Sherwood number defined by Equation (6.1.22)
N_{St}	Stokes number defined by Equation (6.1.3) (−)
$N_{St_{eff}}$	effective Stokes number defined by Equation (6.3.12a)
n	unit normal vector to collector surface (−)
p	pressure (Pa) or a quantity defined by Equation (5.4.11b)
R_f	fiber radius (m)
S	collector surface area (m^2)
S_i	off-center distance from a particle trajectory at the edge of Happel's cell (see Fig. 6.7b) (m)
St_{eff}	a parameter defined by Equation (6.1.56)
s	collector projected area
T	temperature (K)
t	time (s)
t^*	dimensionless time defined as ta_c/U (−)
$\overline{U}$	velocity of a particle moving through fluid at rest (m s^{-1})
U	characteristic velocity (m s^{-1})
u_{pi}, u_{pr}	particle incident and rebound velocity (m s^{-1})

u_s	superficial velocity (m s^{-1})
u_∞	approach velocity (m s^{-1})
$\underline{u}$	fluid velocity vector (m s^{-1})
u_p^*	capture limit velocity (m s^{-1})
$\underline{v}$	particle velocity vector (m s^{-1})
V_t	sedimentation velocity defined by Equation (6.1.14) (m s^{-1})
u^*	dimensionless fluid velocity vector defined as u/U (–)
V_s	first impact velocity of a particle given by Equation (6.1.53) (m s^{-1})
$\overline{V}_s$	a quantity defined as $1/(k_s\rho_s)^{1/2}$
$\overline{v}_i$	Poisson's ratio of substance i
$\underline{v}^*$	dimensionless particle velocity defined as v/U (–)
w	a quantity defined by Equation (5.4.11a)
Y	defined by Equation (6.4.9a)
Y_i	Young's modulus of substance i
z_i	valency of the i-th ionic species (–)

Greek Letters

α	filter coefficient ratio defined by Equation (6.5.6) (–) or coefficient of Equation (6.3.7)
α_1, α_2	coefficient and exponent of Equation (6.3.17) (–)
$\overline{\alpha}_1, \overline{\alpha}_2$	coefficient and exponent of Equation (6.2.6) (–)
γ	adhesion probability (–)
Δc	concentration difference [see Equation (6.1.22)]
δ	separation distance (m)
δ^+	dimensionless separation distance defined as δ/a_p (–)
ε	filter medium porosity
$\hat{\varepsilon}$	permittivity of the liquid medium, equal to $\varepsilon_r\hat{\varepsilon}_0$ where ε_r is the dielectric constant (dimensionless) and $\hat{\varepsilon}_0$ permittivity of vacuum ($=8.854 \times 10^{-12}\,kg^{-1}\,m^{-3}\,s^4\,A^2$)
ε_s	medium solidosity ($= 1 - \varepsilon$)
ζ_c, ζ_p	collector and particle surface potential (V)
η	collector efficiency (–)
η_c	particle contact (collision) efficiency (–)
η_0	initial collector efficiency (–)
η_s	single-collector efficiency
$\eta_{BM}, \eta_E, \eta_G, \eta_I, \eta_i$	collector efficiency due to diffusion, electrostatic forces, gravity, interception, and inertial impaction
η_{f_s}	single fiber efficiency (–)
θ	corrected time defined as $t - \int_0^z \dfrac{\varepsilon dz}{u_s}$ (s) or angular coordinate (–)
θ_c	limiting angular position for particle adhesion
κ	reciprocal of the double-layer thickness (m^{-1})
λ	filter coefficient (m^{-1})
λ_e	wave length of electron oscillation, see Figs. 6.12a and 6.12b
λ_0	initial filter coefficient (m^{-1})
$\overline{\lambda}$	filter coefficient evaluated according to Equation (6.2.5b) or an elastic parameter given by Equation (6.1.50)
$(\lambda)_{fav}$	value of λ_0 with favorable surface interaction (m^{-1})
μ	fluid viscosity (Pa.s)

ρ	fluid density $(\text{kg}\,\text{m}^{-3})$
ρ_{p}	particle density $(\text{kg}\,\text{m}^{-3})$
ρ_{s}	filter grain density $(\text{kg}\,\text{m}^{-1})$
σ	specific deposit (vol/vol)
$\bar{\sigma}$	value of σ determined according to Equation (6.2.5a)
τ	surface shear stress given by Equation (6.1.36) $(\text{N}\,\text{m}^{-2})$
$\phi_{\text{DL}}, \phi_{\text{LO}}$	double-layer and London–van der Waals force potential (N m)
ψ	stream function $(\text{m}^2\,\text{s}^{-1})$ or the objective function of optimization or a quantity given by Equation (6.1.57)

In this chapter, we begin our presentation by first discussing the mechanisms of particle deposition in deep bed filtration and introducing a number of methods, which may be used for predicting cylindrical and spherical collector efficiencies followed by a brief discussion on the determination of the filter coefficients of aerosols and hydrosols from experimental data. Finally, a summary of a number of empirical correlations of the filter coefficients based on either calculations or experimental data are given and their use in predicting filter performance demonstrated.

6.1   Deposition Mechanisms and Prediction of Collector Efficiency based on Individual Transport Mechanism

By describing a filter medium as an assembly of collection, the process of filtration may be viewed as the flow of suspension through or past a number of collectors and deposition of the suspended particles onto the collector surfaces. Filtration performance, therefore, can be determined from the knowledge of the extent of particle deposition.

On physical grounds, one may view filtration as a two-step process, transport of suspended particles from a flowing suspension to collector surfaces followed by their attachment (adhesion) to the surfaces. The first step, in principle, can be determined by tracking particle trajectories based on the forces involved. Particle adhesion, however, depends on the interaction between contacting particles and collector surfaces, which, in turn, is determined by the force acting on the contacting particles including the particle-collector interaction forces as well as collector surface geometry and material characteristics. Particle collection efficiency (which may be expressed as the collector efficiency, fiber collector efficiency, single-collector efficiency, or single fiber efficiency mentioned in previous sections), therefore, can be decomposed into two parts, the collision (or contact) efficiency and the adhesion probability as

$$\eta = \eta_{\text{c}} \cdot \gamma \tag{6.1.1}$$

where η may be any one of the collector efficiencies mentioned before, and η_{c} the corresponding contact efficiency. γ is the adhesion probability defined as the fraction of contacting particle which becomes deposited onto the collector; η and η_{c} become the same if contact leads to adhesion automatically (i.e., $\gamma = 1.0$). As this is often the case in many practical situations, the difference between η and η_c is often ignored by filtration workers. For the sake of not creating further complication, the conventional but, strictly speaking, incorrect practice will be followed in this text. Adhesion, however, will be mentioned whenever it becomes important (see 6.3.2).

6.1.1   Mechanism of Particle Transport

As stated previously, the extent of particle transport (from the main body of a suspension flowing past a collector to the collector surface) may be based on particle trajectories, which, for a given particle and collector geometry and fluid flow field, are determined by a number of variables as elaborated below. For the purpose of illustrating the physical significance of the variables, it is useful to analyze their effect in isolation (individually) and obtain expressions of the extent of contact (contact or collision efficiency). The results so obtained are commonly referred to as the collector efficiency due to a particular mechanism with the unstated assumption that the adhesion probability is unity.

A.   By Inertial Impaction

One of the major deposition mechanisms for aerosols of micron size, in the absence of significant external forces, is inertial impaction. The inertial effect is shown schematically in Fig. 6.1, which depicts the flow of suspension over a spherical or cylindrical collector. Assuming that the direction of the flow coincides with the axis of symmetry of the collector body, both fluid and particles move rectilinearly (or fluid streamline and particle trajectory coincide) at distances remote from the collector. However, closer to the collector, fluid streamlines begin to change directions, turning away from the collector at the forward stagnation point in order to conform to the no-slip condition at the collector surface. On

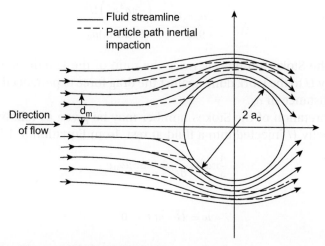

FIGURE 6.1 Particle deposition caused by inertial impaction.

the other hand, particles, because of their inertia, change trajectories differently from the way fluid does. As they deviate from the corresponding streamlines, some of the particle trajectories may intersect the collector surface, leading to particle deposition.

Accordingly, the extent of particle deposition and, therefore, the collector efficiency (or the filter coefficient) can be determined from the knowledge of particle trajectories. The procedure used is the so-called trajectory analysis[1]. Presently, we will restrict our attention to the simple case of a spherical particle moving with velocity $\underline{v}$ in a fluid stream moving with velocity $\underline{u}$. In the absence of any external forces, and assuming that the drag force acting on the particle is given by the Stokes law, the equation of particle motion may be written as

$$\frac{4}{3}\pi a_p^3 \rho_p \frac{d}{dt}\underline{v} = 6\pi\mu a_p(\underline{u} - \underline{v}) \tag{6.1.2}$$

The above expression can be put into dimensionless form by using U, the characteristic velocity of the flow, as the normalizing factor for velocity and a_c/U for time. Equation (6.1.2) becomes

$$N_{St}\frac{d}{dt}\underline{v}^* = \underline{u}^* - \underline{v}^* \tag{6.1.3}$$

and

$$N_{St} = \frac{2\rho_p U a_p^2}{9\mu a_c} \tag{6.1.4.a}$$

where a_p is the particle radius, a_c the collector radius, and U the characteristic velocity of the flow. ρ_p and μ are the particle density and fluid viscosity, respectively, and the superscript * denotes the dimensionless quantities.

The physical meaning of the Stokes number can be explained in several ways. For example, from its definition in Equation (6.1.3), N_{St} can be expressed as

$$N_{St} = \frac{2a_p^2\rho_p U}{9\mu a_c} = \frac{\left(\frac{4}{3}\pi a_p^3 \rho_p\right)U^2}{(6\pi\mu U a_p)a_c} \tag{6.1.4.b}$$

In other words, the Stokes number is twice the ratio of the kinetic energy of a particle moving at velocity U to the work done against the drag force experienced by the particles, $6\pi\mu U a_p$, over a distance of a_c.

Another interpretation of the Stokes number may be given as follows: For a particle moving with velocity $\overline{U}$ injected into a fluid at rest, from Equation (6.1.2), one has

$$\frac{4}{3}\pi a_p^3 \rho_p \frac{dv}{dt} = -6\pi\mu a_p v \tag{6.1.5}$$

and

$$v = \overline{U} \quad \text{at } t = 0 \tag{6.1.6}$$

[1]For a detailed discussion, see Tien and Ramarao (2007).

The solution of Equation (6.1.5) with the initial condition of Equation (6.1.6) is

$$v = \overline{U}e^{-\dfrac{9\mu}{2a_p^2\rho_p}t} \tag{6.1.7}$$

The particle velocity, v, may be written as dx/dt, where x is the particle position. Assuming $x = x_0$ at $t = 0$, Equation (6.1.7) can be further integrated to give

$$x = x_0 + \frac{2a_p^2\rho_p\overline{U}}{9\mu}\left(1 - e^{-\dfrac{9\mu}{2a_p^2\rho_p}t}\right) \tag{6.1.8}$$

The total distance traveled by the particle before it comes to a stop, namely, the value $x - x_0$ at $t \to \infty$, is

$$\frac{2a_p^2\rho_p\overline{U}}{9\mu} = \frac{2a_p^2\rho_p\overline{U}}{9\mu a_c}a_c = N_{St}a_c \tag{6.1.9}$$

In other words, N_{St} can be considered the particle's stopping distance (expressed in multiples of the characteristic length of the collector). By either interpretation, N_{St} is an indicator of the importance of the inertial effect.

In the above discussion, the drag force on the particle is assumed to be given by the Stoke's law, which holds true if the relative fluid/particle motion is slight and the no-slip condition is satisfied at the particle surface. The former condition is likely satisfied in deep bed filtration. On the other hand, significant velocity slip at the particle surface occurs if the particle size is comparable to the mean free path of the entraining gas molecules as discussed previously. To account for this effect, the Cunningham correction factor, c_s, is introduced. The Stokes number is then modified as

$$N_{St} = \frac{2}{9}c_s\frac{\rho_p\overline{U}a_p^2}{\mu a_c} = \frac{c_s\rho_p\overline{U}d_p^2}{9\mu d_c} \tag{6.1.10}$$

According to Millikan's formula (Millikan, 1923), c_s is given as

$$c_s = 1 + \frac{\overline{\ell}}{a_p}\left[1.23 + 0.41\exp\left(-0.88a_p/\overline{\ell}\right)\right] \tag{6.1.11}$$

where $\overline{\ell}$ is the mean free path of the entraining gas molecules and is given as

$$\overline{\ell} = \frac{\mu}{\sqrt{2\rho p/\pi}} \tag{6.1.12}$$

where p is the pressure and ρ the gas density. Generally speaking, for particles of 1 μm diameter at normal temperature and pressure, $c_s \sim 1.16$. On the other hand, since $\overline{\ell}$ is inversely proportional to the square root of the fluid density, c_s may differ significantly from unity at low pressure and high temperature.

A large number of studies on the determination of the collector/fiber efficiency due to inertial impaction, η_i, including the development of correlations, which can be used for calculating η_i have been reported in the literature. The differences among the results of these studies were found to be significant (see Davis, 1973, Brown, 1993). Examples of these results are shown in Figs. 6.2 and 6.3. Fig. 6.2 gives $(\eta_s)_i$ vs. N_{St} for

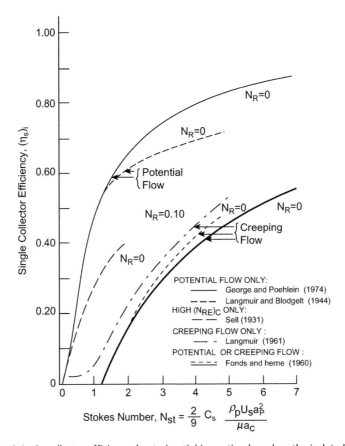

FIGURE 6.2 Calculated single-collector efficiency due to inertial impaction based on the isolated-sphere model.

spherical collectors obtained using different flow fields. Fig. 6.3(a) and (b) present the results on the single fiber efficiency due to the combined effect of inertial impaction and interception, [η_s vs. N_{St} Fig. 6.3(a) and $(\eta_s)_i$ vs. N_{St} Fig. 6.3(b)]. In spite of the differences of these results, some common trend can be easily discerned. $(\eta_s)_i$, generally speaking, is negligible for small N_{St}. Once N_{St} exceeds a certain value, $(\eta_s)_i$ increases rapidly with N_{St} and approaches to its asymptotic value. The significant difference shown in these figures casts doubts about the validity of trajectory calculation results (Davis, 1973).

Approximate expression of single fiber efficiency due to inertial impaction has been given by earlier investigations, including, as summarized by, Brown (1993).

For High Stokes Number

$$(\eta_{f,s})_i = 1 - \frac{1.61}{N_{St}} \quad \text{for } \varepsilon_s = 0.05 \tag{6.1.13a}$$

For Low Stokes Number

$$(\eta_{f,s})_i = \left[(29.6 - 28\varepsilon_s^{0.62})N_R - 27.5N_R^{2.8}\right]\frac{N_{St}}{8(Ku)^2} \qquad (6.1.13b)$$

for $0.0035 < \varepsilon_s < 0.111$

$$0.01 < N_R < 0.4$$

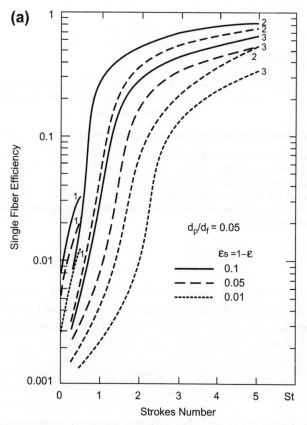

FIGURE 6.3(a) Single fiber efficiency due to combined inertial impaction and interception, $(\eta_{f,s})_{RI}$ vs. the stokes number, N_{St}.

1. Analytical Approximation by Stechkina et al. (1970)
2. Stepwise Calculation by Harrop using Happel's Model (1969)
3. Stepwise Calculation of Dawson (1969) using Spielman–Goren Flow Model
Reprinted from C.N. Davis, "Air Filtration", Academic Press, 1973, with permission of Elsevier.

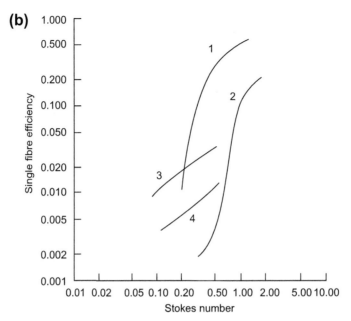

FIGURE 6.3(b) Calculated single efficiencies due to inertial impaction vs. the stokes number, N_{St}.

1. Results of Harrop and Stenhouse (1969) with $\varepsilon_s = 0.11$
2. Results of Harrop and Stenhouse (1969), $\varepsilon_s = 0.01$
3. Results of Stechkina et al. (1969), $\varepsilon_s = 0.1$, $N_R = 0.05$
4. Results of Stechkina et al. (1969), $\varepsilon_s = 0.01$, $N_R = 0.05$
Reprinted from R.C. Brown, "Air Filtration: An integrated approach to the theory and application of fibrous filters",
Pergamon Press, 1993, with permission of Elsevier.

B. By Gravity[2]

If the particle density is different from that of the fluid (and it is assumed to be greater), then particles may settle out in the direction of the gravitational force. The sedimentation velocity of small particles in dilute suspensions, V_t, can be approximated by the Stokes law [see Equation (5.6.8)]:

$$V_t = \frac{2}{9} \frac{a_p^2 g (\rho_p - \rho)}{\mu}$$ (6.1.14)

Accordingly, the single collector/fiber efficiency due to sedimentation, $(\eta_s)_G$, is

$$(\eta_s)_G = \frac{(V_t)(s)c_\infty}{(s)u_\infty c_\infty} = \frac{2a_p^2 g(\rho_p - \rho)}{9\mu u_\infty} = N_G$$ (6.1.15)

[2]The treatment given is for the case when the direction of the main flow coincides with the direction of the gravitational force.

where s is the projected collector area on a plane normal to the direction of flow and c_∞ the particle concentration. Note that this expression is applicable to both spherical and cylindrical collectors.

C. By Interception

Particle deposition by interception occurs because particles are finite in size. If all the forces acting on particles are negligible, particle trajectories coincide with fluid stream-lines. Following the discussion in 6.1.1, the extent of deposition can be estimated by identifying the streamlines, which come within a distance of $d_p/2$ away from collector surfaces. The physical situation is depicted in Fig. 6.4 for either spherical or cylindrical collectors.

a. Spherical Collector (Granular Media). The spherical collector case will be discussed first. As shown in Fig. 6.4, for the upper half of the collector, the farthest point away from the front stagnation point, where deposition may take place, is the one with coordinates $r = a_c + a_p$, $\theta = \pi/2$. The stream function value corresponding to the streamline passing through $r = a_c + a_p$, $\theta = \pi/2$ is $\psi|_{r=a_c+a_p, \theta=\pi/2}$ where ψ is given by Equation (5.4.8) if Happel's model is used to represent filter medium. The collector efficiency due to interception $(\eta)_I$, by definition, is

$$(\eta)_I = \frac{2\pi\psi|_{r=a_c+a_p, \theta=\pi/2}}{\pi b^2 u_\infty} \tag{6.1.16}$$

The numerator of the above expression represents the volumetric flow rate of the feed enclosed by the streamline passing through point $r = a_c + a_p$, $\theta = \pi/2$, $0 < \phi < 2\pi$ and the numerator is the total volumetric flow rate into the Happel cell.

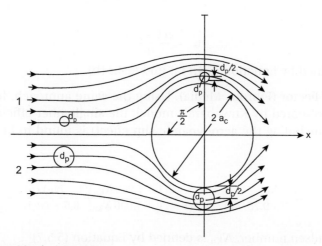

FIGURE 6.4 Particle deposition by interception.

The value of $\psi|_{r=a_c+a_p, \theta=\pi/2}$ can be found from Equation (5.4.8) as

$$\psi|_{r=a_c+a_p} = (u_s/2)a_c^2\left[\frac{K_1}{1+N_R} + K_2(1+N_R) + K_3(1+N_R)^2 + K_4(1+N_R)^4\right] \tag{6.1.17a}$$

and

$$N_R = (a_p/a_c) \tag{6.1.17b}$$

Substituting Equation (6.1.17a) into Equation (6.1.16) and writing

$$\frac{1}{1+N_R} = 1 - N_R + N_R^2 - N_R^3 + \dots \tag{6.1.18}$$

one has

$$\eta_I = \left(\frac{a_c}{b}\right)^2\left[(K_1 + K_2 + K_3 + K_4) + (-K_1 + K_2 + 2K_3 + 4K_4)N_R + (K_1 + K_3 + 6K_4)N_R^2 + O(N_R^3)\right]$$

The coefficients K_1, K_2, K_3, and K_4 are given by Equations (5.4.10b)–(5.4.10e). It is simple to show that

$$K_1 + K_2 + K_3 + K_4 = 0$$

$$-K_1 + K_2 + 2K_3 + 4K_4 = 0$$

$$K_1 + K_3 + 6K_4 = \frac{3(1-p^5)}{w}$$

Since in granular filtration, a_c, in many cases, can be expected to be at least two orders of magnitude greater than a_p, therefore, if only the leading term is retained, $(\eta)_I$ is found to be

$$\eta_I = 1.5 A_s(1-\varepsilon)^{2/3}N_R^2 \tag{6.1.19a}$$

and

$$A_s = \frac{2(1-p^5)}{w} \tag{6.1.19b}$$

with w and p defined by Equation (5.4.11a) and (5.4.11b).

b. Cylindrical Collector (Fibrous Media). Using the same approach, for cylindrical collectors represented by Kuwabara's model, the single fiber efficient due to interception (including the aerodynamic slip effect) is found to be (Brown, 1993, p. 79)

$$(\eta_s)_I = \frac{(1+N_R)^{-1} - (1+N_R) + 2(1+1.966\,N_{Kn})(1+N_R)\,\ell n(1+N_R)}{2[-0.75 - (1/2)\ell n(1-\varepsilon)] + 1.996\,N_{Kn}[-0.5 - \ell n(1-\varepsilon)]} \tag{6.1.20}$$

where the Knudsen number, N_{Kn} is defined by Equation (5.5.7)

■ ■ ■ ▬▬▬▬▬▬▬▬▬▬▬▬▬▬▬▬▬▬▬▬▬▬▬▬▬▬▬▬▬▬▬▬▬

Illustrative Example 6.1

Obtain an expression of the collector efficiency due to interception, η_I, assuming that filter medium may be represented as a bundle of capillaries of radii R_c (namely, the capillaric model). The flow through the capillaries is laminar and fully developed. The particle concentration across the capillary cross section is uniform. If the radius R_c may be assumed to be proportional to the grain radius, a_g, or $R_c = \beta\, a_g$, what is the value of β?

Solution

The flow through a capillary of radii R_c is

$$u = 2u_{av}\left[1 - \left(\frac{r}{R_c}\right)^2\right]$$

As particles present in the fluid flowing through a capillary with radial position equal to $R_c - a_c$ are to be collected, η_I is given as

$$\eta_I = \frac{\displaystyle\int_{R_c-a_p}^{R_c} r\,u\,dr}{\displaystyle\int_0^{R_c} r\,u\,dr} = \frac{\displaystyle\int_{1-\frac{a_p}{R_c}}^{1} r^+(1-r^{+2})\,dr^+}{\displaystyle\int_0^1 r^+(1-r^{+2})\,dr^+}$$

$$= 4\left(\frac{a_p}{R_c}\right)^2 - 4\left(\frac{a_p}{R_c}\right)^3 + \left(\frac{a_p}{R_c}\right)^4$$

$$\simeq 4\left(\frac{a_p}{R_c}\right)^2 \quad \text{if } (a_p/R_c) \ll 1 \tag{i}$$

If R_c is assumed to be proportional to the grain radius,

$$\eta_I = (4/\beta^2).N_R^2 \tag{ii}$$

To estimate β, recall that η_I based on Happel's model is given by Equation (6.1.18), or

$$\eta_I = 1.5\,A_s(1-\varepsilon)^{2/3}N_R^2 \tag{iii}$$

Comparing Equation (ii) with Equation (iii) yields

$$\beta = \sqrt{\frac{8}{3\,A_s}}(1-\varepsilon)^{-1/3} \tag{iv}$$

▬▬▬▬▬▬▬▬▬▬▬▬▬▬▬▬▬▬▬▬▬▬▬▬▬▬▬▬▬▬▬▬▬ ■ ■ ■

D.   By Brownian Diffusion

For particles of submicron size and with the absence of electrostatic forces, the Brownian diffusion force is the dominant factor in determining particle deposition. The problem of particle deposition may then be considered as a mass transfer problem with the Brownian diffusivity replacing the ordinary binary diffusion coefficient [for a detailed discussion, see Tien and Ramarao (2007)]. The Brownian diffusivity, D_{BM}, is given as

$$D_{BM} = \frac{c_s kT}{3\pi\mu d_p} \qquad (6.1.21)$$

where c_s and k are the Cunningham corrector factor and the Boltzmann constant, respectively.

a. Spherical Collector (Granular Media). We will first consider the granular media case. For mass transfer in granular packed beds, the Sherwood number, N_{Sh}, is defined as

$$N_{Sh} = \frac{d_g}{D} \frac{I}{(\Delta c)S} \qquad (6.1.22)$$

where I is the mass flux over a granule. D, S, and Δc are, respectively, the diffusivity, the surface area of the grain, and the mass transfer driving force (that is, concentration difference). As filter grains are assumed to be spherical,

$$S = \pi d_g^2 \qquad (6.1.23)$$

The driving force for mass transfer may be taken as the difference between the concentration in the bulk of the fluid, c, and that at the fluid–grain interface. Using the perfect sink assumption that all particles transported to collector surfaces are deposited, the surface concentration vanishes. The driving force Δc therefore equals c_∞.

For particle deposition, the single-collector efficiency, $(\eta_s)_{BM}$, may be written as

$$(\eta_s)_{BM} = \frac{4I_p}{(\pi d_g^2)u_s c} \qquad (6.1.24)$$

where I_p is the particle flux over a filter grain. For deposition of Brownian particles, $I = I_p$ and $D = D_{BM}$. Comparing Equations (6.1.22) with (6.1.24), one has

$$N_{Sh} = \frac{1}{4}N_{Pe}(\eta_s)_{BM} \qquad (6.1.25)$$

and the Peclet number N_{Pe} is defined as

$$N_{Pe} = \frac{d_g u_\infty}{D_{BM}} \qquad (6.1.26)$$

For mass transfer in packed beds, the Sherwood number is found to be (Pfeffer and Happel 1964)

$$N_{Sh} = A_s^{1/3} N_{Pe}^{1/3} \qquad (6.1.27)$$

where A_s is given by Equation (6.1.19b). By combining Equations (6.1.26) and (6.1.25), one has

$$(\eta_s)_{BM} = 4A_s^{1/3} N_{Pe}^{-2/3} \qquad (6.1.28)$$

b. Cylindrical Collector (Fibrous Media). Similarly, for fibrous media and using Kuwabara's model, Lee and Liu (1982) obtained the single fiber efficiency expression as

$$(\eta_{f,s})_{BM} = 1.6\left(\frac{\varepsilon}{Ku/u_s}\right)^{1/3} N_{Pe}^{-2/3}$$

(6.1.29)[3]

with N_{Pe} defined as $d_f u_\infty / D_{BM}$ and Ku, the Kuwabara hydrodynamic factor, is given by Equation (5.4.27e).

The results given above [namely, Equations (6.1.28) and (6.1.29)] do not allow for the presence of repulsive force barriers between particles and collectors, which may be present in hydrosol filtration or the aerodynamic slip effect in the aerosol case. Correlations which account for the presence of repulsive force barrier for hydrosol filtration in granular media have been developed by several investigators and a summary of these studies can be found in the monograph of Tien and Ramarao (2007). The monograph of Davis (1973) provides the results on the aerodynamic slip effect in fibrous media.

■ ■ ■ ▬▬▬▬▬▬▬▬▬▬▬▬▬▬▬▬▬▬▬▬▬▬▬▬▬▬▬▬▬▬▬

Illustrative Example 6.2

For aerosol filtration in fibrous media under the following conditions:

Aerosol Particle Diameter	1 μm
Aerosol Particle Density	1200 kg m^{-3}
Fiber Diameter	20 μm
Fiber Packing Density	$\varepsilon_s = 0.05$
Gas Velocity	10 cm s^{-1} (0.1 ms^{-1})
Temperature	20 °C or 293 K
Pressure	1 atm or 1.013×10^5 Pa

(a) Calculate the single fiber efficiencies due to gravity, interception, and diffusion by Equations (6.1.14), (6.1.20), and (6.1.29), respectively.
(b) Estimate approximately the single fiber efficiency due to inertial impaction based on the results shown in Figs. 6.3a and 6.3b. Comment on the results obtained.
(c) Ignore the contribution due to inertial impaction; calculate the single fiber efficiency due to the combined effect of gravity, interception, and diffusion.

Solution

(a) From Equation (6.1.14),

$$(\eta_{sf})_G = N_G$$

(i)

[3]Equation (6.1.29) is obtained from Equation (6.3.4) by ignoring the contribution due to interception.

and

$$N_G = \frac{2\,a_p^2 g(\rho_p - \rho)}{9\,\mu\,u_\infty} \tag{ii}$$

At 20 °C and 1 atm

$$\mu = 0.0165 \text{ cp} \quad \text{or} \quad 1.65 \times 10^{-5} \text{ Pa s}$$

For air density, assume that the ideal gas law is valid, the volume occupied by 1 g-mol ideal gas is

$$22.4 \times \frac{293}{273} = 24.04 \text{ } \ell \quad \text{or} \quad 2.404 \times 10^{-2} \text{ m}^3$$

and air density $p = 0.029/(2.404 \times 10^{-2}) = 1.208 \text{ kg m}^{-3}$.

$$N_G = \frac{(2)[(1 \times 10^{-6})/2]^2(9.8)(1200 - 1.208)}{(9)(1.65 \times 10^{-5})(0.1)} = \frac{(5.875) \times 10^{-9}}{(1.485) \times 10^{-5}} = 3.956 \times 10^{-4}$$

and

$$(\eta_{f,s})_g = 3.956 \times 10^{-4}$$

For calculating $(\eta_{f,s})_I$, from Equation (6.1.20)

$$(\eta_{f,s})_I = \frac{(1 + N_R)^{-1} - (1 + N_R) + 2(1 + 1.966 N_{Kn})(1 + N_R)\ell n(1 + N_R)}{2\left[-0.75 - \frac{1}{2}\ell n(1 - \varepsilon)\right] + 1.966 N_{Kn}[-0.5 - \ell n(1 - \varepsilon)]} \tag{iii}$$

$$N_R = 1/20 = 0.05$$

From Equation (5.5.7)

$$N_{Kn} = \frac{2\bar{\ell}}{d_f} \tag{iv}$$

The gas molecule mean free path, $\bar{\ell}$, is given by Equation (5.5.8). In Illustrative Example 5.5, λ is found to be 6.521×10^{-8} m at $T = 298$. Since $\bar{\ell}$ is proportional to the square root of T, $\bar{\ell}$ at 293 K is

$$\bar{\ell} = 6.521 \times 10^{-8}\sqrt{293/298} = 6.466 \times 10^{-8} \text{ m}$$

and

$$N_{Kn} = \frac{2 \times 6.466 \times 10^{-8}}{(20) \times 10^{-6}} = 6.466 \times 10^{-3}$$

The single fiber efficiency due to interception is

$$\left(\eta_{f,s}\right)_I = \frac{(1.05)^{-1} - (1.05) + 2[1 + (1.966)(6.466 \times 10^{-3})](1.05)\ell n(1.05)}{2\left[-0.75 - \frac{1}{2}\ell n(0.05)\right] + (1.966)(6.466 \times 10^{-3})[-0.5 - \ell n(0.05)]}$$

$$= \frac{0.00125}{4.462} = 2.8 \times 10^{-4}$$

For calculating $(\eta_{s,f})_D$, according to Equation (6.1.29)

$$(\eta_{f,s})_{BM} = (1.6)\left(\frac{\varepsilon}{Kn}\right)^{1/3} N_{Pe}^{-2/3} \tag{v}$$

and

$$N_{Pe} = \left(\frac{d_f u_s}{D_{BM}}\right) \tag{vi}$$

The Brownian diffusivity, D_{BM}, is given as [see Equation (6.1.21)]

$$D_{BM} = \frac{c_s.kT}{3\pi\mu d_p} \tag{vii}$$

where k is the Boltzmann constant ($k = 1.3805 \times 10^{-23}$ J K^{-1}). The Cunningham correction factor is given by Equation (6.1.11) or

$$c_s = 1 + \frac{\bar{\ell}}{a_p}[1.23 + 0.41\exp(-0.88\, a_p/\bar{\ell})]$$

$$= 1 + (6.466 \times 10^{-8}/5 \times 10^{-7})[1.23 + 0.41 \exp(44/6.466)] = 1.16$$

and

$$D_{BM} = \frac{(1.16)(1.3805 \times 10^{-23})(293)}{(3\pi)(1.65 \times 10^{-5})(1 \times 10^{-6})} = 3.017 \times 10^{-11}\, m^2\, s^{-1}$$

$$N_{Pe} = \frac{(2 \times 10^{-5})(0.1)}{3.017 \times 10^{-11}} = 6.629 \times 10^4$$

The Kuwabara hydrodynamic factor, (Ku/u_s), is given by Equation (5.4.27e)

$$Ku = -\frac{1}{2}\ell n\,(\varepsilon_s) - (3/4) + \varepsilon_s - (\varepsilon_s^2/4)$$

$$= -\frac{1}{2}\ell n(0.05) - (3/4) + (0.05) - \frac{1}{4}(0.05)^2 = 0.7973$$

The single fiber efficiency due to diffusion, according to Equation (v), is

$$(\eta_{f,s})_{BM} = (1.6)\left(\frac{0.95}{0.7973}\right)^{1/3}\left(6.629 \times 10^4\right)^{-2/3}$$

$$= 1.035 \times 10^{-3}$$

(b) It is difficult to obtain a reasonable estimate of $(\eta_{s,f})_i$ from these figures. As shown in Fig. 6.3a, for the three different curves, $(\eta_{s,f})_I$ ranges from 0.002 to 0.005. However, according to Fig. 6.3b, $(\eta_{f,s})_i$ is negligible. This uncertainty underscores the contention of Brown's that the inertial impaction contribution cannot be accurately estimated based on existing knowledge.

(c) Assuming that the combined effect on deposition can be expressed as the sum of the individual effect,

$$\eta_{f,s} = (\eta_{f,s})_G + (\eta_{f,s})_I + (\eta_{f,s})_{BM}$$
$$= 3.556 \times 10^{-4} + 2.8 \times 10^{-4} + 1.035 \times 10^{-3} = 1.671 \times 10^{-3}$$

Alternatively, one may apply Equations (6.1.33b)

$$(1 - \eta_{f,s}) = \left[1 - (\eta_{f,s})_G\right]\left[1 - (\eta_{f,s})_I\right]\left[1 - (\eta_{f,s})_{BM}\right]$$

or

$$\eta_{f,s} = 1 - (1 - 3.956 \times 10^{-4})(1 - 2.8 \times 10^{-4})(1 - 1.891 \times 10^{-3})$$
$$= 1 - (0.9996)(99972)(0.9981)$$
$$= 0.00257 \text{ or } 2.57 \times 10^{-3}$$

$$\eta_{f,s} = 1 - (1 - 3.956 \times 10^{-4})(1 - 2.8 \times 10^{-4})(1 - 1.035 \times 10^{-3})$$
$$= 1 - (0.9996)(99972)(0.9989)$$
$$= 0.00178 \text{ or } 1.78 \times 10^{-3}$$

As expected, the difference is insignificant since all $\eta_{f,s}$'s are small. This subject will be further discussed in 6.1.3.

■ ■ ■

E.   By Electrostatic Forces

In aerosol filtration, particles and filter media often become electrically charged by natural process (such as raindrops acquire charges from the atmosphere) or by design (e.g., manufactured charged media). The effect of the electric charge may be significant as shown by Lundgren and Whitby (1965).

The electrostatic force between a particle and a collecting body is complicated because of the presence of a variety of interactions. Using a two-body assumption, the electrostatic force between an aerosol particle (charged or uncharged) and a collecting body, Fe, can be expressed according to Kraemer and Johnstone (1955) as a sum of several forces including:

1. The Coulombic force between charged particles and charged collectors, F_{EC}.
2. The electric image force between charged particles and neutral collectors, F_{EI}.
3. The electric image force between neutral particles and charged collectors, F_{EM}.
4. The force on charged particles in the presence of a neutral collector by a uniform external electric field, F_{ex}.
5. The particle charged in the same sign produces a repulsive force among themselves, F_{es}.
6. The electric dipole interaction force between an uncharged particle and an uncharged collector both being polarized by an external electric field, F_{icp}.

or

$$Fe = F_{EC} + F_{EI} + F_{EM} + F_{EX} + F_{ES} + F_{icp} \qquad (6.1.30)$$

The various electrostatic force expressions are given in Table 6.1.

Deposition due to the electrostatic forces can be estimated by tracking particle trajectories as described previously (see Section 6.1.1). Equation (6.1.2) may be generalized to include the electrostatic forces given by Equation (6.1.30), or,

$$\frac{3}{4}\pi a_{p}^{3}\rho_{p}\frac{dv}{dt} = 6\pi\mu\,a_{p}(u-v) + F_{e} \qquad (6.1.31)$$

Equation (6.1.31) can be put into dimensionless form by using U, the characteristic velocity, a_{c}, the collector radius, and a_{c}/U as normalizing factors for velocity, coordinate distance, and time. Equation (6.1.31) becomes

$$N_{St}\frac{dv^{*}}{dt^{*}} = (u^{*} - v^{*}) + \sum_{j} K_{Ej}f_{Ej} \qquad (6.1.32)$$

where the subscript Ej denotes EC, EI, EM, EX, ES, and icp. Expressions of K_{Ei}'s and f_{Ei}'s are listed in Table 6.2. Single collector efficiencies due to some of the above-mentioned electrostatic forces based on trajectory tracking obtained by Nielson and Hill (1976) for spherical collector with Stokes flow field are shown in Fig. 6.5.

Similarly, for cylindrical collectors, the single fiber efficiency due to the electrostatic forces can also be estimated by the trajectory analysis. Some of the results cited by Brown (1993) are given in Table 6.3.

Table 6-1 Electrostatic Forces Between a Charged Aerosol Particle and a Spherical Collector

Force	Radial Component	Angular Component
F_{EC}, Coulombic Attraction Force	$\dfrac{Q_c Q_p}{4\pi\,\varepsilon_f a_c^2}\dfrac{1}{r^{*2}}$	0
F_{EI}, Charged-Collector Image force	$\dfrac{\gamma_p Q_c^2 a_p^3}{2\pi\,\varepsilon_f a_c^5}\dfrac{1}{r^{*5}}$	0
F_{EM}, Charged-Particle Image force	$\dfrac{\gamma_p Q_p^2}{4\pi\,\varepsilon_f a_c^2}\left[\dfrac{1}{r^{*3}} - \dfrac{r^*}{(r^{*2}-1)^2}\right]$	0
F_{ES}, Charged-Particle	$-\dfrac{\gamma_c Q_p^2 a_c c}{3\,\varepsilon_f}\dfrac{1}{r^{*2}}$	0
F_{EX}, External Electric Field force	$Q_p E_0\left(1+\dfrac{2\gamma_c}{r^{*3}}\right)\cos\theta$	$-Q_p E_0\left(1-\dfrac{\gamma_c}{r^{*3}}\right)\sin\theta$
F_{ICP}, Electric dipole Interaction force	$\left(\dfrac{-12\pi\,\gamma_c\gamma_p\varepsilon_f a_p^3 E_0^2}{a_c}\right)$ $\left[-2\left(1+\dfrac{2\gamma_c}{r^{*3}}\right)\cos^2\theta - \left(1-\dfrac{\gamma_c}{r^{*3}}\right)\sin^2\theta\right]\Big/r^{*4}$	$\left(\dfrac{-12\pi\,\gamma_c\gamma_p\varepsilon_f a_p^3 E_0^2}{a_c}\right)$ $\left(2+\dfrac{\gamma_c}{r^{*3}}\right)\dfrac{\sin\theta\cos\theta}{r^{*3}}$

Q_c: charge on collector Q_p: charge on particle ε_f: dielectric constant of fluid ε_p: dielectric constant of particle ε_c: dielectric constant of collector c_s: Cunningham's corrector factor c: particle concentration (number of particle/m³) γ_p: $(\varepsilon_p - \varepsilon_f)/(\varepsilon_p + 2\varepsilon_f)$ γ_c: $(\varepsilon_c - \varepsilon_f)/(\varepsilon_c + 2\varepsilon_f)$ E_0: uniform external electric field strength

Table 6.2   Expressions of K_{Ej} and f_{Ej}

	Radial Component	Angular Component
f_{EC}	$\dfrac{1}{r^{*2}}$	0
f_{Ei}	$\dfrac{1}{r^{*5}}$	0
f_{EM}	$\dfrac{1}{r^{*3}} - \dfrac{r^*}{(r^{*2}-1)^2}$	0
f_{ES}	$-\dfrac{1}{r^{*2}}$	0
f_{EX}	$\left(1 + \dfrac{2\gamma_C}{r^{*3}}\right)\cos\theta$	$\left(1 - \dfrac{\gamma_C}{r^{*3}}\right)\sin\theta$
f_{icp}	$\dfrac{\left[-2\left(1+\dfrac{2\gamma_c}{r^{*3}}\right)\cos^3\theta - \left(1-\dfrac{\gamma_c}{r^{*3}}\right)\sin^2\theta\right]}{r^*}$	$\left(2 + \dfrac{\gamma_C}{r^{*3}}\right)\dfrac{\sin\theta\cos\theta}{r^{*3}}$

Parameter

$$K_{EC} = \frac{c_s Q_c Q_p}{24\pi^2 \varepsilon_f a_c^2 a_p \mu U} \qquad\qquad K_{ES} = \frac{\gamma_c c_s Q_p^2 a_c c}{18\pi\varepsilon_f \mu U a_p}$$

$$K_{EI} = \frac{\gamma_p Q_c^2 a_p^2}{12\pi^2 \varepsilon_f a_c^5 \mu U} \qquad\qquad K_{EX} = \frac{2\gamma_c \gamma_p \varepsilon_f c_s a_p^2 E_0^2}{a_c \mu U}$$

$$K_{EM} = \frac{\gamma_c c_s Q_p^2}{24\pi^2 \varepsilon_f a_c^2 \mu U a_p} \qquad\qquad K_{ICP} = \frac{2\gamma_c \gamma_p \varepsilon_f c_s \theta_p^2 E_0^2}{a_c \mu U}$$

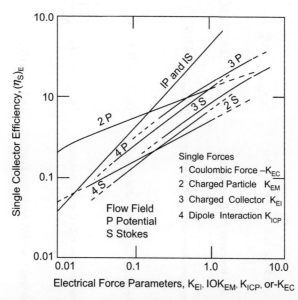

FIGURE 6.5 Single collector efficiencies due to electrostatic forces. *(Reprinted from K.A. Nielson and James C. Hill, "Collection of inertialess particles on spheres with electric forces" Ind. Eng. Chem. Fundam., 15, 145–156, 1976, American Chemical Society).*

Table 6.3   Single Fiber Efficiencies due to Electrostatic Forces

Mechanism	Single Fiber Efficiency*	Parameter
Charged fiber with charged particles	$(\eta_s)_{Qq} = \pi N_{Qq}$ $\qquad N_{Qq} = \dfrac{Qq}{12\pi^2 \varepsilon_0 \mu a_p a_f u_\infty}$	Q: charge per unit length of fiber
Charged fiber with neutral particles	$(\eta_s)_{Qo} = \pi N_{Qo}^2$ $\qquad N_{Qo} = \dfrac{Q^2 a_p^2}{6\pi^2 \varepsilon_0 \mu a_f^3 u_\infty}$	q: charge held by a particle $\;E$: electron field s strength
Neutral fiber with charged particles	$(\eta_s)_{oq} = (2/\zeta^{1/2})(N_{oq}^{1/2})$ $\qquad N_{oq} = \left(\dfrac{\varepsilon_f - 1}{\varepsilon_f + 1}\right)\dfrac{q^2}{96\pi^2 \mu u_\infty \varepsilon_0 a_p a_f^2}$	a_p: particle radius $\;a_f$: fiber radius
Line-dipole fiber with charged particles	$(\eta_s)_{oq} = 0.57(N_{\sigma q}^{0.83})$ $\qquad N_{\sigma q} = \dfrac{\sigma q}{6\pi\mu(\varepsilon_f + 1)u_\infty \varepsilon_0 a_p}$	u_∞: fluid velocity $\;\varepsilon_o$: permittivity of free space
Line-dipole fiber with neutral particles	$(\eta_s)_{\sigma 0} = 0.47(N_{\sigma 0}^{0.40})$ $\qquad N_{\sigma 0} = \dfrac{2}{3}\left(\dfrac{\varepsilon_p - 1}{\varepsilon_p + 2}\right)\dfrac{2\sigma^2 a_p^2}{\varepsilon_0(1+\varepsilon_f)^2 a_f \mu u_\infty}$	ε_f: dielectric constant of fiber
Polarized fiber with charged particles	$(\eta_s)_{pq} = N_{pq}\left[\dfrac{\frac{\varepsilon_f-1}{\varepsilon_f+1}+1}{N_{pq}+1}\right]$ $\qquad N_{pq} = \dfrac{Eq}{6\pi\mu a_p u_\infty}$	ε_p: dielectric constant of particle
Polarized fiber with neutral particles	$(\eta_s)_{p0} = (N_{p0}/2)$ $\qquad N_{p0} = \dfrac{2}{3}\left(\dfrac{\varepsilon_p - 1}{\varepsilon_p + 2}\right)\left(\dfrac{\varepsilon_f - 1}{\varepsilon_f + 1}\right)\dfrac{2 a_p^2 \varepsilon_0 E^2}{a_f \mu u_\infty}$	σ: surface charge derived of fiber $\;\mu$: fluid viscosity $\;\zeta$: equal to 0.998

*Taken from Eqs. (6.16), (6.22), (6.49), (6.32), (6.33), (6.61) and (6.62); Brown (1993).

■ ■ ■ ▬▬▬▬▬▬▬▬▬▬▬▬▬▬▬▬▬▬▬▬▬▬▬▬

Illustrative Example 6.3

Obtain the single-collector efficiency expressions due to the combined effect of interception and Coulombic (electrostatic) force in granular media assuming Stokes' flow around the collector.

Solution

Since the Coulombic force is solenoidal, particle trajectories may be expressed in terms of a "particle stream function" given by the sum of the Coulombic force potential and the stream function of the flow field. The dimensionless force potential, ψ^*_{EC}, given in Table 6.1, is

$$\psi^*_{EC} = K_{EC} \cos \theta \tag{i}$$

The stream function (dimensionless) of the Stokes flow is (Nielsen and Hill, 1976)

$$\psi^* = \frac{1}{2} r^{*2} \sin^2\theta \, h(r^*) \tag{iia}$$

$$h(r^*) = 1 - (3/2)(r^*) + (1/2)(1/r^{*3}) \tag{iib}$$

with r^* being r/a_c and $\psi^* = \psi/(a_c^2 u_0)$.

Accordingly, the dimensionless particle stream function ψ^*_p is

$$\psi^*_p = (1/2)r^{*2}(\sin \theta) \, h(r^*) + K_{EC} \cos \theta \tag{iii}$$

On the surface, it may appear that one can obtain the collector efficiency of this case in the same way used to obtain the interception collector efficiency (see Section 6.1.3). Closer examination shows, however, that the two cases are not identical. For the interception case, deposition takes place over the upstream half of the spherical collector. Particles initially situated at the axis of symmetry at $z = -\infty$ will be deposited at the front stagnation point, while the farthest point away from the center of the axisymmetry results in deposition at the angular position, $\theta = \pi/2$. In other words, deposition occurs over $0 < \theta < \pi/2$. With the presence of the Coulombic force and depending upon its nature (repulsive vs. attractive) and magnitude, the angular position range over which deposition takes place is not $0 < \theta < \pi/2$ but $\theta_s < \theta < \theta_i$. The values of θ_s and θ_i vary from 0 to π subject to the condition $\theta_s < \theta_i$. In particular, a particle initially located at the axis of symmetry at a large distance upstream from the collector may be deposited at an angular position other than the front stagnation point.

Determining the collector efficiency therefore requires knowing θ_i and θ_s. First, recall that the value of ψ^*_p along a given particle trajectory is constant. Consequently, one can identify a trajectory by its ψ^*_p value or, alternatively, by its off-center distance, x^*_0 at $z^* = -\infty$, since

$$\lim_{z^* \to -\infty} r^{*2} \sin^2 \theta = x^*_0$$

$$\lim_{z^* \to -\infty} \cos \theta = 1$$

The value of ψ_p^* corresponding to a trajectory which has an off-center distance x_0^* upstream from the collector is

$$(\psi_p^*)_0 = \frac{1}{2}x_0^{*2} + K_{EC}$$

The trajectory of a particle with an initial off-center distance, x_0^{*2}, is then given by the expression

$$\frac{1}{2}x_0^{*2} + K_{EC} = \frac{1}{2}r^{*2}\sin^2\theta\, h(r^*) + K_{EC}\cos\theta$$

or

$$x_0^{*2} = (\sin^2\theta)\, r^{*2}h(r^*) - 2\, K_{EC}(1 - \cos\theta) \tag{iv}$$

The value of θ_s can be found from the above equation with $x_0^* = 0$, $\theta = \theta_s$, and $r^* = 1 + N_R$, or

$$A\,\sin^2\theta_s - 2\,K_{EC}(1 - \cos\theta_s) = 0 \tag{va}$$

where

$$A = r^{*2}h(r^*)\Big|_{1+N_R} \tag{vb}$$

To obtain θ_i, first note that θ_i is the angular position of a particle trajectory which makes contact with the collector and has the largest off-center distance, x_0^*. Accordingly, we can find θ_i by solving $(dx_0^*/d\theta) = 0$ of Equation (iv), or

$$\sin\theta_i(A\cos\theta_i - K_{EC}) = 0 \tag{vi}$$

One may define a quantity $\hat{\eta}$ which gives the particle flux over the region $\theta_s \leq \theta \leq \theta.\,\hat{\eta}$ can, therefore, be expressed as

$$\hat{\eta} = \int_{\theta_s}^{\theta} [2\,\pi\,\sin\theta]\zeta(\theta)d\theta \tag{vii}$$

and

$$(\eta_s)_E = \hat{\eta}\Big|_{\theta=\theta_i} = \int_{\theta_s}^{\theta_i} [2\,\pi\,\sin\theta]\zeta(\theta).d\theta \tag{viii}$$

where $\zeta(\theta)$ may be considered as the local deposition flux (normalized by $\pi\,a_c^2 U_\infty c_\infty$). On the other hand, by the definition of the single-collector efficiency, one has

$$(\eta_s)_E = x_0^{*2}\Big|_{\substack{x_n^* \text{ (or } \theta_i) \\ 0 \text{ (or } \theta_s)}}$$

Similarly, one may write

$$\hat{\eta} = x_0^{*2}\Big|_{\substack{x_0^* \text{ (or } \theta) \\ 0 \text{ (or } \theta_s)}}$$

Accordingly

$$\frac{\partial}{\partial\theta}\left[\int_{\theta_s}^{\theta}[2\,\pi\,\sin\,\theta]\zeta(\theta)\mathrm{d}\theta\right] = \frac{\mathrm{d}}{\mathrm{d}\theta}x_0^{*2}$$

The value of $\dfrac{\mathrm{d}}{\mathrm{d}\theta}x_0^{*2}$ can be obtained from Equation (iv). Therefore,

$$\zeta(\theta) = \frac{1}{2\pi\,\sin\,\theta}[(2\,\sin\,\theta)(\cos\,\theta)A - 2\,K_{EC}\,\sin\,\theta] = \frac{A\cos\,\theta - K_{EC}}{\pi} \tag{ix}$$

Substituting Equation (vi) into Equation (viii) and carrying out the integration, we find $(\eta_s)_E$ to be

$$(\eta_s)_E = (\cos\,\theta_s - \cos\,\theta_i)\;[A(\cos\,\theta_s + \cos\,\theta_i) - 2\,K_{EC}] \tag{x}$$

The above expression together with Equations (va) and (vi) can be used to calculate the single-collector efficiency due to the combined effect of interception and the Coulombic force. First, consider the two limiting cases:

(a) No interception effect: If there is no interception effect, we may consider $N_R = 0$, and

$$h(r^*)|_{r^*=1} = 1 - (3/2) + (1/2) = 0 \quad \text{or} \quad A = 0$$

Equations (va) and (vi) give

$$2\,K_{EC}(1 - \cos\,\theta_s) = 0$$

$$-K_{EC}\,\sin\,\theta_i = 0$$

or

$$\theta_s = 0,\,\pi$$
$$\theta_i = 0,\,\pi$$

Since $\theta_s < \theta_i$, one has

$$\theta_s = 0,\;\theta_i = \pi$$

Therefore, from Equation (x), one has

$$(\eta_s)_{EC} = -2\,K_{EC}.[1 - (-1)] = -4\,K_{EC}$$

Note that by definition, K_{EC} is negative for attractive Coulombic force. K_{EC} is positive if the force is repulsive. On the other hand, the single-collector efficiency cannot be negative. Therefore,

$$(\eta_s)_{EC} = -4\,K_{EC} \quad \text{if } K_{EC} < 0$$
$$(\eta_s)_{EC} = 0 \qquad\qquad \text{if } K_{EC} > 0$$

(b) No Coulombic force. With $K_{EC} = 0$, Equations (va) and (vi) gives

$$A \sin^2 \theta_s = 0$$

$$\cos \theta_i \sin \theta_i = 0$$

or

$$\theta_s = 0, \ \pi$$

$$2 \theta_i = 0, \ \pi$$

Since $\theta_i > \theta_s$, the above result implies

$$\theta_s = 0, \quad \theta_i = \pi/2$$

According to Equation (x)

$$(\eta_s)_E = A = r^{*2} h(r^*)\big|_{1+N_R}$$

$$= (1 + N_R)^2 \left[1 - (3/2)\frac{1}{1 + N_R} + \frac{1}{2} \frac{1}{(1 + N_R)} \right]$$

$$= (3/2)N_R^2 + \text{terms of } N_R^i \ \ i \geq 3$$

Since $N_R \ll 1$

$$(\eta_s)_E = (3/2)N_R^2$$

which is the single collector efficiency due to interception given in most texts (for example, see Equation (4.20), Tien and Ramarao, 2007).

(c) Combination of interception and the Coulombic force. Here, we have two possibilities:
(i) Attractive Coulombic force or $K_{EC} < 0$. From Equations (ix) and (viii), $(\eta_s)_E$ is given as

$$(\eta_s)_E = \int_{\theta_s}^{\theta_i} [2\pi \sin \theta] \left[\frac{A \cos \theta - K_{EC}}{\pi} \right] . d\theta$$

$$= \int_{\theta_s}^{\theta_i} [2\pi \sin \theta] \frac{A \cos \theta}{\pi} . d\theta - \int_{\theta_s}^{\theta_i} [2\pi \sin \theta] \frac{K_{EC}}{\pi} . d\theta$$

For the first integral, $\theta_s = 0$, $\theta_i = \pi/2$ since interception is operative over $0 < \theta < \pi/2$. For the second integral, $\theta_s = 0$, $\theta_i = \pi$. The above expression gives

$$\eta_s = (3/2)N_R^2 - 4K_{EC}$$

namely, the two effects are additive.

(ii) Repulsive Coulombic force, or $K_{EC} > 0$. For this case, from Equation (va),

$$\sin \theta_s = 0, \quad \cos \theta_s = 1, \quad \text{or} \quad \theta_s = 0$$

From Equation (vi), if $\theta_i > \theta_s$, one has

$$A \cos \theta_i - K_{EC} = 0$$

As shown before, with $N_R \ll 1$, $A \to (3/2) N_R^2$. The above expression is

$$\frac{3}{2} N_R^2 \cos \theta_i - K_{EC} = 0$$

or

$$\theta_i = \cos^{-1} \frac{2K_{EC}}{3N_R^2}$$

subject to the condition $\dfrac{2K_{EC}}{2N_R^2} \le 1$

According to Equation (x), $(\eta_s)_E$ is

$$(\eta_s)_E = A\left(\cos^2 \theta_s - \cos^2 \theta_i\right) - 2K_{EC}(\cos \theta_s - \cos \theta_i)$$

$$= (3/2)N_R^2\left[1 - \left(\frac{2K_{EC}}{3N_R^2}\right)^2\right] - 2K_{EC}\left(1 - \frac{2K_{EC}}{3N_R^2}\right) \quad \text{if} \quad \frac{2K_{EC}}{3N_R^2} < 1$$

$$(\eta_s)_E = 0 \qquad\qquad\qquad\qquad\qquad\qquad \text{if} \quad \frac{2K_{EC}}{3N_R^2} > 1$$

■ ■ ■

6.1.2   Criteria of Particle Adhesion

With an incoming particle making contact with a collector, deposition takes place if the particle becomes attached to or immobilized in the close proximity of the collector surface. On the other hand, there will be no adhesion if the particles, upon contact, slide along, roll over, or bounce-off the collector surface. The conditions leading to the presence of these different situations are discussed below.

A. Adhesion Criteria based on Sliding or Rolling

For an incoming particle, if the size of the collector is significantly greater than that of the particle, the interaction between the particle and the collector may be considered as that between a sphere and a flat surface. Let F_x and F_y denote the tangential and normal forces acting on the particle (see Fig. 6.6a); the criterion of particle sliding is

$$F_x > \mu_f F_y \tag{6.1.33}$$

where μ_f is the friction coefficient between the particle and the collector.

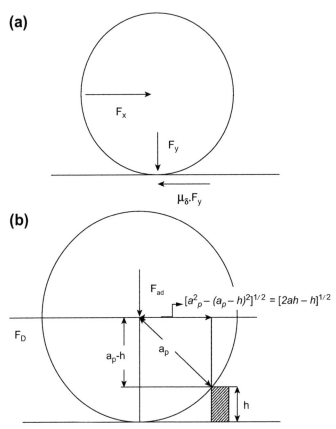

FIGURE 6.6 **(a)** Adhesion criterion based on particle sliding. **(b)** Adhesion criterion based on particle rolling.

For the case of particle rolling, if the roughness of the collector surface is represented by the presence of a number of protrusions of finite height, h, an incoming particle making contact with the collector will eventually reach such a protrusion leading to its deposition if the particle fails to roll over the protrusion (see Fig. 6.6b).

The condition of a particle's rolling over a protrusion can be obtained by taking a moment balance at point a (see Fig. 6.6b). Assuming that both F_x and F_y act through the center of this particle, F_x is at a distance $a_p - h$ from point "a" and causes a moment in the clockwise direction. On the other hand, F_y acts at a distance of $[2a_p h - h^2]^{1/2}$ from point "a", resulting in a moment in the counter clockwise direction. In addition, a moment, M_D, in the clockwise direction due to fluid drag can be expected. The condition of a particle's rolling over the protrusion based on moment balance is

$$F_s(a_p - h) + M_D > F_y[20\, a_p\, h - h^2]^{1/2} \tag{6.1.34a}$$

The above expression may be rewritten as

$$\left(F_x + \frac{M_D}{a_p - h}\right) > \frac{(2a_p h - h^2)^{1/2}}{a_p - h} F_y \tag{6.1.34b}$$

Equation (6.1.34a) may be considered equivalent to Equation (6.1.33) if $F_x + M_D/(a_p - k)$ of Equation (6.1.34b) is viewed as equivalent to F_x of Equation (6.1.33) and $(2\,a_p h - h^2)^{1/2}/a_p - h$ is the same as μ_f. Accordingly, only the rolling case will be discussed in the following paragraphs.

For the evaluation of the adhesion probability, the forces involved and the flow field near the collector surface must be specified. Corresponding to the common conditions found in deep bed filtration (especially for hydrosol particles), the dominant contributions to F_x and F_y of Equations (6.1.33) and (6.1.34a) are, respectively, the hydrodynamic drag force, F_D, and the London–van der Waals force,[4] F_{LO}. According to the results of Goldman et al. (1967), F_D and M_D are given as

$$F_D = 10.205\,\pi\,a_p^2\tau \tag{6.1.35a}$$

$$M_D = 3.776\,\pi\,a_p^3\tau \tag{6.1.35b}$$

where τ is the surface shear stress. If the filter medium is described by Happel's model, τ is given as

$$\tau = 3\,\mu\,A_s u_s \sin\theta/d_g \tag{6.1.36}$$

where A_s is the Happel parameter defined by Equation (6.1.19).

The London–van der Waals force for particle-collection interaction is given as

$$F_{LO} = -Ha_p/6\delta^2 \tag{6.1.37}[5]$$

where H is the Hamaker constant and δ the separation distance between the particle and collector.

Substituting Equations (6.1.35a), (6.1.35b), and (6.1.37) into (6.1.34a) yields

$$\frac{3\,\mu\,A_s u_s}{d_g}\,\pi\,a_p^3\left[10.205\left(1 - \frac{h}{a_p}\right) + 3.776\right]\sin\theta > \frac{Ha_p^2}{6\delta^2}\left[2\frac{h}{a_p} - \left(\frac{h}{a_p}\right)^2\right]^{1/2}$$

for particle rolling over.

Alternatively, the adhesion criterion may be written as

$$\frac{\left[2\left(\dfrac{h}{a_p}\right) - \left(\dfrac{h}{a_p}\right)^2\right]^{1/2}}{10.205\left(1 - \dfrac{h}{a_p}\right) + 3.776} \le B\sin\theta \tag{6.1.38}$$

---

[4]The double-layer force is not considered here. However, it can be included in estimating F_y.
[5]For more information, see 6.5.1. Equation (6.1.37) is obtained from Equation (6.5.2) under the conditions $\delta^+ \ll 2$.

with

$$B = 103.79 \, \mu \, A_s u_s (a_p/a_g)(\delta^2/H) \tag{6.1.39}$$

For a given situation, B is known. One may determine a critical value of h_c, as a function of the angular position from Equation (6.1.38) (with the equal sign). Particle adhesion may take place if the actual protrusion height is greater than h_c.

As an example, the values of h_c as a function of θ for $B = 0.2$, 0.5, and 1.0 are shown in Fig. 6.7a. The results are given in the form of (h_c/a_p) vs. θ for different values of B. If the height of protrusion over the collector is $h = 0.2 \, a_p$, the results suggest that adhesion takes place over $0 < \theta < 10.95°$ and by symmetry, $169.05° < \theta < 180°$.

The particle size effect on adhesion can be seen from Fig. 6.7a. Consider two cases with $a_p = (a_p)_1$ for case (1) and $a_p = (a_p)_2 = 2(a_p)_1$ for case (2). With B of case (1) being 0.5, then B of case (2) is 1.0 (Note that B is proportional to a_p) and with adhesion taking place over

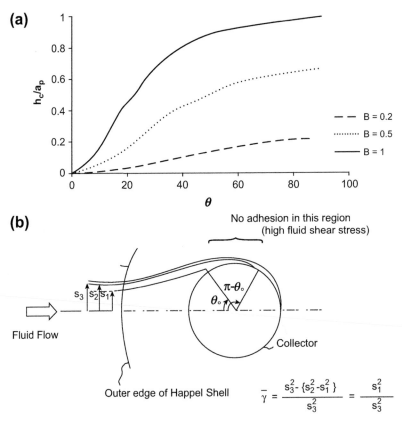

FIGURE 6.7 a(h_c/a_p) vs. θ for different values of B according to Equation (6.1.38). b Estimation of γ according to Equation (6.1.40).

a smaller area, namely, $0 < \theta < 7.39°$ and $172.61° < \theta < 180°$. In other words, the likelihood of deposition decreases with the increase of particle size.

Evaluation of γ can be made as follows: From a collector with protrusions of height, h_1, adhesion takes place over $0 < \theta < \theta_a$ and $\pi - \theta_a < \theta < \pi$ where θ_a is the angular position corresponding to $h_c = h_1$. Fig. 6.7b which gives upper half of a spherical collector composed of regions where adhesion may or may not take place. Considering those particle trajectories which reach the collector surface at $\theta = \theta_a$, $\theta = \pi - \theta_a$, and $\theta = \pi$. Let the off-center distance of these trajectories at the outer edge of the Happel cell be S_1, S_2, and S_3, respectively. By definition, the overall adhesion probability, γ is

$$\gamma = \frac{S_3^2 - S_2^2 + S_1^2}{S_3^2} \cong \frac{S_1^2}{S_3^2} \tag{6.1.40}$$

since, in general, S_2 and S_3 are essentially the same.

The above treatment was first suggested by Gimbel and Sontheimer (1978) for hydrosol filtration. Based on their experimental data, correction for the adhesion effect due to particle rolling (or sliding) becomes necessary for large particles and high suspension flow rate.

B. Adhesion Based on Particle Bouncing Off

As shown in 6.1.1. A., aerosol deposition is expected to increase with the increase of particle inertia. Experimentally, aerosol deposition in the inertia-dominated region is found to increase with the increase of the gas velocity initially. However, at sufficiently high velocity, deposition actually decreases, suggesting the failure of adhesion of a portion of the impending particles.

Investigation of adhesion of impacting particles with high inertia has been made by a number of investigators. As an example of demonstrating the basic principle used, a brief account of Dahneke's work (1971, 1995) is presented below.

The collision of a particle and a surface may be characterized in terms of the energy of the particle-surface system. The total energy of a particle, E_t, may be written as

$$E_t = KE + E \tag{6.1.41}$$

where KE is the kinetic energy of the particle and E, the potential energy, namely, the energy due to the particle-surface interaction (or their relative position).

Considering an incoming particle colliding with a surface, let the incident velocity (in the normal direction) be $(u_p)_i$ and the total energy of the incident particle, $(E_t)_i$ is

$$(E_t)_i = (KE)_i + E_i = \frac{m_p}{2}(u_p)_i^2 + E_i \tag{6.1.42}$$

where m_p is the particle mass.

Upon colliding with the surface, in the event that the particle rebounds, the total energy of the rebounding particle, $(E_t)_r$, with a rebounding velocity $(u_p)_r$, may be written as

$$(E_t)_r = (KE)_r + E_r = \frac{m_p}{2}(u_p)_r^2 + E_r \qquad (6.1.43)$$

The relationship between $(E_t)_i$ and $(E_t)_r$ is given as

$$(E_t)_r = e^2(E_t)_i \qquad (6.1.44)$$

where e is the coefficient of restitution. Combining Equations (6.1.42), (6.1.43), and (6.1.44), after rearrangement, one has

$$\frac{(u_p)_r}{(u_p)_r} = \left[e^2 - \frac{E_r - e^2 E_i}{(m_p/2)(u_p)_i^2}\right]^{1/2} \qquad (6.1.45)$$

Specifically, the condition of particle adhesion can be determined from Equation (6.1.45) under the condition $(u_p)_r$ vanishes namely, particle fails to rebound. The value of $(u_p)_i$ corresponding to $(u_p)_r = 0$, termed the capture limit velocity, $(u_{p_i}^*)$, can be found from the expression

$$E_r = e^2 \left[\frac{m_p}{e}(u_{p_i}^*)^2 + E_i\right]$$

or

$$(u_{p_i}^*) = \sqrt{2\frac{E_r - e^2 E_i}{m_p e^2}} \qquad (6.1.46)$$

The capture limit velocity, $u_{p_i}^*$, is the threshold value for the adhesion of impacting particles, namely, particle adhesion takes place if the incident velocity u_{p_i} is less than $u_{p_i}^*$. For $u_{p_i} > u_{p_i}^*$, the impacting particle rebounds and moves away from the collector surface.[6]

Two special cases of Equations (6.1.45) may be distinguished. If $E_i = E_r = E$, Equation (6.1.46) becomes

$$u_{p_i}^* = \left[\frac{2E}{m_p}\frac{1 - e^2}{e^2}\right]^{1/2} \qquad (6.1.47a)$$

and Equation (6.1.45) reduces to

$$(u_p^*)_r = (u_p)_i\left[e^2 - \frac{2E(1 - e^2)}{m_p(u_p)_i^2}\right]^{1/2} \qquad (6.1.47b)$$

[6]That a particle rebounds does not imply that it will escape deposition. Rather, its fate should be determined by following its subsequent trajectory, which is influenced by its possible collisions with other rebounding and/or incoming particles. For a discussion, see Wang (1986).

On the other hand, for $E_i \ll E_r$, one has

$$u^*_{p_i} = \left[\frac{2E_r}{m_p e^2} \right]^{1/2} \tag{6.1.48a}$$

$$(u_p)_r = u_p \left[e^2 - \frac{2E_r}{m_p (u_p)_i^2} \right]^{1/2} \tag{6.1.48b}$$

For applying the results given above, the values of e, E_i, and E_r must be known. According to Dahneke (1995), e may be estimated according to

$$e = [e_0^2 + \exp(-3.4\,\bar{\lambda}) - 1]^{1/2} \tag{6.1.49}$$

where e_0 is the value of e corresponding to the case of no flexural work and $\bar{\lambda}$ the inelasticity parameter. For granular filtration with filter grains of diameter d_p, Dahneke's result for the sphere–plate collision case may be applied. $\bar{\lambda}$ is given as

$$\bar{\lambda} = \frac{2}{3\pi^{2/5}} \left(\frac{d_p}{d_g} \right)^2 \left[\frac{1}{1 + (d_p/d_g)} \right]^{1/10} [u_{p_i}/\bar{V}_s]^{1/5} \left(\rho_p/\rho_s \right)^{3/5} \left(\frac{k_s}{k_p + k_s} \right)^{2/5} \tag{6.1.50}$$

where $\bar{V}_s$ is defined to be $1/(k_s \rho_s)^{1/2}$. The subscripts s and p stand for filter grain and particle, respectively. $k_i = (1 - \bar{v}_i^2) Y_i$ where $\bar{v}_i$ and Y_i are Poisson's ratio and Young's modulus for material with $i =$ p for particle and $i =$ s for filter grain.

The results given above can be readily applied in determining whether an impacting particle becomes deposited or not. However, to determine the adhesion efficiency, it requires the tracking of rebound particles. In fact, in many cases, deposition of a particle can be ascertained only after its repeated impacts with a collector. Such an approach was used by Wang (1986). An empirical expression of the adhesion probability results from trajectory calculations based on trajectory calculations of cylindrical collector, and the potential flow field was obtained by Wang (1986). His results may be summarized as follows:

$$\gamma = 1 \quad \text{for } \frac{u^*_{p_i}}{V_s} \gg 1 \tag{6.1.51a}$$

$$\gamma = 1 - \left[1 - \left(\frac{u^*_p}{V_s} \right)^2 \right]^{1/2} + 0.054\sqrt{e} \left[1 - \left(\frac{u^*_p}{V_s} \right)^2 \right]^{1/4} \quad \text{for } 0.27\, e^{0.85} \le u^*_{p_i}/V_s < 1 \tag{6.1.51b}$$

where $u^*_{p_i}$ is the capture limit velocity discussed before (see Equation (6.1.46)) with E for the cylinder–sphere case given by

$$E = \frac{H}{12\,\delta} \frac{d_p}{(1 + d_p/d_f)^{1/2}} \tag{6.1.52}$$

where H and δ are the Hamaker constant and the separation distance as before. d_p and d_f are the particle and fiber (cylinder) diameter. V_s is the first impact velocity of a particle at the front stagnation point. According to Wang's calculation, V_s can be approximated as

$$V_s/u_\infty = (2/\pi)\tan^{-1}\left[1.15\left(\text{St}_{\text{eff}} - \frac{1}{8}\right)^{0.85}\right] \tag{6.1.53}$$

where u_∞ is the fluid velocity over the cylindrical collector. St_{eff} is defined as

$$\text{St}_{\text{eff}} = N_{\text{St}}\psi \tag{6.1.56}$$

where N_{St} is defined according to Equation (6.1.10) and ψ is given as

$$\psi = 18.99 N_{\text{Re}}^{-2/3} - 47.77 \tan^{-1}(03975\, N_{\text{Re}}^{1/3})/N_{\text{Re}} \tag{6.1.57}$$

and the Reynolds number, N_{Re}, is based on particle diameter ($N_{\text{Re}} = d_p U_\infty \rho/\mu$).

The results given above can be readily applied in estimating the adhesion probability of hydrosol particles [Equation (6.1.40)] and aerosol particles [Equation (6.1.51)].

However, one should bear in mind that the values of the various parameters (such as the Hamaker's constant, the coefficient of restitution, friction coefficient, etc.) and quantities (such as the particle/collector separation distance and the protrusion height) are known approximately or can only be assumed. Caution must be exercised in assessing the accuracy of the calculated adhesion probability before their use.

■ ■ ■ ▬▬▬▬▬▬▬▬▬▬▬▬▬▬▬▬▬▬▬▬▬▬▬▬▬▬▬▬▬▬

Illustrative Example 6.4

From the results given, 6.1.2 shows that the adhesion criterion based on particle sliding is equivalent to that based on particle rolling. For a particular case with the friction coefficients [see Equation (6.1.33)] to be 0.2, what is the equivalent protrusion height if the particle size is 5 μm?

Solution

From Equations (6.1.33) and (6.1.34b), the respective adhesion criteria are

$$\text{Based on Particle Sliding} \quad F_x < \mu_f F_y \tag{i}$$

$$\text{Based on Particle Rolling} \quad F_x + \frac{M_D}{a_p - h} < \frac{(2ah - h^2)^{1/2}}{(a_p - h)} \tag{ii}$$

With F_x and M_D given by Equations (6.1.35a) and (6.1.35b), Equation (ii) may be rewritten as

$$F_x < \frac{\left[2\dfrac{h}{a_p} - \left(\dfrac{h}{a_p}\right)^2\right]^{1/2}}{\left[1 + 0.37\left(1 - \dfrac{h}{a_p}\right)^{-1}\right]\left[1 - \dfrac{h}{a_p}\right]} F_y = \frac{\left[2\dfrac{h}{a_p} - \left(\dfrac{h}{a_p}\right)^2\right]^{1/2}}{1.37 - (h/a_p)} F_y \tag{iii}$$

Comparing Equations (i) with (iii), one has

$$\mu_f = \frac{\left[2\left(\dfrac{h}{a_p}\right) - \left(\dfrac{h}{a_p}\right)^2\right]^{1/2}}{1.37 - \dfrac{h}{a_p}} = \frac{(2h/a_p)^{1/2}}{1.37} \quad \text{if } h/a_p \ll 1$$

$$= 0.2$$

Solving for h/a_p, one has

$$(2\,h/a_p)^{1/2} = 0.274$$
$$h/a_p = (0.274)^2/2 = 0.0375$$
$$h = (0.0375)(5) = 0.1877 \ \ \mu\text{m}$$

■ ■ ■

Illustrative Example 6.5

Estimate the adhesion probability of particles of diameter 10 μm in deep bed filtration under the following conditions.

Filtrate Medium	$\varepsilon = 0.5$
	$d_g = 1{,}000 \ \mu\text{m}$
	$h = 2 \ \mu\text{m}$
Operating Condition	$u_s = 1 \text{ cm s}^{-1} = 1 \times 10^{-2} \text{ ms}^{-1}$
Relevant Physical Properties	$\mu = 10^{-3} \text{ Pa.s}$
	$H = 10^{-20} \text{ J}$

Particle deposition is assumed to be by interception. The particle/collector separation distance, δ, may be taken as 4 nm (4×10^{-9} m). Happel's model may be used to describe the flow field outside filter grains.

Solution

The relevant information about Happel's model are (see 5.4.1):

(a) Dimension of Happel's cell $b = (d_g/2)(\varepsilon)^{-1/3} = 1.2599 \ (d_g/2)$

(b) Expression of the stream function, ψ. From Equation (5.4.8),

$$\psi = A\left[\frac{K_1}{r^*} + K_2 r^* + K_3 r^{*^2} + K_4 r^{*^4}\right]\sin^2\theta \qquad 1 < r^* < 1.2599 \qquad \text{(i)}$$

$$K_1 = 1/w$$

$$K_2 = -\frac{3 + 2p^5}{w}$$

$$K_3 = \frac{2 + 3p^5}{w} \tag{ii}$$

$$K_4 = \frac{-p^5}{w}$$

$$p = (1 - \varepsilon)^{1/3}$$

$$w = 2 - 3p + 3p^5 - 2p^6$$

Based on the given conditions, the values of these parameters are

$$p = (1 - \varepsilon)^{1/3} = (0.5)^{1/3} = 0.7937$$

$$w = 2 - 3(0.7937) + 3(0.315) - 2(0.25) = 0.0639$$

$$K_1 = 1/0.0639 = 15.6495$$

$$K_2 = -\frac{3 + 2(0.315)}{0.0639} = -56.8075$$

$$K_3 = -\frac{2 + 3(0.315)}{0.0639} = -46.0875$$

$$K_4 = -\frac{0.315}{0.0639} = -4.9296$$

For the calculation of the adhesion probability, the part of the collector surface over which adhesion occurs should be identified. From Equation (6.1.38), the condition of adhesion is

$$\frac{\left[2\left(\frac{h}{a_p}\right) - \left(\frac{h}{a_p}\right)^2\right]^{1/2}}{10.205\left(1 - \frac{h}{a_p}\right) + 3.776} = B \sin \theta_c \tag{iii}$$

$$B = 103.79 \, \mu \, A_s u_s (a_p/a_g)(\delta^2/H)$$

$$A_s = 2(1 - p^5)/w = 2[1 - (0.7937)^5]/0.0639 = 21.44$$

$$(a_p/a_g) = 10/1000 = 10^{-2}$$

$$B = (103.79)(10^{-3})(21.44)(10^{-2})(10^{-2})\frac{(4 \times 10^{-9})^2}{10^{-20}} = 0.356$$

and

$$\frac{[2(2/5) - (2/5)^2]^{1/2}}{10.205\left(1 - \dfrac{2}{5}\right) + 3.776} = (0.356)\sin\theta_a$$

and

$$\sin\theta_a = 0.227$$

If one assumes that deposition is by interception alone, only the front half of the filter grain functions as particle collector; namely, the part of the collector surfaces over which adhesion takes place is given by $0 < \theta < \theta_a$ (see Fig. 6.7b). Equation (6.1.40) becomes

$$\gamma = \frac{S_1^2}{S_2^2} \tag{iv}$$

where S_1 is the off-center distance from the outer edge of the Happel cell of the trajectory (in the present case, the stream line) which passes through the point $r^* = 1 + 0.01$, $\theta = \theta_a$ and S_2 is the off-center distance value of the trajectory passing through the point $r^* = 1 + 0.01$, $\theta = \pi/2$. To obtain the value of S_2, the stream function value of the latter trajectory, from Equation (i), ψ_2 is

$$\psi_2 = \psi\Big|_{\substack{r^* = 1.01 \\ \theta = \pi/2}} = A\left[\frac{15.6445}{1.01} - 56.8075\,(1.01) + 48.0846\,(1.01)^2 - 4.9296\,(1.01)^4\right] \tag{1}$$

$$= 0.0033\,A$$

The value of S_2 can be found as follows. The angular position of the stream line at $r^* = 1.2599$, $\theta = \theta_2$ is given by

$$0.0033\,A = A\left[\frac{15.6495}{1.2599} - 56.8075(1.2599)^2 + 48.0876(1.2599)^2 - 4.9296(1.2599)^2\right]\sin^2\theta_2$$

Solve for $\sin\theta_2$

$$\sin\theta_2 = 0.0456$$

and

$$S_2 = (1.2599)(a_g)0.0456 = 0.0575.a_g$$

Similarly, S_1 is found to be

$$S_1 = 0.01035\,a_g$$

and

$$\gamma = \left(\frac{0.01035}{0.0575}\right)^2 = 0.051$$

■ ■ ■

■ ■ ■ ━━━━━━━━━━━━━━━━━━━━━━━━━━━━━

Illustrative Example 6.6[7]

For deep bed filtration of nanoparticle suspensions, as the Brownian diffusive force is the dominant factor of particle transport, the incident velocity of a particle making contact with collector may be approximated by its thermal velocity, which follows the Maxwell–Boltzman distribution function.

(a) Based on the principles outlined in 6.1.2B, obtain a particle adhesion probability expression using Equation (6.1.48a) with $e^2 = 1$ for the capture limit velocity and E taken to be $E = -Hd_p/(12\delta)$.

(b) For a particle of diameter $d_p = 10$ nm, what are the adhesion probabilities in air filtration? And in water filtration? From the calculation results, comment on the relative likelihood of nanoparticle deposition in air and water filtration.

The values of the various relevant properties are

$$\rho_p = 1200 \text{ kg m}^{-3}$$

$$k = 1.38 \times 10^{-23} \text{ J K}^{-1}$$

$$T = 298 \text{ K}$$

For air filtration $H = 10^{-19}$ J, $\delta = 0.4$ nm

For water filtration $H = 10^{-23}$ J, $\delta = 3$ nm

Solution

(a) The probability density function of the particle thermal velocity, u_p, of mass m_p is

$$f(u_p) = 4\pi u_p^2 \left(\frac{m_p}{2\pi kT}\right)^{3/2} \exp\left[-\frac{m_p u_p^2}{2kT}\right] \tag{i}$$

where k is the Boltzmann constant. The mean value of u_p, $\bar{u}_p$, is

$$\bar{u}_p = \left[\frac{8kT}{\pi m_p}\right]^{1/2} \tag{ii}$$

In 6.1.2b, adhesion is based on comparing the particles incident velocity with the capture limit velocity, $u_{p_i}^*$. From Equation (6.1.48a), with $e = 1$, $u_{p_i}^*$ is given as

$$u_{p_i}^* = \left(\frac{2E}{m_p}\right)^{1/2} \tag{iii}$$

and

$$E = -H\,d_p/(12\,\delta) \tag{iv}$$

Substituting Equation (iv) (absolute value) into (iii) and with $m_p = (\pi/6)d_p^3\rho_p$, $u_{p_i}^*$ is found to be

$$u_{p_i}^* = \left[\frac{H}{\pi\,\rho_p\delta\,d_p^2}\right]^{1/2} \tag{v}$$

The adhesion probability, by definition, is

$$\gamma = \frac{\int_0^{u_p^*} f(u_p).du_p}{\int_0^{\infty} f(u_p)du_p} \tag{vi}$$

Substituting Equation (i) into (vi), after rearrangement, one has

$$\gamma = \int_0^{u_{p_i}^*/\bar{u}_p} \left(\frac{32}{\pi^2}\right) X^2 \exp\left(\frac{-4X^2}{\pi}\right) dX$$

In other words, γ is dependent on $u_{p_i}^*/\bar{u}_p$. The results are shown in Fig. i in which γ is shown as a function of the reciprocal of $u_{p_i}^*/\bar{u}_p$.

(b) For a 10 nm (diameter) particle, the mean thermal velocity is

$$\bar{u}_p = \left(\frac{8kT}{m_p\pi}\right)^{1/2} = \left(\frac{48kT}{\pi^2\rho_p d_p^3}\right)^{1/2} = \frac{(48)(1.38\times10^{-23})(298)}{\pi^2(1200)(10^{-2})^3} = 16.32\ \text{m s}^{-1}$$

The capture limit velocity of air filtration, $u_{p_i}^*$ is

$$u_{p_i}^* = \left[\frac{H}{\pi\,\rho_p\delta\,d_p^2}\right]^{1/2} = \frac{10^{-19}}{\pi(1200)(4\times10^{-10})(10^{-3})^2} = 25.75\ \text{m s}^{-1}$$

The capture limit velocity of water filtration $u_{p_i}^*$ is

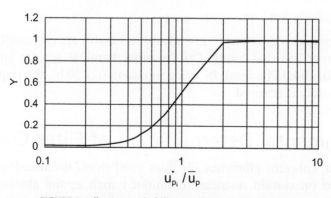

FIGURE I Adhesion probability vs. capture limit velocity.

$$u_{p_i}^* = \left[\frac{10^{-20}}{\pi(1200)(3 \times 10^{-9})(10^{-8})^2} \right]^{1/2} = 2.97 \text{ m s}^{-1}$$

The ratio $\bar{u}_p / u_{p_i}^*$ for air and water filtration is

$$\bar{u}_p / u_{p_i}^* = \begin{matrix} 16.22/25.75 = 0.63 & \text{air filtration} \\ 16.32/2.97 = 5.49 & \text{water filtration} \end{matrix}$$

The corresponding γ values for Fig. i are

$\gamma = 1.0$ for air filtration

$\gamma < 0.01$ for water filtration

The results suggest that adhesion failure of 10 nm particles by their thermal velocity is more pronounced in water filtration than air filtration. However, one should bear in mind that the calculations were made on the basis of monodispersed suspensions, although in aqueous media nanoparticles are most likely to form flocs.

---

[7]For a detailed account in nano-size aerosol filtration, see Wang and Kasper (1991).

■ ■ ■

6.1.3 Prediction of Collector Efficiency

The results given above provide particle collection information due to individual mechanisms with the assumption that particle contact leads to deposition automatically. In actual situations, particle transport is likely to be caused by more than a single mechanism. As an approximation, if one assumes that these mechanisms act independently, the collector efficiency may be expressed as (Stairmand, 1950)

$$\eta = \eta_i + \eta_I + \eta_G + \eta_{BM} + \eta_E \tag{6.1.50.a}$$

Alternatively, $(1 - \eta)$ may be viewed as the escape probability from deposition of a particle. If the actions due to the individual mechanism are independent of one another, one may write (Strauss, 1966)

$$1 - \eta = (1 - \eta_I)(1 - \eta_i)(1 - \eta_G)(1 - \eta_{BM})(1 - \eta_E)$$

$$\eta = 1 - (1 - \eta_I)(1 - \eta_i)(1 - \eta_G)(1 - \eta_{BM})(1 - \eta_E) \tag{6.1.50.b}$$

The two expressions [Equations (6.1.50.a) and (6.1.50.b)] are essentially the same if all η_j's are significantly less than unity. On the other hand, if η_j's are sufficiently large, Equation (6.1.50.b) should be used. By using Equation (6.1.50.b), the requirement that η does not exceed unity is ensured.

6.2 Experimental Determination of Filter Coefficient

Predictions of the collector efficiency (or filter coefficient) discussed in the preceding section are based on certain restricted conditions such as the absence of deposited particles over collector surface. Equally important, some of the results may not have sufficient accuracy. Accordingly, experimental determination of filter coefficient from filtration data is inevitably required to provide sufficiently accurate information for validation of analyses and in design calculations. In the following sections, the basic principles used for determining filter coefficient (and therefore collector efficiency) from experimental data are outlined.

6.2.1   Determination of the Initial (or clean) Filter Coefficient, λ_0

As discussed in Chapter 5, the initial filter coefficient, λ_0, provides a benchmark in assessing deep bed filtration performance. For a clean homogeneous medium devoid of any deposited particles, if the depth of the media is L, from Equations (5.1.4), (5.2.2), (5.2.3a), and with $F(\alpha, \sigma) = 1$, one has

$$\frac{c_{\text{eff}}}{c_{\text{in}}} = \exp[-\lambda_0 L]$$

or

$$\lambda_0 = -\frac{1}{L}\ell n(c_{\text{eff}}/c_{\text{in}}) \tag{6.2.1}$$

where c_{in} and c_{eff} are the influent and effluent particle concentrations. Strictly speaking, if Equation (6.2.1) is to be used for determining λ_0 from experimental data, c_{eff} should be the value of effluent concentration at $\theta = 0$. Such a value, however, is difficult if not impossible to measure. Instead, tests may be conducted over a period of time to obtain the effluent particle concentration history (c_{eff} vs. θ). λ_0 can be determined according to

$$\lambda_0 = \frac{1}{L}\lim_{\theta \to 0}\left[\ell n\frac{c_{\text{in}}}{c_{\text{eff}}}\right] \tag{6.2.2a}$$

or

$$\lambda_0 = \frac{1}{L}\ell n\left[\lim_{\theta \to 0}\frac{c_{\text{in}}}{c_{\text{eff}}}\right] \tag{6.2.2b}$$

In other words, one may obtain the limiting values of $\ell n(c_{\text{in}}/c_{\text{eff}})$ or $(c_{\text{in}}/c_{\text{eff}})$ at $\theta \to 0$ from effluent particle concentration history for the determination of λ_0. The results obtained may not be the same but the difference, if any, can be expected to be insignificant. Practically speaking, using either limiting value in determining λ_0 may not be necessary if the value of c_{eff} remains constant (or nearly constant) for a period of time.

There are two major sources of errors in the determination of the effluent particle concentration history, c_{eff} vs. θ, namely, (i) the experimental filter used may not be homogeneously packed and (ii) the inherent inaccuracies associated with particle concentration determination. The former may be significant for shallow filters. On the other hand, increasing filter depth tends to give a lower effluent particle concentration, which makes accurate particle concentration determination more difficult. To minimize such errors, λ_0 may be determined from effluent concentration data obtained using filters of different depths. An objective function, $\psi(\lambda_0)$ is defined as

$$\psi(\lambda_0) = \sum_{m=1}^{M} \left[(c_{in}/c_{eff})_{L_m} - \exp(\lambda_0 L_m) \right]^2 \tag{6.2.3}$$

where $(c_{in}/c_{eff})_{L_m}$ is the limiting value ($\theta \to 0$) of the influent/effluent concentration ratio obtained with filter of depth L_m, and filters of depths $L_1, L_w,..., L_m$ are used in measurements. λ_0 is determined based on the minimization of ψ.

6.2.2   Determination of Deposition Effect on Filter Coefficient

As stated before, λ, in general, is a function of the extent of particle deposition. The effect of deposition can be expressed by the ratio of λ to its initial value, λ_0 or

$$\frac{\lambda}{\lambda_0} = F(\alpha, \sigma) \tag{6.2.4}$$

A number of methods may be applied to determine λ (or F) as a function of the specific deposit, σ including

(i) Based on uniform deposition assumption. If one assumes that particle deposition within a filter of depth L is uniform, the specific deposit throughout the filter is the same as the average specific deposit, $\bar{\sigma}$, which can be, at a given instant, θ, determined from overall mass balance as

$$\bar{\sigma} = \frac{1}{L} \int_0^L u_s (c_{in} - c_{eff}) dz \tag{6.2.5a}$$

where L is the filter depth.
The corresponding filter coefficient, $\bar{\lambda}$, is

$$\bar{\lambda} = \frac{1}{L} \ell n (c_{in}/c_{eff}) \tag{6.2.5b}$$

In other words, from experimental data of c_{eff} vs. θ, values of $\bar{\lambda}$ vs. $\bar{\sigma}$ can be obtained from the above two expressions. These results may be assumed to be an approximation of the relationship of λ vs. σ.
The procedure has often been used in describing the effect of deposition on filter performance, although it is difficult to assess its accuracy.

(ii) Based on extrapolation of results using uniform deposition assumption. Physically speaking, the uniform deposition assumption becomes more valid as the filter height decreases. Using this argument, Walata et al. (1986), in their aerosol study, proposed a limiting procedure in order to obtain accurate relationship of λ vs. σ, or the function, F.

From their experimental data, it was found that $\bar{\lambda}/\lambda_0$, in many cases, can be related to $\bar{\sigma}$ by the power-law expression

$$\frac{\bar{\lambda}}{\lambda_0} = 1 + \bar{\alpha}_1 \bar{\sigma}^{\bar{\alpha}_2} \tag{6.2.6}$$

By evaluating $\bar{\alpha}_1$ and $\bar{\alpha}_2$ from data obtained using filters of different heights and then by extrapolating $\bar{\alpha}_1$ (and $\bar{\alpha}_2$) against filter height, the respective limiting values of $\bar{\alpha}_1$ and $\bar{\alpha}_2$ as $L \to 0$ can be determined, which were then taken to be those of the correct expression of F.

(iii) Based on the assumption that F is a linear function of σ. In Illustrative Example 5.7, we have shown that if F is a linear function of σ or $F = 1 - k\sigma$, the effluent concentration history is given as

$$\frac{(c_{in}/c_{eff}) - 1}{[\exp(\lambda_0 L)] - 1} = \exp[-u_s \lambda_0 c_{in} k\theta] \tag{6.2.7}$$

In other words, a plot of $\ln [(c_{in}/c_{eff}) - 1]$ vs. θ yields a straight line with an intercept of $[\exp(\lambda_0 L)] - 1$ and a slope of $-u_s \lambda_0 c_{in} k$. Thus, with effluent concentration history known, Equation (6.2.7) can then be used to obtain the values of λ_0 and k (therefore F). On the other hand, if such a linearity is not observed, one may surmise that F is not a linear function of σ and cannot be represented as $1 - k\sigma$. As an approximation, one may take the slope of the tangent to the curve, $\ln[c_{in}/c_{eff}) - 1]$ vs. σ, as the instantaneous value of $-u_s \lambda_0 c_{in} k$ and to obtain a relationship between k vs. θ. As $\bar{\sigma}$ can be shown to be a function of θ [see Equation (6.2.5)], a relationship between k and $\bar{\sigma}$ can be established and may be taken as an approximation of k as a function of σ.

(iv) Determination of F as a search-optimization problem. Determination of F from experimental data can be treated as a problem of search and optimization. First, a particular expression of F is selected based on the considerations stated before. Search and optimization procedure can then be used to obtain the values of the parameters present in F. The required procedure has the following elements:

(a) A method for integrating the macroscopic equations of granular filtration with the assumed expressions for $F(\alpha, \sigma)$ to yield concentration profiles at various times;

(b) A search-optimization technique to determine the values of the constants of $F(\alpha, \sigma)$ in order to obtain the best agreement between predictions and experiments.

This search-optimization technique has been applied by a number of investigators in the past (for example, see Payatakes et al., 1975, Bai and Tien, 2000). Its advantage resides mainly in its conceptual simplicity and rigor. However, it is computationally demanding and its use may not be justified in preliminary calculations.

■ ■ ■ ▬▬▬▬▬▬▬▬▬▬▬▬▬▬▬

Illustrative Example 6.7

Determine the initial filter coefficient from the following data obtained by Vaidyanathan (1986). The experiments were conducted under the following conditions:

Filter Bed	Barium titanate glass spheres packed beds of $L = 2$ cm, 4 cm,	
	Glass sphere diameter	$d_g = 345$ μm
	Glass sphere density	$\rho_g = 4{,}400$ kg m^{-3}
	Bed Porosity	$\varepsilon = 0.38$
	Filter grain zeta potential	$\psi_p = -3$ mV
Suspensions	Aqueous solutions of latex particles	$d_p = 11.4$ μm,
		$\rho_p = 1050$ kg m^{-3}
	Electrolyte concentration	0.181 mol/ℓ
	Sodium acetate	0.01 mol/ℓ
	Acetic Acid	0.01 mol/ℓ
	NaCl	0.161 mol/ℓ
	Solution pH	4.5
	$u_s = 0.2$ cm s^{-1}	

The effluent concentration history expressed as effluent to influent concentration ratio vs. the total volume of particle introduced, $V_t = (u_s)(A)(c_{in})t$ [u_s, superficial velocity, A, filter cross section area, c_{in}, influent particle (volume) concentration] are:

$L = 2$ cm			$L = 4$ cm		
$V_t = (u_s)(A)$ $(c_{in})(t) \times 10^3$	c_{eff}/c_{in}	$\frac{1}{L}\ell n(c_{in}/c_{eff})$	$V_t = (u_s)(A)$ $(c_{in})(t) \times 10^3$	c_{eff}/c_{in}	$\frac{1}{L}\ell n$ (c_{in}/c_{eff})
7.15	0.619	0.2398	6.11	0.355	0.2589
14.8	0.664	0.2047	14.2	0.479	0.1840
22.1	0.710	0.1712	22.4	0.498	0.1743
31.1	0.730	0.1574	25.6	0.536	0.1559
39.7	0.789	0.1185	34.7	0.563	0.1436
48.6	0.802	0.1103	40.3	0.612	0.1269
55.8	0.803	0.1097	47.4	0.581	0.1358
62.5	0.800	0.116			

with V_t given in cm^3 and $\frac{1}{L}\ell n(c_{in}/c_{eff})$ in cm^{-1}.

Solution

Several methods can be used to determine λ_0 as discussed in the text.

(1) Determination of λ_0 based on the limiting value of $\frac{1}{L}\ell n(c_{in}/c_{eff})$ as $t \to 0$. According to Equation (6.2.2a),

$$\lambda_0 = \frac{1}{L}\lim_{\theta \to 0}\left[\ell n \frac{c_{in}}{c_{eff}}\right] \qquad (i)$$

(2) Determination of λ_0 based on the limiting value of c_{in}/c_{eff} as $t \to 0$. According to Equation (6.2.2b),

$$\lambda_0 = \frac{1}{L}\ell n \left[\lim_{\theta \to 0} \frac{c_{in}}{c_{eff}} \right] \tag{ii}$$

(3) Optimizing the objective function ψ to search for λ_0. ψ is defined as [see Equation (6.2.3)]

$$\psi = \sum_{m=1}^{M} \left[(c_{in}/c_{eff})_{exp} - \exp(\lambda_0 L_m) \right]^2 \tag{iii}$$

(4) Analogous to (3), ψ may be defined as

$$\psi = \sum_{m=1}^{\infty} \left[\left(\ell n \frac{c_{in}}{c_{eff}} \right)_m - \lambda_0 L_M \right]^2$$

For (1) and (2), the limiting values $[(c_{in}/c_{eff})$ or $\ell n(c_{in}/c_{eff})]$ are those corresponding to $\theta \to 0$. The data presented are those at various values of V_t. However, since A, c_{in}, and u_s are constants and the difference between θ and t is small, $V_t \to 0$ implies $\theta \to 0$.

(a) By fitting $y = \frac{1}{L} \ell n \frac{c_{in}}{c_{eff}}$ vs. $x = V_t$ as a second-order polynomial, we have

For $L = 2$ cm

$$\frac{1}{L} \ell n \frac{c_{in}}{c_{eff}} = 0.2806 - 0.0593 \ V_t + 0.0051 \ V_t^2$$

Therefore, $\lambda_0 = 0.2806$ cm^{-1}

Following the same procedure, with the results obtained with $L = 4$ cm, the initial filter coefficient is found to be

$$\lambda_0 = 0.296 \text{ cm}^{-1}$$

(b) If the data of c_{eff}/c_{in} are fitted as a second-order polynomial of V_t, the limiting value of c_{eff}/c_{in} as $V_t \to 0$ are found to be

$$c_{eff}/c_{in} = 0.561 \quad L = 2 \text{ cm}$$
$$= 0.288 \quad L = 4 \text{ cm}$$

the corresponding values of λ_0 are [from Equation (ii)]

$$\lambda_0 = 0.289 \quad L = 2 \text{ cm}$$
$$= 0.311 \quad L = 4 \text{ cm}$$

The results may be summarized as

Bed Length (cm)	$(c_{eff}/c_{in})_{\theta \to 0}$	$\left[\frac{1}{L} \ln(c_{in}/c_{eff}) \right]_{\theta=0}$	$\lambda_0 (\text{cm}^{-1})$	
			Equation (i)	Equation (ii)
2	0.561	0.2806	0.289	0.281
4	0.288	0.296	0.311	0.296

(c) From Equation (iii), the optimum value of λ_0 can be found from

$$\frac{\partial \psi}{\partial \lambda_0} = 0$$

or

$$\sum_{m=1}^{2} \left[(c_{in}/c_{eff}) - \exp(\lambda_0 L_m) \right] (L_m) \exp(\lambda_0 L_m) = 0$$

From the above expression and with the experimental data given, one has

$$[(1/0.561) - \exp(2\lambda_0)]\exp(2\lambda_0) + 2[(1/0.288) - \exp(4\lambda_0)]\exp(4\lambda_0) = 0$$

By trial and error, $\lambda_0 = 0.31$

(d) With ψ given by Equation (iv), $\dfrac{\partial \psi}{\partial \lambda_0} = 0$ gives

$$\sum_{m=1}^{2} \left[\left(\ell n \frac{c_{in}}{c_{eff}} \right)_m - \lambda_0 L_m \right] L_m = 0$$

with the data given, one has $2(0.5812 - 2\lambda_0) + 4(1.184 - 4\lambda_0) = 0$
and $\lambda_0 = 0.2949$.
The λ_0 values obtained above range from 0.281 to 0.311 or a maximum difference of $0.311 - 0.281 = 0.03$, which is within the error associated with the experimental determination of λ_0.

■ ■ ■

6.3   Correlations of Filter Coefficient/Collector Efficiency of Aerosols

As discussions of the preceding sections show, accurate quantitative information of the filter coefficient/collector efficiency is essential to predict and assess particle deposition phenomenon and to design calculations of filtration systems. The results given in 6.1, to a degree, serve these purposes, but not completely, because of the restricted condition under which they may be applied and their limited accuracy.

Over the past three decades, a large number of studies, mostly experimental, have yielded results upon which empirical correlations of the filter coefficient/collector efficiency were developed. A summary of these correlations for aerosols will be given in this section followed by those of hydrosols in 6.4.

6.3.1   Single Fiber Efficiency of Aerosols in Fibrous Media

A.   Initial Single Fiber Efficiency in the Inertial Impaction-Dominated Region

For relatively large particles with $N_{St} > 0.5$, $N_R > 1$, based on experimental data from three different sources, the following empirical correlation of (η_{f_s}) was obtained using real and model filters (Nguyen and Beeckmans, 1975):

$$(\eta_{f_s})_0 = \frac{N_{St}^3 f^3}{N_{St}^3 f^3 + 1.54\left(15k_3 N_{Re}^{-1/2} + k_4 N_{Re}^{-1}\right)N_{St}^2 f^2 + 4.64} \tag{6.3.1}$$

and

$$f = 1 + k_1 \varepsilon_s + k_2 \varepsilon_s^2 \tag{6.3.2}$$

The values of k_i's, $i = 1$, 2, 3, and 4, are given in Table 6.3. The agreement between Equation (6.3.2) and the data upon which Equation (6.3.2) was developed is shown in Fig. 6.8. Although the agreement may appear acceptable, the scattering of data points, in fact, reveals the degree of inconsistency of experimental data used in developing the correlation. This lack of agreement among experimentally determined η_{f_s} obtained

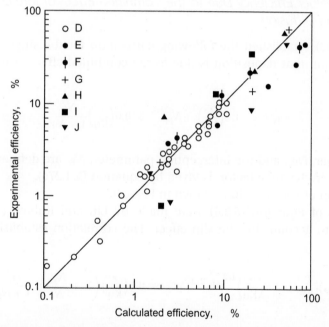

FIGURE 6.8 Experimental Single-Collector Efficiency vs. Calculated Results according to the Correlation of Nguyen and Beeckman. Sources of Data:

D	Fibrous Filter Results of Nguyen and Beeckmans, (1975)
E	Paper Mat Results of Stern et al. (1960)
F	20 oz Felt;
G	Permeable Aluminum "A";
H	Coarse Foam
I	Fine Foam
J	Glass Fibers, Results of Whitby (1965)

Reprinted from x. Nguyen and J.M. Beeckman, "Single fiber capture efficiencies of aerosol particles in real and model filters in the inertia-interception domain", J. Aerosol Science, 6, 205–212, 1975, with permission of Elsevier.

under comparable conditions from various sources was first highlighted by Brown (1993)[8] leading to the conclusion that using single fiber efficiency approach in describing aerosol collection in the inertial impaction dominant region may not be appropriate.

Numerical Values of Coefficients k_1, k_2, k_3, and k_4 of the Correlation of Nguyen and Beeckmans

	k_1	k_2	k_3	k_4
Model Filters	37	91	12	60
Real Filters	4	2250	4	65

B.   Initial Single Fiber Efficiency Due to the Combined Effect of Interception and Brownian Diffusion

Lee and Liu (1982) established the following correlation for the single fiber efficiency under the condition that deposition is due to the combined effect of interception and diffusion:

$$(\eta_{s_f}) = 1.6\left(\frac{\varepsilon}{Ku}\right)^{1/3} N_{Pe}^{-2/3} + 0.6\left(\frac{\varepsilon}{Ku}\right)\frac{N_R^2}{1 + N_R} \tag{6.3.4}$$

The Peclet number, N_{Pe}, and the interception parameter, N_R, are defined as before. Ku, the Kuwabara hydrodynamic factor, is given by Equation (5.4.27e).

The correlation of Lee and Liu is shown in Fig. 6.9.

Modifications of Equation (6.3.4) were made by Liu and Rubow (1990) and later by Payet (1992) to account for the slip effect. The correlation established by Payet is given as

$$(\eta_{st})_{I,BM} = 1.6\left(\frac{\varepsilon}{Ku}\right)^{1/3} N_{Pe}^{-2/3} C_1 C_2 + 0.6\left(\frac{\varepsilon}{Ku}\right)\left(1 + \frac{N_{kn}}{N_R}\right)\frac{N_R^2}{1 + N_R} \tag{6.3.5}$$

where

$$C_1 = 1 + 0.388 N_{kn}\left(\frac{\varepsilon N_{Pe}}{Ku}\right)^{1/3} \tag{6.3.6a}$$

$$C_2 = \frac{1}{1 + 1.6\left(\frac{\varepsilon}{Ku}\right)^{1/3} N_{Pe}^{-2/3} C_1} \tag{6.3.6b}$$

[8]See Figs. 4.7–4.10, Brown, 1993.

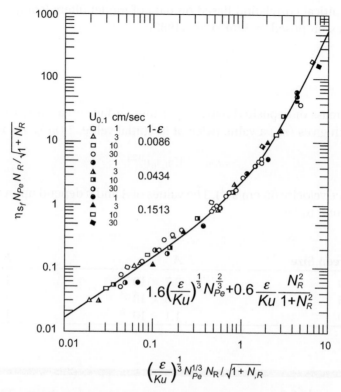

FIGURE 6.9 Comparison of single fiber efficiency experimental results with Lee-Liu correlation, Aerosol Science and Technology, K.W. Lee and B.Y.H. Liu, "Theoretical study of aerosol filtration by fibrous filters, 1, 147–161, Mount Lauret, NJ, 1982, Reprint with permission.

C.   Effect of Deposition on Single Fiber Efficiency

It is well known experimentally that single fiber efficiency increases with the increase of deposition (the so-called loading effect). The experimental results of Myojo et al. (1984) yield the following empirical correlation

$$\frac{\eta_{S_f}}{(\eta_{S_f})_0} = 1 + \alpha M \tag{6.3.7}$$

where $(\eta_{S_f})_0$ is the initial single fiber efficiency, M is the mass of deposit per unit volume of filter material, and α is an empirical constant. M is related to the specific deposit as

$$\sigma = (M/\rho_p)\big/\varepsilon_s \tag{6.3.8}$$

and the coefficient, α, is specific to the filter used. Qualitatively speaking, α varies from 0.05 to 1.0 $m^3\,kg^{-1}$ as the Stokes number ranges from 0.025 to 0.5.

A similar empirical correlation based on data of model filters (composed of layers of wire screens) was proposed by Emi et al., (1982):

$$\frac{\eta_{f_s}}{(\eta_{f_s})_0} = 1 + \left(\frac{m}{m_0}\right)^b \tag{6.3.9}$$

where m is the mass of deposited particle per unit surface area (in $kg\,m^{-2}$) and m_0, the value of m which gives (η_{f_s}) a value twice of its initial value, $2\,(\eta_{f_s})_0$. m_0 is given as

$$m_0 = A(\eta_{f_s})_0 u_\infty^{0.25} \tag{6.3.10}$$

and u_∞ is the face velocity (in $cm\,s^{-1}$). The values of A and b depend upon the wire screen used and are found to be

Screen Size	A	b
200	2.7×10^{-3}	1.15
325	1.5×10^{-3}	1.23
500	1.1×10^{-3}	1.34

■ ■ ■ ▬▬▬▬▬▬▬▬▬▬▬▬▬▬▬▬▬▬▬▬▬▬▬▬▬▬▬▬▬▬

Illustrative Example 6.8

(a) Calculate the single fiber efficiency, η_s, as a function of particle size according to Equation (6.3.4) for the following conditions:

Particle diameter	0.1, 0.5, 1.0, 2.0, 3.0, 4.0 μm
Fiber diameter	10 μm, Fiber packing density (or solidosity), $\varepsilon_s = 0.05$
Temperature	298 K
Gas Density	1.1853 kg m^{-3}, Gas Viscosity 1.8×10^{-5} Pa s
Gas Velocity	10 cm s^{-1}

The Cunningham correction factor may be assumed to be unity.

(b) Based on the results obtained, what is your conclusion about the effect of particle size on particle collection?

Solution

Equation (6.3.4) is given as

$$\eta_{S_f} = 1.6 \left(\frac{\varepsilon}{Ku}\right)^{1/3} N_{Pe}^{-2/3} + 0.6\left(\frac{\varepsilon}{Ku}\right) \frac{N_R^2}{1 + N_R} \tag{i}$$

and

$$N_{Pe} = \frac{d_f u_s}{D_{BM}}$$

$$N_R = d_p/d_f$$

$$Ku = -(1/2)\,\ell n\,\varepsilon_s - (3/4) + \varepsilon_s - \varepsilon_s^2/4$$

$$D_{BM} = \frac{c_s.kT}{3\,\pi\,\mu\,d_p}$$

The Kuwabara hydrodynamic factor is

$$Ku = (1/2)\ell n(0.05) - (3/4) + 0.05 - (0.05)^2/4$$

$$=-(1/2)(-2.9958) - 0.75 + 0.05 - 0.006 = 0.7973$$

For particles of diameter 1 μm, D_{BM} is

$$D_{BM} = \frac{(1.0)(1.3805 \times 10^{-23})(298)}{(3\pi)(1.8 \times 10^{-5})10^{-6}} = 2.425 \times 10^{-11}\,m^2\,s^{-1}$$

$$N_{Pe} = \frac{(1 \times 10^{-5})(0.1)}{2.425 \times 10^{-11}} = 4.12 \times 10^4$$

$$N_R = 1/10 = 0.1$$

Note that $D_{BM}\alpha(1/d_p)$, $N_R\alpha\,d_p$ and $N_{Pe}\alpha\,(1/D_{BM})$ or $N_{Pe}\,\alpha\,d_p$; therefore, the following results were obtained:

d_p(μm)	N_R	D_{BM}	N_{Pe}	$N_{Pe}^{2/3}$	$N_{Pe}^{-2/3}$
0.1	1×10^{-2}	2.425×10^{-10}	4.12×10^3	257	3.891×10^{-3}
0.5	5×10^{-2}	4.85×10^{-11}	2.06×10^4	752	1.329×10^{-3}
1.0	0.1	2.425×10^{-11}	4.12×10^4	1193	8.38×10^{-4}
2.0	0.2	1.213×10^{-11}	8.24×10^4	1894	5.28×10^{-4}
3.0	0.3	8.08×10^{-12}	1.236×10^5	2481	4.03×10^{-4}
4.0	0.4	6.06×10^{-12}	1.648×10^5	3006	3.33×10^{-4}

The single fiber efficiency, for particle size d_p, is

$$\eta_{s_f} = (1.6)\frac{0.95}{0.7973}\,N_{Pe}^{-2/3} + (0.6)\frac{0.95}{0.7973}\frac{N_R^2}{1 + N_R}$$

$$= 1.9064\,N_{Pe}^{-2/3} + 0.7149\,\frac{N_R^2}{1 + N_R}$$

The results are

$d_p(\mu m)$	$1.6\,\dfrac{\varepsilon}{Ku}\,N_{Pe}^{-2/3}$	$0.6\left(\dfrac{\varepsilon}{Ku}\right)\dfrac{N_R^2}{1+N_R}$	η_{s_f}
0.1	7.418×10^{-3}	7.009×10^{-5}	7.49×10^{-3}
0.5	2.534×10^{-3}	1.702×10^{-3}	4.24×10^{-3}
1.0	1.598×10^{-3}	6.499×10^{-3}	8.10×10^{-3}
2.0	1.007×10^{-3}	0.0238	2.48×10^{-2}
3.0	7.683×10^{-4}	0.0495	5.72×10^{-2}
4.0	6.348×10^{-4}	0.0817	8.23×10^{-2}

The above results show that at small values of d_p, η_{s_f} increases with the decrease of d_p. However, at large values of d_p, η_{s_f} increases with the increase of d_p. Therefore, there is a d_p value at which η_{s_f} is a minimum. This d_p value, $(d_p)_{min}$, is commonly known as the most penetrating size. For the present case, further calculation gives the following:

$d_f(\mu m)$	η_{s_f}
0.35	4.064×10^{-3}
0.39	4.041×10^{-3}
0.40	4.05×10^{-3}
0.41	

As an approximation, $d_f = 0.39\ \mu m$ may be taken as the most penetrating size.

■ ■ ■

6.3.2   Collector Efficiency of Aerosols in Granular Media

A.   Initial Collector Efficiency in the Inertial Impaction-Dominated Region

A number of initial collector efficiency correlations in the inertial impaction-dominated region have been reported in the literature (see Tien and Ramarao, 2007, for detailed information). Two of the more recent ones based on substantial amounts of data are:

(1) Correlation of D'Ottavio and Goren (1983)

$$(\eta_s)_0 = \frac{N_{St_{eff}}^{3.55}}{1.67 + N_{St_{eff}}^{3.55}} \tag{6.3.11}$$

where

$$N_{St_{eff}} = \left[A_s + 1.14\,N_{Re_s}^{1/2}(\varepsilon)^{-3/2}\right]\frac{N_{St}}{2} \tag{6.3.12a}$$

and

$$A_s = \frac{2(1-p^5)}{w} \tag{6.3.12b}$$

$$N_{Re_s} = \frac{d_c u_s \rho}{\mu} \tag{6.3.12c}$$

with N_{St} defined by Equation (6.1.10). p and w are given by Equations (5.4.11b) and (5.4.11c), respectively. Equation (6.3.11) was based on both liquid and solid aerosol filtration data. The ranges of variables covered are: $0.017 < N_{St_{eff}} < 9.5$, $13 < N_{Re_s} < 500$ and $3.4 \times 10^{-4} < N_R < 2.27 \times 10^{-3}$. For solid aerosols, Equation (6.3.11) may be applied for $N_{St_{eff}} < 1$ without accounting for possible particle bouncing off (see 6.3.3).

(2) Correlation of Jung et al. (1989)

$$(\eta_s)_0 = 0.2589\, N_{St_{eff}}^{1.3437} N_R^{0.23} \quad \text{for } N_{St_{eff}} < 1.2 \tag{6.3.13}$$

The above expression was based on data obtained under the following conditions: $2.15 \times 10^{-3} < N_R < 4.56 \times 10^{-2}$, $0.399 < N_{Re_s} < 19.96$ and $0.036 < N_{St_{eff}} < 27.5$. Equation (6.3.13) is applicable for $N_{St_{eff}} < 1.0$. For $N_{St_{eff}} > 1.0$, particle bounce-off becomes significant and the results obtained from Equation (6.3.11) require correction, see later discussion.

Equation (6.3.11) suggests that $(\eta_s)_0$ is a function of $N_{St_{eff}}$ only. On the other hand, Equation (6.3.13) indicates that $(\eta_s)_0$ depends upon N_R as well and $(\eta_s)_0/N_R^{0.23}$ is a function of $N_{St_{eff}}$. In Figs. 6.10.a and 6.10.b, the data of Ottavio and Goren and those of Jung et al. were plotted as η_{s_0} *vs.* $N_{St_{eff}}$ and $\eta_{s_0}/N_R^{0.23}$ *vs.* $N_{St_{eff}}$, respectively. As seen from these figures, the inclusion of N_R indeed reduces the scattering of data points. In addition, two important observations can be made from these figures: First, the initial single-collector efficiency increases monotonically with $N_{St_{eff}}$ until a certain threshold value of $N_{St_{eff}}$ is reached. The same phenomena were also found by investigators in the past and were generally attributed to the bounce-off of impacting particles with sufficiently high particle inertia. To account for this effect, the adhesion probability of impacting particles should be considered (see next section). Second, the relatively large scattering of data points shown in these two figures underscores the difficulty of obtaining collector efficiency data with a high degree of accuracy. A major factor for the scattering can be attributed to the inherent problem of particle counting as mentioned by Tien and Ramarao previously (2007).

B.   Estimation of the Onset of Particle Bounce-Off and Correlation of Particle Adhesion Probability

For aerosol filtration, the correlation of Equation (6.3.13) may be assumed to be valid in the absence of particle bounce-off (or automatic deposition of particles making contact

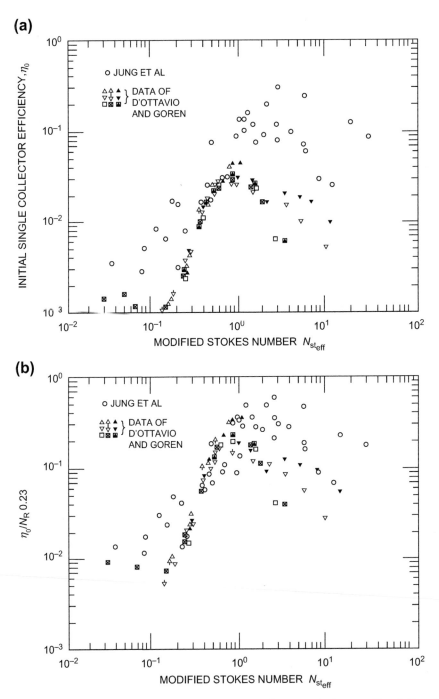

FIGURE 6.10 a Plot of Experimental Data of D'Ottavio and Goren and Jung et al. of η_{s_0} vs. $N_{St_{eff}}$. **b** Plot of Experimental Data of D'Ottovia and Goren and Jung et al. of $\eta_{s_0}/N_R^{0.23}$ vs. $N_{St_{eff}}$ (Jung, 1991).

with collectors); the predicted collector efficiency from Equation (6.3.13), in reality, is the collision efficiency. Accordingly, one may define an adhesion probability, γ, to be the ratio of the experimentally determined collector efficiency to the collision efficiency with the latter given by Equation (6.3.13), or

$$\gamma = \frac{(\eta_s)_{0,\exp}}{(\eta_s)_0 \text{ from Eq. (6.3.13)}} \tag{6.3.14}$$

A plot of γ vs. $N_{St_{eff}}$ for $N_{St_{eff}} > 1.0$ based on the data of Ottavio and Goren and Jung et al. is shown in Fig. 6.11. In spite of the scattering of data points, it is clear that γ is a monotonically decreasing function of $N_{St_{eff}}$ and is given by the expression

$$\gamma = 1.4315 \, N_{St_{eff}}^{-1.968} \quad \text{for } N_{St_{eff}} > 1.0 \tag{6.3.15}$$

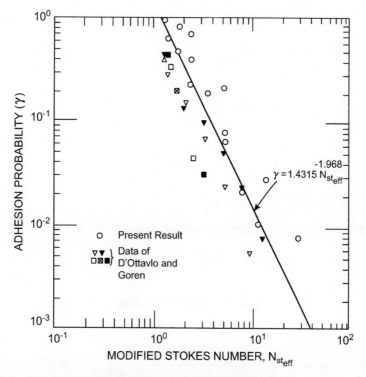

FIGURE 6.11 Adhesion Probability as a Function of $N_{St_{eff}}$ (Jung, 1991).

■ ■ ■ ━━━━━━━━━━━━━━━━━━━━━━━━━━━━━━━

Illustrative Example 6.9

Calculate the initial collector efficiency of aerosol filtration in granular bed under the following conditions:

Particle Diameter:	$d_p = 4\ \mu m\ (4 \times 10^{-6}\ m)$
Grain Diameter:	$d_g = 500\ \mu m\ (5 \times 10^{-4}\ m)$
Superficial gas velocity:	$u_s = 0.2\ m\,s^{-1}$
Filter Media Porosity:	$\varepsilon_0 = 0.34$
Particle Density:	$\rho_p = 1050\ kg\,m^{-3}$
Gas Viscosity and Density:	$\mu = 0.018\ cp$ or $1.8 \times 10^{-5}\ Pa\,s$
	$\rho = 0.074\ lb\,ft^{-3}$ or $1.1853\ kg\,m^{-3}$

Solution

The relatively large particle size and high gas velocity suggest that deposition is controlled by inertial impaction. Equations (6.3.11) or (6.3.13) may be applied.

The various dimensionless parameters present in these correlations are:

$$N_{St_{eff}} = \left[A_s + 1.14\ N_{Re_s}^{1/2}(\varepsilon)^{-3/2} \right](N_{St}/2)$$

$$N_{St} = (\rho_p\ d_p^2\ u_s\ c_s)/(9\ \mu\ d_g)$$

$$N_R = d_p/d_g$$

$$A_s = 2(1 - p^5)/w$$

$$w = 2 - 3p + 3p^5 - 2p^6$$

$$p = (1 - \varepsilon_0)^{1/3}$$

With the given conditions, the various dimensionless parameters are:

$$N_R = 4/500 = 8 \times 10^{-3}$$

$$p = (1 - 0.34)^{1/3} = 0.8707$$

$$A_s = \frac{2\left[1 - (0.8707)^5\right]}{2 - 3(0.8707) + 3(0.8707)^5 - 2(0.8707)^6} = 56.45$$

$$N_{St} = \frac{(1050)(4 \times 10^{-6})^2(0.2)}{(9)(1.8 \times 10^{-5})(5 \times 10^{-4})} = 0.4148 \quad \text{with } c_s = 1.0$$

$$N_{St_{eff}} = \left[56.45 + 1.14(0.6585)^{1/2} \left(\frac{1}{0.34}\right)^{3/2} \right] \frac{0.4148}{2} = 12.68$$

For the calculation of η_0

(a) Apply Equation (6.3.11). The initial single-collector efficiency, η_{s_0}, is found to be

$$\eta_{s_0} = \frac{(12.68)^{3.55}}{1.67 + (12.68)^{3.55}} = 0.9998$$

In terms of the collector efficiency, η_0,

$$\eta_0 = (\eta_{s_0})(1 - \varepsilon)^{2/3} = (0.9998)(0.66)^{2/3} = 0.7579$$

(b) Apply Equation (6.3.13). Since $N_{St_{eff}} > 1.0$, the initial collector efficiency obtained from Equation (6.3.11) should be modified by the adhesion probability given by Equation (6.3.15), is therefore

$$\gamma = 1.4315 \, (12.68)^{-1.968} = 9.657 \times 10^{-3}$$

and

$$\eta_{s_0} = 0.2598(12.68)^{1.3437}(8 \times 10^{-3})^{0.23}(9.657 \times 10^{-3}) = 2.5089 \times 10^{-2}$$

and

$$\eta_0 = (2.5089 \times 10^{-2})(0.66)^{2/3} = 1.9019 \times 10^{-2}$$

As shown in Fig. 6.8a, the experimentally determined η_0 varies significantly ranging approximately from 5×10^{-3} to 10^{-1} with $N_{St_{eff}} \simeq 10$. This is comparable to the value obtained from Equation (6.3.13) modified by γ. The value obtained from Equation (6.3.9) is much higher. However, as stated in the text, Equation (6.3.9) is not valid for $N_{St_{eff}} > 0.5$. The result based on Equations (6.3.13) corrected by (6.3.15) is therefore more reasonable.

■ ■ ■

C.   Initial Collector Efficiency in the Diffusion Dominated Regime

The single-collector efficiency of Equation (6.1.28) can be readily applied. By combining Equations (5.4.19) and (6.1.28), the initial collector efficiency due to diffusion is found to be:

$$(\eta_0)_{BM} = 4(1 - \varepsilon)^{2/3} A_s^{1/3} N_{Pe}^{-2/3} \tag{6.3.16a}$$

The above expression was based on the assumption of creeping flow over a spherical collector and is applicable for $N_{Re_s} < 30$ [Yoshida and Tien (1985)]. For higher values of N_{Re_s}, the results of Tardos et al. (1978) may be used or

$$(\eta_0)_{BM} = 4(1 - \varepsilon)^{2/3} \left(\frac{1.31}{\varepsilon^{1/2}}\right) N_{Pe}^{-1/2} \quad \text{for } N_{Re_s} > 30 \qquad (6.3.16b)$$

D.   Effect of Deposition on the Collector Efficiency

Following the procedure described previously (procedure (ii) of 6.2.2), Jung and Tien (1991) established the following empirical expressions for the increase in collector (or filter coefficient) efficiency due to deposition:

$$F = \frac{\eta}{\eta_0} = \frac{\lambda}{\lambda_0} = 1 + \alpha_1 \sigma^{\alpha_2} \qquad (6.3.17)$$

The coefficient α_1 and exponent α_2 depend upon the Stokes number N_{St} and the interception parameter N_R. α_1 and α_2 are found to be

$$\alpha_2 = 0.4416 \ N_{St}^{-0.3649} N_R^{0.2397} \qquad (6.3.18a)$$

$$\alpha_1 = (0.0954 \ 10^{3\alpha_2}) \ N_{St}^{-1.478} N_R^{0.4322} \qquad (6.3.18b)$$

The above expressions are applicable for $1.7 \times 10^{-3} < N_{St} < 3.8 \times 10^{-2}$ and $1.7 \times 10^{-3} < N_R < 8.0 \times 10^{-3}$. The conditions used in experiments are those under which the inertial impaction is the dominant mechanism for deposition.

6.4   Filter Coefficient Correlations of Hydrosols

6.4.1   Filter Coefficient of Fibrous Media

A.   Initial Filter Coefficient Correlation based on Trajectory Calculation Results

Assuming that deposition is due to the combined effects of interception, gravitation, and the London–van der Waals force, Guzy et al. (1983) and later Choo and Tien (1992) conducted trajectory calculations to obtain the initial filter coefficient under the condition of favorable surface interactions[9] using several cylinder-in-cell models. Based on the results obtained with the use of Kuwabara's model, Choo and Tien established the following empirical correlations:

$$\lambda_0 = \left(\frac{6}{\pi}\right) \frac{1 - \varepsilon_0}{d_f} A_s \left[0.216 \times 10^{-0.41\varepsilon_0} N_R^{1.55} N_{LO}^{0.1542} + 2.99 \times 10^{-4} \times 10^{3\varepsilon_0} N_G^{1.1} N_R^{-0.3}\right] \qquad (6.4.1)$$

[9]A discussion on the nature of surface interaction between particles and collectors is given in Section 6.5.

for

$$10^{-3} < N_R < 10^{-1}$$

$$10^{-4} < N_G < 10^{-1}$$

$$0.01 < \varepsilon_s < 0.65$$

$$10^{-8} < N_{LO} < 10^{-3}$$

The dimensionless parameters N_R and N_G are the same as defined before (see Equations (6.1.16b) and (6.1.14) with $u_\infty = u_s$). The London force parameter N_{LO} is defined as

$$N_{LO} = \frac{H}{9 \pi \mu a_p^2 u_s} \tag{6.4.2}$$

where H is the Hamaker constant.

The various hydrodynamic parameters are[10]

$$A_s = \frac{(2/3)(4 c_1 + c_4)}{c_1[(1/\varepsilon_s) - 2 + \varepsilon_s] + (c_4/2)(\varepsilon_s - 1 - \ell n \, \varepsilon_s)} \tag{6.4.3a}$$

$$c_1 = -\varepsilon_s c_4/4 \tag{6.4.3b}$$

$$c_2 = -c_1 - c_3 \tag{6.4.3c}$$

$$c_3 = c_1 + (c_4/2) \tag{6.4.3d}$$

$$c_4 = -4/(2 \, \ell n \, \varepsilon_s + 3 - 4\varepsilon_s + \varepsilon_s^2) \tag{6.4.3e}$$

B.   Filter Coefficient due to the Brownian Diffusion

The correlation of Equation (6.4.1) does not include the Brownian diffusion contribution. Based on the results from the solution of the convective diffusion equation using Lighthill's formula, $(\lambda_0)_{BM}$ is found to be (Choo and Tien, 1992)

$$(\lambda_0)_{BM} = (9.2/\pi)(c_1 + c_3)^{1/3}[(1 - \varepsilon_0)/d_f]N_{Pe}^{-2/3} \tag{6.4.4}$$

---

[10]These parameters are equivalent to those of Equations (5.4.27a)–(5.4.27e).

6.4.2   Filter Coefficient of Granular Media

A. Initial Filter Coefficient, Favorable surface interactions case. Rajagopalan and Tien (1976) presented the following initial filter coefficient correlation:

$$\lambda_0 = A_s\left(\frac{1-\varepsilon_0}{d_g}\right)\left[1.5\,N_{LO}^{1/8}\,N_R^{15/8} + 5.06\times 10^{-3}N_G^{1.2}N_R^{-0.4} + 6(A_sN_{Pe})^{-2/3}\right] \quad \text{for } N_R \leq 0.18 \quad (6.4.5)$$

This correlation was developed based on trajectory calculation results (using Happel's model for media representation and with favorable surface interaction; i.e., no repulsive force barrier between particles and collectors) and supplemented with contributions from diffusion according to Equation (6.1.28). It assumes that deposition is due to the combined effects of interception, gravitation, and the Brownian diffusion. The hydrodynamic retardation effect was included in the trajectory calculations.

Since the development of Equation (6.4.5), a number of similar correlations have appeared in the literature including those by Yoshimura (1980), Cushing and Lawler (1998), and Tufenkji and Elimelech (2004). While all these correlations supposedly present new features different from the work of Rajagopalan and Tien, in terms of practical utility, they provide essentially the same information, namely, approximate predictions of the filter coefficient with uncertain degrees of accuracy.

Agreements between these correlations with selected experimental data have been used to justify the claim of superiority of one correlation over others. What is often overlooked is that the degree of agreement inevitably depends upon the experiments selected for comparisons. The inherent problems of obtaining consistent filtration data (for example, see Table 6.2, Tien and Ramarao, 2007) and using selected experimental data for comparisons make it difficult to make any unequivocal conclusion regarding the relative merits of these comparisons. (For a more detailed discussion, see pp. 301–307, Tien and Ramarao, 2007.) It should also be mentioned that the difference between these correlations, on the whole, is not significant and within typical errors of experimentally determined collector efficiencies.

■ ■ ■ ▬▬▬▬▬▬▬▬▬▬▬▬▬▬▬▬▬▬▬▬▬▬▬▬▬

Illustrative Example 6.10

Estimate the initial filter coefficient for the filtration of aqueous solution in granular beds under the following conditions:

$$d_p = 6.1\,\mu\text{m}\ (6.1\times 10^{-6}\,\text{m})$$

$$d_g = 345\,\mu\text{m}\ (3.45\times 10^{-4}\,\text{m})$$

$$u_s = 0.2 \text{ cm s}^{-1}(2 \times 10^{-3} \text{ m s}^{-1})$$

$$\varepsilon_0 = 0.38$$

$$\mu = 1 \text{ } cp(10^{-3} \text{ Pa s})$$

$$\rho_p = 1050 \text{ kg m}^{-3}$$

$$\rho = 1005.8 \text{ kg m}^{-3}$$

$$H = 1.1 \times 10^{-20} \text{ J}$$

$$\varepsilon_r = 80$$

The particle-collector surface interactions may be assumed favorable.

Solution

With the given conditions, the initial filter coefficient, λ_0, can be estimated from the correlation of Equation (6.4.5) which expresses λ_0 in terms of a number of dimensionless parameters. The values of the constants present in these parameters are:

Boltzmann's Constant $k_B = 1.3805 \times 10^{-23} \text{ J K}^{-1}$
Avagadro's Number $N_A = 6.0225 \times 10^{23} \text{ mol}^{-1}$

The various dimensionless groups present in the correlation of λ_0 [i.e., Equation (6.4.5)] are

$$N_R = d_p/d_g = 6.1/345 = 1.768 \times 10^{-2}$$

$$N_{LO} = \frac{H}{9\pi\mu a_p u_s} = \frac{1.1 \times 10^{-20}}{9\pi(10^{-3})(3.05 \times 10^{-6})^2(2 \times 10^{-3})} = 2.091 \times 10^{-5}$$

$$N_G = \frac{(2)(\Delta\rho)a_p^2 g}{9\mu u_s} = \frac{(2)(44.2)(3.05 \times 10^{-6})^2 9.8}{(9)(10^{-3})(2 \times 10^{-3})} = 4.477 \times 10^{-4}$$

$$(D_{BM})_\infty = \frac{k_B T}{3\pi\mu d_p} = \frac{(1.385 \times 10^{-23})(298)}{3\pi(10^{-3})(6.1 \times 10^{-6})} = 7.156 \times 10^{-14}$$

$$N_{Pe} = \frac{u_s d_g}{(D_{BM})_\infty} = \frac{(2 \times 10^{-3})(6.1 \times 10^{-6})^2}{7.156 \times 10^{-14}} = 1.705 \times 10^{-5}$$

The various Happel parameters are

$$p = (1 - \varepsilon)^{1/3} = (1 - 0.38)^{1/3} = 0.8527$$

$$w = 2 - 3p + 3p^5 - 2p^6 = 0.0255$$

$$A_s = \frac{2(1 - p^5)}{w} = 43.07$$

Therefore, λ_0 according to the correlation of Equation (6.4.5) is

$$\lambda_0 = A_s \frac{1 - \varepsilon_0}{d_g} \left[1.5(N_{L0})^{1/8}(N_R)^{15/8} + 5.06(10^{-3})N_G^{1.2}N_R^{-0.4} + 6(A_sN_{Pe})^{-2/3} \right]$$

$$= (43.07)\frac{0.62}{345 \times 10^{-6}}\left[(1.5)(2.091 \times 10^{-5})^{1/8}(1.768 \times 10^{-2})^{15/8} \right.$$

$$\left. + 5.06 \times 10^{-3}(4.477 \times 10^{-4})^{1.2}(1.768 \times 10^{-2})^{-0.4} + 6(43.7 \times 1.705 \times 10^5)^{-2/3} \right]$$

$$= 28.1 \text{ m}^{-1}$$

■ ■ ■

B.   Effect of Deposition on Filtration Coefficient

Chiang and Tien (1985a,b), based on simulation results and comparisons with experiments, established an expression which gives the increase of filter coefficient (in terms of $F = \lambda/\lambda_0$) as a function of the specific deposit, σ. The constricted tube model of porous media was used in their simulations carried out under two limiting conditions. In the first case (limiting condition A), deposition was assumed to result in the formation of nonuniform deposit layers. For the second case (limiting condition B), deposition leads to particle dendrite growth. By comparing simulation results with experiments, it was found that in order to obtain good agreement between simulations and experiments, particle deposition should be assumed to proceed in a combination of these two limiting situations with a proper weighting factor for the combination. On this basis, the following empirical expression was established:

$$F = \frac{\lambda}{\lambda_0} = \frac{\eta}{\eta_0} = 1 + \sigma^{0.755}\left[492 - (1.6 \times 10^4)N_R + (1.46 \times 10^5)N_R^2 \right] \qquad (6.4.6)$$

A different expression of F was later obtained by Choo and Tien (1995) based on their nonuniform permeable deposit model and the assumption that interception is the main mechanism of deposition. An empirical correlation of their numerical results was found to be

$$F = \frac{\lambda}{\lambda_0} = \frac{\eta}{\eta_0} = Y\left(\frac{\lambda}{\lambda_0}\right)_{k_d=0} + (1 - Y)\left(\frac{\lambda}{\lambda_0}\right)_{k_d \to \infty} \qquad (6.4.7)$$

and

$$\left(\frac{\lambda}{\lambda_0}\right)_{k_d \to 0} = 1 + 9.61(1 - \varepsilon_0)^{2/3}\left(\frac{\sigma}{1 - \varepsilon_d}\right) \qquad (6.4.8a)$$

$$\left(\frac{\lambda}{\lambda_0}\right)_{k_d \to \infty} = 1 + \frac{0.6794}{1 - \varepsilon_0}\left(\frac{1}{N_R} - 0.921\right)\left(\frac{\sigma}{1 - \varepsilon_d}\right) + \frac{0.1731}{(1 - \varepsilon_0)}\left(\frac{1}{N_R^2} + \frac{3}{N_R} - 1.171 \times 10^2\right)$$
$$\times \left(\frac{\sigma}{1 - \varepsilon_d}\right)^2$$

(6.4.8b)

$$Y = \frac{f_1}{1 + f_1}$$

(6.4.9a)

$$f_1 = 0.598 k_d^{-0.8}(1 - \varepsilon_0)^{-2}\left(1 + \frac{0.0128}{N_R}\right)\left(\frac{\sigma}{(1 - \varepsilon_d)}\right)^{(1.63 + 5.5 \times 10^{-4}/N_R)}$$

(6.4.9b)

where ε_d is the deposit porosity and k_d the dimensionless deposit permeability (normalized by a_c^2).

Similar to Equation (6.4.6), F, given by Equation (6.4.7), is a function of σ and N_R. Furthermore, in applying Equation (6.4.7), the values of ε_d and k_d must be known. In the absence of experimentally determined values, ε_d may be assumed to be 0.7–0.8 and k_d can be estimated according to the Kozeny–Carman equation.

A few words about the above two correlations [Equations (6.4.6) and (6.4.7)] may be in order. Both correlations were developed with limited data. More important, as mentioned in 5.2, the effect of deposition may not always be positive. While present knowledge does not provide us with precise information about the conditions under which deposition enhances subsequent deposition, experimental evidence indicates that the enhancement of filter performance due to deposition (the so-called ripening period of filtration) is likely to be found in direct filtration of chemically pretreated or flocculated feed streams. This is the fact one should bear in mind in applying Equations (6.4.6) or (6.4.7) in calculations.

Illustrative Example 6.11

Estimate the effluent particle concentration history of granular bed filtration of aqueous suspension for the following conditions:

Filter grain diameter	$d_g = 350\ \mu m$
Filter bed height	$L = 30$ cm
Bed porosity	$\varepsilon = 0.38$
Suspension flow rate	$u_s = 0.2$ cm s^{-1} or 2×10^{-3} m s^{-1}
Suspension concentration	$c_{in} = 100$ PPM (by vol) or 10^{-4} vol/vol
Particle diameter	$d_p = 5\ \mu m$
Initial filter coefficient	$\lambda_0 = 5$ m^{-1}

The increase of the filter coefficient due to deposition can be estimated according to the correlation of Equation (6.4.6). To estimate the effluent concentration history, the procedure outlined in Section 5.7 may be applied.

Solution

According to Equation (6.4.6), one has

$$F = \frac{\lambda}{\lambda_0} = 1 + \left[492 - 1.6 \times 10^4\, N_R + 1.46 \times 10^5 N_R^2\right]\sigma^{0.755}$$

$$N_R = 5/350 = 1.43 \times 10^{-2}$$

$$F = 1 + \left[492 - 1.6 \times 10^4(1.43 \times 10^{-2}) + (1.46 \times 10^5)(1.43 \times 10^{-2})^2\right]\sigma^{0.755}$$

$$= 1 + 293\,\sigma^{0.755}$$

According to the procedure outlined in 5.7, one has [see Equation (5.7.20)]

$$\frac{c_{in}}{c_{eff}} = \frac{\sigma_{in}}{\sigma_L} \tag{i}$$

where c_{in} and c_{eff} are the influent and effluent particle concentrations of the suspension. σ_{in} is the value of the specific deposit at inlet $(z=0)$ and σ_L the value at exit $(z=L)$.
  The relationship between σ_i and θ (corrected time) is [see Equation (5.7.11)]

$$\int_0^{\sigma_{in}} \frac{d\sigma}{F(\sigma)}\, d\sigma = u_s \lambda_0 c_{in} \theta$$

or

$$\int_0^{\sigma_{in}} \frac{d\sigma}{1 + 293\sigma^{0.755}} = (2 \times 10^{-3})(5)(10^{-4})\theta = (10^{-6})\theta \tag{ii}$$

For a given σ_{in}, the corresponding θ can be found from the above equation. The results are

σ_{in}	$\int_0^{\sigma_{in}} \dfrac{d\sigma}{1+293\sigma^{0.755}}$	$\theta(s)$
2.56×10^{-5}	2.445×10^{-5}	24.45
6.41×10^{-5}	5.798×10^{-5}	57.98
1.097×10^{-4}	9.45×10^{-5}	94.5
1.605×10^{-4}	1.322×10^{-4}	132
2.159×10^{-4}	1.705×10^{-4}	169
5.402×10^{-4}	2.564×10^{-4}	296
1.353×10^{-3}	6.353×10^{-4}	635
3.389×10^{-3}	1.179×10^{-3}	1179
6.447×10^{-3}	1.689×10^{-3}	1689

Once σ_{in} is known, the corresponding σ at $z = L$, σ_L can be calculated according to Equation (5.7.17)

$$\int_{\sigma_L}^{\sigma_{in}} \frac{d\sigma}{\sigma(1 + 293\sigma^{0.755})} = \lambda_0 L = (5)(0.3) = 1.5 \qquad \text{(iii)}$$

Therefore, for a specified σ_{in} (corresponding to a given θ), the above expression can be used to determine σ_L. Once σ_L is known, the effluent particle concentration, c_{eff}, can be found according to Equation (i). As an approximation, Equation (iii) may be written as

$$\frac{1}{1 + 293\bar{\sigma}^{0.755}} \, \ell n \, \frac{\sigma_{in}}{\sigma_L} = 1.5 \qquad \text{(iv)}$$

where $\bar{\sigma}$ is a value between σ_L and σ_{in}. If one assumes

$$\bar{\sigma} = (1/2)(\sigma_{in} + \sigma_L) \qquad \text{(v)}$$

By trial and error, one can easily determine σ_L vs. σ_i according to Equation (iv). The results are

$\theta(s)$	σ_{in}	$\sigma_L/\sigma_{in} = c_{eff}/c_{in}$
0	0	0.233
24.45	2.56×10^{-5}	0.2014
57.98	5.998×10^{-5}	0.183
296	5.402×10^{-4}	0.0863
635	1.353×10^{-3}	0.036

■ ■ ■

6.5   Particle-Collector Surface Interactions Effect on Hydrosol Deposition in Granular Media

6.5.1   Surface Interaction Forces

An important factor in determining the extent of hydrosol deposition is the particle-collector surface interaction forces operating in the immediate neighborhood of collectors. Depending upon the nature of the surface interactions, particle deposition may be significantly reduced. Under such circumstances, the results obtained from correlation such as those given by Equations (6.4.1) or (6.4.5) are no longer valid and corrections are required.

There are a number of particle-collector surface interaction forces for hydrosol deposition considerations. Two of the most important ones are:

A.   London–van der Waals Force

The London–van der Waals force arises from the instantaneous dipole moments generated by the temporary asymmetrical distribution of electrons around atomic nuclei. This force is largely responsible for particle adhesion. In fact, without it, the hydrodynamic retardation effect would prevent particles from reaching any collector surfaces.

As filter grains are often two orders of magnitude larger than deposited particles, we can approximate the London–van der Waals force between a collector and a particle by thinking of it as that between a flat surface and a particle of radius a_p. The force potential, ϕ_{LO}, and the force itself, F_{LO}, are given as

$$\phi_{LO} = \frac{-H}{6}\left[\frac{2(\delta^+ + 1)}{\delta^+(\delta^+ + 2)} - \ell n\left(\frac{\delta^+ + 2}{\delta^+}\right)\right] \tag{6.5.1}$$

$$F_{LO} = -\frac{2}{3}\left[\frac{(H/a_p)}{\delta^{+2}(\delta^+ + 2)^2}\right]\underline{n} \tag{6.5.2}$$

where H is the Hamaker constant and is of the order of 10^{-14} ergs (or 10^{-21} Joules) for particles suspended in aqueous media; δ^+, the dimensionless separation distance using a_p as the normalizing quantity; and n, the unit normal vector to the surface. Note that the London–van der Waals force acts along the normal direction of the collector surface and is attractive.

B.   Double-Layer Force

The double-layer force arises from the fact that solid materials placed in an aqueous environment may acquire surface charges from preferential adsorption of ions or dissociations of surface groups. This surface charge is balanced by countercharged ions present in the solution. Thus, a double-layer of charge is established, characterized by an electrical potential between the outer portion of the double-layer and the bulk solution. This potential, known as the zeta potential, is used to approximate the potential difference between the material surface and the bulk solution. Derjaguin and Landau (1941) and Verwey and Overbeek (1945) developed a theory (the DVLO theory) describing the interaction of the two double layers. Their idea was extended to the sphere–plate system (for constant surface potential) by Hogg et al. (1966). The double-layer potential, ϕ_{DL}, is given as

$$\phi_{DL} = \pi a_p \hat{\varepsilon}\left(\zeta_p^2 + \zeta_c^2\right)\left[\frac{2\zeta_p\zeta_c}{\zeta_p^2 + \zeta_c^2}\ell n\left(\frac{1 + e^{-\kappa\delta}}{1 - e^{-\kappa\delta}}\right) + \ell n\left(1 - e^{-\kappa\delta}\right)\right] \tag{6.5.3}[11]$$

where $\hat{\varepsilon} = \hat{\varepsilon}_0 \cdot \varepsilon_r$, $\hat{\varepsilon}$ and $\hat{\varepsilon}_0$ are the permittivities of the liquid and vacuum, respectively, ε_r the permittivity ratio, is also known as the dielectric constant (dimensionless); ζ_p and ζ_c are the respective particles' and collectors' surface potentials (which are often

[11]The numerical value of $\hat{\varepsilon}_0$ is 8.8542×10^{-12} kg^{-1} m^{-3} s^4 A^2 or 0.07965 (esu)^2(erg)$^{-1}$(cm)$^{-1}$. As the value of $4\pi\hat{\varepsilon}_0$ is unity in c.g.s. units, therefore, Equation (6.5.3) becomes

$$\phi_{DL} = (\varepsilon_r a_p/4)(\zeta_p^2 + \zeta_0^2)\left[\frac{2\zeta_p\zeta_c}{\zeta_p^2 + \zeta_c^2}\ell n\frac{1 + e^{-\kappa\delta}}{1 - e^{-\kappa\delta}} + \ell n\left(1 - e^{-2\kappa\delta}\right)\right]$$

with all quantities expressed in c.g.s. units.

approximated by the zeta potentials obtained from electrophoretic measurements); and κ is the Debye–Huckel reciprocal double-layer thickness, defined as

$$\kappa = \sqrt{\frac{e^2}{\hat{\varepsilon}kT} \sum z_j^2 m_j} \qquad (6.5.4)$$

where e is the charge of the electron, κ is the Boltzmann constant, T is the absolute temperature, and m_j is the number concentration of the jth ion species present in the solution with variance z_j. Equation (6.5.3) is valid for $|\zeta| < 60$ mV and $\kappa\, a_p > 10$.

The double-layer force, F_{DL}, is given as

$$F_{DL} = \frac{2\pi\hat{\varepsilon}a_p(\zeta_p^2 + \zeta_c^2)\kappa e^{-\kappa\delta}}{(1 - e^{-2\kappa\delta})} \left[2\frac{\zeta_p\zeta_c}{\zeta_p^2 + \zeta_c^2} - e^{-\kappa\delta} \right] \underline{n} \qquad (6.5.5)$$

As Equation (6.5.5) shows, the double-layer force acts along the normal direction to the collector surface and is attractive (or F_{DL} being negative) whenever the surface potentials of the particle and the collector have opposite signs, and repulsive when they share the same sign. The net surface potential (or force) is the algebraic sum of the London and double-layer potentials (forces). Whether a repulsive force barrier exists or not depends upon the combination of these two forces. Examples illustrating the absence or presence of the repulsive force barrier are shown in Figs. 6.12a and 6.12b.

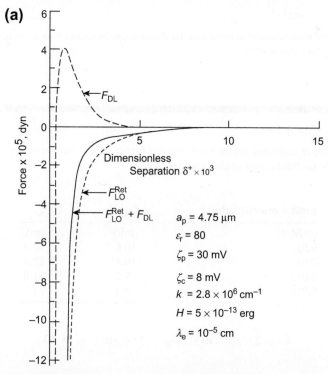

FIGURE 6.12a Plot of the retarded London Force and the double-layer force and their sums vs. separation: The case where the net force is always attractive.

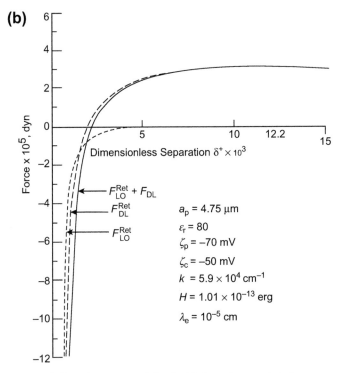

FIGURE 6.12b Plot of the retarded London Force and the double-layer force and their sums vs. separation: The case where a repulsive force barrier is present.

■ ■ ■ ━━━━━━━━━━━━━━━━━━━━━━━━━━━━━━━━━━━

Illustrative Example 6.12

Calculate the particle-collector surface interaction forces as a function of separation distance corresponding to the following conditions:

Ionic Concentration	ζ_p	ζ_c
(mM/ℓ)	(mV)	(mV)
0.011	10.6	12.6
0.012	10.2	12.2
0.013	9.8	11.8
0.014	9.5	11.5

$$u_s = 2 \times 10^{-3} \text{ m s}^{-1}; d_p = 11.4 \text{ μm} \qquad \varepsilon_r = 80$$

Solution

The combined surface interaction force is the sum of the double-layer force (F_{DL}) and the London–van der Waals (F_L) or

$$\frac{F_{LO} + F_{DL}}{6\pi\mu a_p u_s} = \frac{-N_{LO}}{\delta^{+2}(\delta^+ + 2)^2} + 4\pi N_{E_1} \left[N_{E_2} - e^{-N_{DL}\delta^+}\right] \left[\frac{e^{-N_{DL}\delta^+}}{1 - e^{-2N_{DL}\delta^+}}\right]$$

The dimensionless separation distance is δ/a_p and the various parameters are:

$$N_{LO} = H/(9\pi\mu a_p^2 u_s)$$

$$N_{E_1} = \varepsilon_r \hat{\varepsilon}_0 (\zeta_p^2 + \zeta_c^2)\kappa/(3\mu u_s)^{12}$$

$$N_{E_2} = 2(\zeta_p\zeta_c/\zeta_p^2 + \zeta_c^2)$$

$$N_{DL} = \kappa.a_p$$

The reciprocal double-layer thickness κ is [see Equation (6.5.4)]

$$\kappa = \left[\frac{e^2}{\hat{\varepsilon}k_B T} \sum_L z_j^2 m_j\right]^{1/2}$$

The value of the quantities of the above expression are:
$e = 1.6021 \times 10^{-19}$, $\hat{\varepsilon} = \varepsilon_r.\hat{\varepsilon}_0$ with $\varepsilon_r = 80$, $\hat{\varepsilon}_0 = 8.8542 \times 10^{-12}$, $k_B = 1.3805 \times 10^{-23}$, $T = 298$ K, the number concentration of ions per m³ can be related to the molar concentration M and the Avagadro's number, N_{Av} or

$$m = (1000) N_{Av}.M$$

For 1:1 electrolyte

$$\sum z_i^2 m_i = (1000)(6.023 \times 10^{23})(2M)$$

For the case of max concentration being 0.011, κ is

$$\kappa = \left[\frac{(1.602 \times 10^{-19})^2(1000)(6.023 \times 10^{23})(0.022)}{(80)(8.8542 \times 10^{-12})(1.3805 \times 10^{-23})(298)}\right]^{1/2} = 3.4163 \times 10^8 \text{ m}^{-1}$$

and

$$N_{LO} = \frac{10^{-20}}{(9\pi)(10^{-3})(5.7 \times 10^{-6})^2(2 \times 10^{-3})} = 5.4429 \times 10^{-6}$$

$$N_{E_1} = \frac{(80)(8.3542 \times 10^{-12})(3.4163 \times 10^8)}{(12)(\pi)(10^{-3})(2 \times 10^{-3})} \left[(1.26 \times 10^{-2})^2 + (1.06 \times 10^{-2})^2 \right] = 0.8695$$

$$N_{E_2} = \frac{(2)(12.6)(10.6)}{(13.6)^2 + (10.6)^2} = 0.9852$$

$$N_{DL} = (3.4163 \times 10^8)(5.7 \times 10^{-6}) = 1.9473 \times 10^3$$

The surface interaction force vs. separation is

$$\frac{F_{DL} + F_{LO}}{6\pi\mu a_p u_s} = \frac{5.4429 \times 10^{-6}}{\delta^{+2}(\delta^+ + 2)^2} + 0.8695 \left[0.9852 - e^{-1.9473 \times 10^3 \delta^+} \right] \left[\frac{e^{-1.9473 \times 10^3 \delta^+}}{1 - e^{-3.8946\delta^+}} \right]$$

The results with $M = 0.011\ M/\ell$ as well as those of the other cases are shown in the following figures. As shown in these figures, the normalized maxima of $F_{DL} + F_{LO}$ were 0.1425, 0.0268, -0.0714, and -0.1302, respectively. By inter-polation, it is found that repulsive force barrier disappears at ionic concentration between 0.0122 and 0.123 M/ℓ.

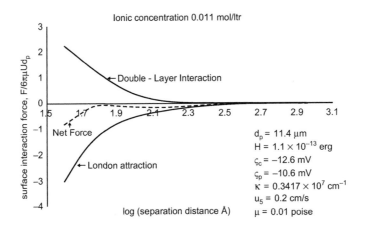

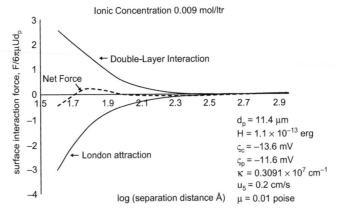

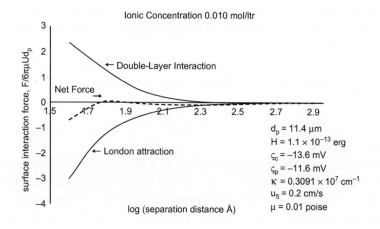

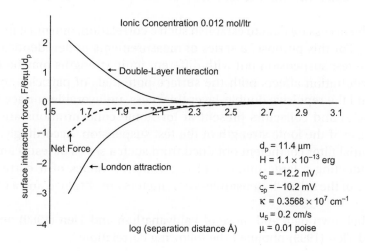

---

[12]Note that this definition differs from that of Equation (6.5.8a) by a factor of $(1/4\pi)$.

■ ■ ■

6.5.2   Initial Filter Coefficient with Unfavorable Surface Interactions

Intuitively, one may expect that the presence of unfavorable surface interactions between particles and collector act as barriers to particle transport leading to a reduction of the filter coefficient. Trajectory analyses based on relatively simple porous media models, however, have yielded results which grossly overestimate the degree of decrease. It predicts that with the onset of unfavorable surface interactions, filter coefficient decreases catastrophically, which is at variances with experiments. To overcome this difficulty, attempts to develop correlations based on experimental data, have been made. The result is given below.

To properly quantify the effect of unfavorable surface interactions on the filter coefficient, one may introduce a filter coefficient ratio, α, defined as

$$\alpha = \frac{\lambda_0}{(\lambda_0)_{\text{fav}}} \tag{6.5.6}$$

where λ_0 is the filter coefficient corresponding to a set of conditions with the presence of unfavorable surface interactions. $(\lambda_0)_{\text{fav}}$ is the value of the filter coefficient corresponding to the same condition except the presence of unfavorable surface interactions. By identifying the relevant dimensionless parameter, which may affect α and with sufficient data, a correlation of α in terms of the relevant parameters may be established and used to predict the filter coefficient in the presence of unfavorable surface interactions.

To obtain the necessary data to establish such a correlation, one must first obtain the value of $(\lambda_0)_{\text{fav}}$. For this purpose, a series of measurements under identical conditions using the same test suspension but with different ionic strengths may be made. Since the ionic concentration affects both the surface potentials of particles and collectors (filter grain) and the repulsive double-layer thickness, unfavorable surface interactions between particles and collectors present at low ionic concentrations are suppressed with the increase of the ionic strength of the test suspension. Accordingly, the highest value of the initial filter coefficient obtained from such a series of tests can be taken as $(\lambda_0)_{\text{fav}}$. By conducting a large number of experiments, values of α corresponding to different values of the select dimensionless parameters are obtained and a correlation of α established.

Based on their own data and those of Vaidyanathan and Tien (1989) and Elimelech (1992), Bai and Tien (1999) obtained the following correlation:

$$\alpha = 2.0354 \times 10^{-3} \, N_{\text{LO}}^{0.7031} \, N_{\text{E1}}^{-0.3132} \, N_{\text{E2}}^{3.5111} \, N_{\text{DL}}^{1.6641} \tag{6.5.7}$$

N_{LO} is the same as given before [see Equation (6.4.2)]. The other dimensionless parameters are defined as

$$N_{\text{E1}} = \hat{\varepsilon}_0 \varepsilon_{\text{r}} (\zeta_{\text{c}}^2 + \zeta_{\text{p}}^2)/(12 \, \pi \, \mu \, u_{\text{s}}) \tag{6.5.8a}[13]$$

$$N_{\text{E2}} = 2\zeta_{\text{p}}\zeta_{\text{c}}/(\zeta_{\text{p}}^2 + \zeta_{\text{c}}^2) \tag{6.5.8b}$$

$$N_{\text{DL}} = \kappa \, a_{\text{p}} \tag{6.5.8c}$$

---

[13] N_{E_1} as defined by Equation (6.5.8a) differ from that given in Illustrative Example 6.9 by a factor of 4π.

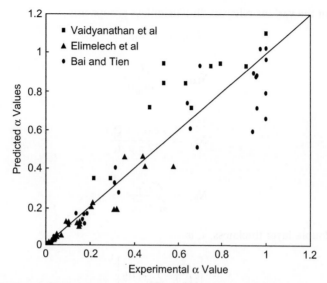

FIGURE 6.13 Predicted Filter Coefficient Ratio vs. Experimental Values (Bai and Tien, 1999).
Reprint from R. Bai and C. Tien, "Particle deposition under unfavorable surface interactions", J. Colloid Interface Sci.,
218, 488–499, 1999, with permission of Elsevier.

A comparison between Equation (6.4.11) with experiments is shown in Fig. 6.13.

■ ■ ■ ▬▬▬▬▬▬▬▬▬▬▬▬▬▬▬▬

Illustrative Example 6.13

Estimate the initial filter coefficient for the filtration of aqueous suspension of particles of
$d_p = 3.063$ μm under the following conditions (unfavorable).

$u_s = 3.7$ m/hr,	$T = 25\,°C$	$\mu = 10^{-3}\,Pa-s$	$H = 1.2 \times 10^{-20}$ J
Ionic concentration	(a) $I = 0.001\,M/\ell (NaC\ell)$		
	(b) $I = 0.03\,M/\ell\ (NaC\ell)$		
Surface potential (a)	$\zeta = -23$ mv;	$\zeta_g = -11$ mv;	
(b)	$\zeta = -10$ mv;	$\zeta_g = -5$ mv;	

The initial filter coefficient of the suspension with $I = 0.2\,M/\ell$, corresponding to favorable
surface interactions, was found to be 8.25 m^{-1}.

Solution

The correlation of Equation (6.5.7) may be used to estimate the filter coefficient ratio,
$\alpha = \lambda_0/(\lambda_0)_F$

$$\alpha = 2.0354 \times 10^{-3}\ N_{LO}^{0.7031} N_E^{-0.3132} N_{E_2}^{3.5111} N_{DL}^{1.6641}$$

The definitions of the various relevant dimensionless groups are

$$N_{DL} = \kappa \cdot a_p$$

$$N_{LO} = \frac{H}{9\pi\mu a_p^2 u_s}$$

$$N_{E_1} = \frac{\hat{\varepsilon}\kappa(\zeta_c^2 + \zeta_p^2)}{12\pi\mu u_s}$$

$$N_{E_2} = \frac{2\zeta_c\zeta_p}{\zeta_c^2 + \zeta_p^2}$$

The reciprocal double-layer thickness, κ, is

$$\kappa = \left[\frac{e^2}{\hat{\varepsilon}k_B T}\sum_i z_i^2 m_i\right]^{1/2}$$

(a) For $I = 0.001\,M$, with 1:1 electrolyte, and with values of e, $\hat{\varepsilon}$ and k_B given before

$$\kappa = \left[\frac{(1.6021\times 10^{-19})^2}{(80\times 8.8542\times 10^{-12})(1.3805\times 10^{-23})\times 298}(10^3)2(0.001)\right]^{1/2} = 1.03\times 10^8\ m^{-1}$$

The various dimensionless groups are

$$N_{DL} = (1.03\times 10^8)(1.5315\times 10^{-6}) = 1.5774\times 10^2$$

$$N_{LO} = \frac{1.2\times 10^{-20}}{9\pi(10^{-3})(1.0278\times 10^{-3})} = 1.7605\times 10^{-4}$$

$$N_{E_1} = \frac{(80\times 8.8542\times 10^{-12})(1.03\times 10^8)}{(12)\pi(10^{-3})(1.0278\times 10^{-3})}[(0.023)^2 + (0.011)^2] = 1.2239$$

$$N_{E_2} = \frac{(2)(0.023)(0.011)}{(0.023)^2 + (0.011)^2} = 0.7785$$

and

$$\alpha = 2.0345\times 10^{-3}(1.7605\times 10^{-4})^{0.7031}(1.2239)^{-0.3132}$$
$$\times (0.7785)^{3.5111}(1.5774\times 10^2)^{1.6641} = 0.0083$$

with $(\lambda_0)_f = 8.25, (\lambda_0) = (8.25)(0.0083) = 0.068 \text{ m}^{-1}$ which is less than 40% of the experimental value (0.1865 m^{-1}) of Bai and Tien.[14]

(b) For $I = 0.03 \, M/\ell$,

$$\kappa = 5.6418 \times 10^8 \text{ m}^{-1}$$

The various dimensionless parameters are:

$$N_{DL} = 8.6404 \times 10^2$$

$$N_{LO} = 1.7605 \times 10^{-4}$$

$$N_{E_1} = 1.2892$$

$$N_{E_2} = 0.8$$

The filter coefficient ratio is

$$\alpha = 0.1516$$

and $\lambda_0 = (8.25)(0.1516) = 1.25 \text{ m}^{-1}$ as compared with the experimental values of 2.5732 m^{-1} reported by Bai and Tien.

---

[14]See Bai and Tien, J. Colloid Interface Sci., 218, 448 (1999), Table 3.

■ ■ ■

Problems

6.1. Applying Equation (6.3.1), estimate the single fiber efficiency of aerosol filtration due to inertial impaction for the following conditions:

$d_p = 1.95 \ \mu\text{m}$
$d_f = 25 \ \mu\text{m}$
$u_s = 1 \ \text{m s}^{-1}$
$\rho_p = 1.2 \ \text{gram cm}^{-3}$
$\varepsilon_s = 0.01$
$T = 298 \text{ K}, \ p = 1 \text{ atm}$
Air may be treated as an ideal gas.

6.2. For the conditions given in Problem 6.1, do interception and/or diffusion contribute significantly to particle collection? Use Equations (6.1.20), (6.1.29) or (6.3.4) for your calculations.

6.3. Based on Equation (6.3.4), obtain an expression for the size of the most penetrating particle.

6.4. Tufenkji and Elimelech (Environ. Sci. Tech. 2004, 38, 529–536) presented a correlation similar to Equation (6.4.5) for the collector efficiency of granular media under favorable surface interaction:

$$(\eta_s)_0 = 2.4\, A_s^{1/3} N_R^{-0.081} N_{Pe}^{0.715} N_{vdw}^{0.052} + 0.55\, A_s N_R^{1.675} N_A^{0.125}$$
$$+ 0.22\, N_R^{-0.24}\, N_G^{1.11}\, N_{vdw}^{0.053}$$

where

$$N_A = \frac{H}{3\pi\mu d_p^2 u_s} \tag{ii}$$

$$N_{vdw} = \frac{H}{kT}$$

and the definition of the other quantities present in Equation (i) are the same as those used in the present text.

(a) Express the results as the initial filter coefficient in terms of the parameters used in the text.

(b) Estimate the value of λ_0 for the condition of Illustrative Example 6.10.

6.5. Assuming that deposition is due to the combined effect of interception and the Brownian diffusion, based on Equation (6.1.18) and (6.1.28), estimate the filter coefficient of hydrosol filtration of particles of $d_p = 1, 2, 5, 10, 30\ \mu m$ under the following conditions

$d_g = 350\ \mu m$
$\varepsilon = 0.4$
$u_s = 4\ cm\ s^{-1}$
$\rho = 1000\ Kg\ m^{-3}$

6.6. The following aerosol granular filtration results were obtained

Time (min)	$c_{in}(nos/cm^3)$	$c_{eff}(nos/cm^3)$
1.25	1554	1524
40	1786	1465
70	1500	907
100	1667	685
130	1675	448
160	1471	161
190	1737	93
220	1890	68

The data were obtained under the following conditions:

Filter grain diameter	$d_g = 505 \ \mu m$
Filter height	$L = 0.84$ cm
Particle size	$d_p = 2.02 \ \mu m$
Air velocity	$u_s = 11.3$ cm s^{-1}
Filter porosity	$\varepsilon = 0.35$

(a) Calculate the initial collector efficiency η_0.
(b) Assuming the deposition is uniform throughout the filter on the basis that the filter bed used is shallow, obtain a relationship on the increase of the collector efficiency (expressed as η/η_0) as a function of the specific deposit, σ.

Particle Size	Fraction (number)
< 1 μm	-
1–2 μm	0.10
2–3 μm	0.30
3–4 μm	0.58
4–5 μm	0.02
5 μm >	-

6.7. For an aqueous suspension to be treated by granular filtration, the size distribution of the suspended particles is
    The total concentration is 100 ppm (by volume) or 10^{-4} vol/vol.
(a) What is the size distribution based on volume?
(b) Estimate the initial filter coefficient in terms of filter coefficients of specific sizes based on total number of particles removed.

6.8. Calculate the reciprocal double-layer thickness, κ for aqueous solutions of 0.01 M 2:1 electrolyte. ε_r for water using may be taken as 80. (Note: $\hat{\varepsilon} = \varepsilon_r.\varepsilon_0$)

6.9. Obtain expressions of the double-layer force and London–van der Waals force for short separation distances. Establish conditions under which the London–van der Waals force is dominant.

6.10. What are the conditions under which the approximate expression of the double-layer force given by Hogg et al. (Trans. Faraday Soc., 62, 1638, 1966) is valid?

References

Bai, R., Tien, C., 1999. J. Colloid Interface Sci. 218, 488.

Bai, R., Tien, C., 2000. Colloids and Surfaces, A 165 95.

Brown, R.C., 1993. Air Filtration: An Integrated Approach to the Theory and Applications of Fibrous Filters. Pergamon Press.

Chiang, H.W., Tien, C., 1985a. AIChE J. 31, 1349.

Chiang, H.W., Tien, C., 1985b. AIChE J 31, 1360.

Choo, C.-U., Tien, C., 1992. Separations Technology 1, 122.

Choo, C.-U., Tien, C., 1995. J. Colloid Interface Sci. 169, 13.

Cushing, R.S., Lawler, D.F., 1998. Environ. Sci. Technol. 32, 2793.

Dahneke, B., 1971. J. Colloid Interface Sci. 37, 347.

Dahneke, B., 1995. Aerosol Sci. Technol. 23, 25.

Davis, C.N., 1973. Air Filtration. Academic Press.

Dawson, S.V., 1969. Theory of Collection of Airborne Particles by Fibrous Filters. Sc.D. Thesis. Harvard School of Public Health.

Derjaguin, B.V., Landau, L.O., 1941. Acta. Physica. Chemico (USSR) 14, 631.

D'Ottavio, T., Goren, S.L., 1983. Aerosol Sci. Technol. 2, 91.

Elimelech, M., 1992. Water Res. 26, 1.

Emi, H., Wang, C.-S., Tien, C., 1982. AIChE J. 28, 397.

Gimbel, R., Sontheimer, H., 1978. "Recent Results of Particle Deposition in Sand Filters" in Deposition and Filtration from Gases and Liquids, The Society of Chemical Industry (London).

Goldman, A.J., Cox, R.G., Brenner, H., 1967. Chem. Eng. Sci. 22, 697.

Guzy, O.J., Bonado, E.J., Davis, E.J., 1983. J., Colloid Interface Sci. 95, 523.

Harrop, J.A., 1969. The Effect of Fibre Configuration on the Efficiency of Aerosol Filtration, Thesis Loughborough University of Technology.

Harrop, J.A., Stenhouse, J.I.T., 1969. Chem. Eng. Sci. 24, 1475.

Hogg, R., Healy, T.W., Fuerstenau, D.W., 1966. Trans. Faraday Soc. 62, 1638.

Jung, Y.-W., Walata, S.A., Tien, C., 1989. Aerosol Sci. Technol. 11, 168.

Jung, Y.-W., 1991. Granular Filtration of Monodispersed and Polydispersed Aerosols. PhD Dissertation. Syracuse Univ., Syracuse.

Jung, Y.-W., Tien, C., 1991. J. Aerosol Sci. 22, 187.

Kraemer, H.F., Johnstone, H.F., 1955. Ind. Eng. Chem. 47, 2426.

Lee, K.W., Liu, B.Y.H., 1982. Aerosol Sci. Technol. 1, 147.

Liu, B.Y.H., Rubow, K.L., 1990. Efficiency, Pressure Drop and Figure of Merit of High Efficiency Fibrous and Membrane Filter Media, in 5th World Filtration Congress, Nice, 3, 112.

Lundgren, D.A., Whitby, K.T., 1965. Ind. Eng. Chem. Proc. Des. Dev. 4, 345.

Millikan, R.A., 1923. Phys. Rev. Series. 22, 22, 1.

Myojo, T., Kanaoka, C., Emi, H., 1984. J. Aerosol Sci. 15, 483.

Nielsen, K.A., Hill, J.C., 1976. Ind. Eng. Chem. Fundam., 15, 149.

Nguyen, X., Beeckmans, J.M., 1975. J. Aerosol Sci. 6, 205.

Payatakes, A.C., Turian, R.M. and Tien, C., 1975. Integration of Filtration Equations and Parameter Optimization Techniques, In: Proc. 2nd World Congress of Water Resources, vol. 241.

Payet, S., 1992. Filtration Stationnaire et dynamique des aerosols liquids submicroniques, Thesi de l'Universite Paris XII Rapport CEA-R-5589 1992, as quoted in J. Aerosol Sci. 23, 723.

Pfeffer, R., Happel, J., 1964. AIChE J. 10, 605.

Rajagopalan, R., Tien, C., 1976. AIChE J. 22, 523.

Stairmand, C.J., 1950. Trans. Instn. Chem. Engrs. 28, 1380.

Stechkina, I.B., Kirsch, A.A., Fuchs, N.A., 1970. Kelloidnyi Zhurnal 32, 467.

Stechkina, I.B., Kirsch, A.A., Fuchs, N.A., 1969. Ann. Occ. Hyg. 12, 1.

Strauss, W., 1966. Industrial Gas Cleaning. Permamon Press.

Tardos, G.I., Abuaf, N., Gutfinger, C., 1978. J. Air Pollution Control Association 28, 354.

Tufenkji, N., Elimelech, M., 2004. Environ. Sci. Technol. 38, 529.

Tien, C., Ramarao, B.V., 2007. Granular Filtration of Aerosols and Hydrosols, second ed. Elsevier.

Vaidyanathan, R., 1986. Hydrosol Filtration in Granular Beds, M.Sc. Thesis, Syracuse University.

Vaidyanathan, R., Tien, C., 1989. Chem. Eng. Comm. 81, 123.

Verwey, E.J.W., Overbeek, J.Th.G., 1945. Theory of Stabillity of Lyophobic Coilloids. Elsevier.

Walata, S.A., Takahashi, T., Tien, C., 1986. Aerosol Sci. Technol. 5, 23.

Wang, H.-C., 1986. J. Aerosol Sci. 17, 827.

Wang, H.-C., Kasper, G., 1991. J. Aerosol Sci. 22, 31.

Yoshida, H., Tien, C., 1985. AIChE J. 31, 1752.

Yoshimura, Y., 1980. Initial, Particle Collection Mechanism in Clean Deep Bed Filtration. D. Eng. Dissertation. Kyoto Univ.

7

Deep Bed Filtration Models

Notation

A	constant defined by Equation (7.2.18) (–)
a_g	filter grain radium (m)
a_p	particle radius (m)
a_s	specific surface area per unit grain mass ($m^2\,kg^{-1}$)
c	suspension particle concentration (vol/vol)
$\tilde{c}$	suspension particle concentration (in number) ($nos\,m^{-3}$)
c_{in}	influent particle concentration (vol/vol)
c_i	effluent concentration of the i-th segment (vol/vol)
c_s	number of deposited particles per unit grain surface area (m^{-2})
$c_{s,sat}$	maximum value of c_s defined by Equation (7.4.27)
c^+	dimensionless suspension particle concentration defined by Equation (7.4.18a)
c_s^+	dimensionless c_s defined by Equation (7.4.18b)
D_{BM}	Brownian diffusivity ($m^2\,s^{-1}$)
d_g	filter grain diameter (m)
d_{go}	initial value of d_g (m)
d_p	particle diameter (m)
$\tilde{d_c}$	constriction diameter (m)
E	maximum of grain–particle surface interaction potential
$E[N(t)]$	expected value of N at time t
F	filter coefficient ratio, equal to λ/λ_o
f	fraction of blocked pore constrictions or fraction of grain surface covered with deposited particles defined by Equation (7.2.17)
f_i	value of f of the i-th segment
f_r^t	a hydrodynamic retardation factor present in Equation (7.4.11)
$q(s,t)$	probability generating function defined by Equation (7.5.13)
$I_0(x)$	modified Bessel function of zeroth order with argument x
I^0, I_s^0	particle flux over a filter grain due to diffusion or surface reaction defined by Equations (7.4.10a) or (7.4.10b), respectively
J	particle flux over a single filter grain (s^{-1})
$J(\alpha,\beta)$	function defined by Equation (7.4.24)
K	virtual rate constant defined by Equation (7.4.11)
k	Boltzmann's constant
k_1	constant in Kozeny–Carman equation [Equation (7.2.1)] or a factor to account for the exclusion effect [see Equation (7.4.36)]
k_f	forward reaction rate constant of Equation (7.4.14) (s^{-1})
k_r	reverse reaction rate constant of Equation (7.4.14) (s^{-1})
L	filter height (m)
ℓ	length of periodicity
ℓ_d, ℓ_f	dendrite length and fiber length defined by Equations (7.3.27b) and (7.3.27a), respectively
N	number of filter grain present in a unit filter medium volume or number of particles deposited directly on filter grain
N_c	number of constricted tubes per unit bed element

Principles of Filtration, DOI: 10.1016/B978-0-444-56366-8.00007-4

N_t	total number of particles attached to a filter grain
N_{max}	maximum number of particles deposited on a filter grain
$\overline{N}$	spatial average of N_t defined by Equation (7.3.16)
n	number of open pores (see Equation (7.5.6))
n_0	initial value of n
$P_n(t)$	probability of n pores being blocked
p	pressure (Pa)
Q	filter grain change density necessary to give an unfavorable grain–particle surface interaction (Coloumb)
q_g	electric charge of a filter grain, (Coloumb)
q_p	electric charge of a single particle, (Coloumb)
$\hat{q}_g$	electric charge density of filter grain (Coulomb m^{-2})
$\hat{q}_p$	electric charge density of particle, (Coulomb m^{-2})
q'_g	equal to $q_g/4\pi a_g^2$ (Coloumb/m^2)
r_f	fiber radius, (m)
r_p	particle radius m
S	specific surface area defined as s_p/v_p (m^{-1}) or the cross sectional area for suspension flow m^2
S_0	initial value of S (m^{-1})
S_{w_i}	fraction of irreducible saturation
s	variable of g(s,t)
s_p	surface area of a single filter grain (m^2)
t	time s
u_s	suspension superficial velocity (m s^{-1})
$(u_s)_e$	effective superficial velocity defined by Equation (7.2.11) (m s^{-1})
v_p	volume of a single filter grain (m^3)
z	axial distance (m)
z^+	dimensionless axial distance defined by Equation (7.4.18d)

Greek Letters

α	packing density of fibrous medium (–) or coefficient of Equation (7.5.6)
α_1	a factor to account for the excluded area effect [see Equation (7.3.2.)]
β	an adjustable parameter introduced to estimate the fraction of blocked constriction [see Equation (7.2.16)] or coefficient of Equation (7.5.7)
β'	an arbitrary parameter introduced to estimate pressure drop increase [see Equation (7.3.15)]
γ	equal to $\varphi''(\delta_m^+)$ [see Equation (7.4.11)]
δ	separation distance (m)
δ^+	dimensionless separation distance defined as δ/a_p
δ_m^+	value of δ^+ which gives ϕ, the surface interaction energy patched its maximum
Δd_g	grain diameter increment, (m)
Δp	pressure drop (Pa)
Δp_0	initial value of Δp Pa
$\Delta\theta$	increment of θ s
ε	filter medium porosity
$\hat{\varepsilon}$	permittivity of a medium, equal to $\varepsilon_r\hat{\varepsilon}_0$ where ε_r is the dielectric constant (–) and $\hat{\varepsilon}_0$ permittivity of vacuum (= 0.854 × 10^{12} kg^{-1} m^{-3}A^{4}A^2)
ε_d	porosity of deposit
ε_{tran}	defined by Equation (7.2.29)
ε_{ult}	value of ε corresponding to σ_{ult}

ε_0	initial value of ε
ε^*	defined by Equation (7.3.6)
η_s	single fiber efficiency
η_{s_0}	initial value of η_s
η_{sp}	single collector efficiency of a deposited particle
η_{f_s}	single fiber efficiency of fibers
η_{p_s}	equivalent single fiber efficiency of particle dendrite
θ	corrected time defined as $t - \int_0^z \dfrac{\varepsilon dz}{u_s}$
θ^+	dimensionless θ defined by Equation (7.4.18c)
κ	reciprocal of the double-layer thickness (m^{-1})
λ	filter coefficient (m^{-1})
λ_i	filter coefficient of the i-th segment
λ_0	initial filter coefficient (m^{-1})
λ_n	a quantity defined by Equation (7.5.6)
λ^1	filter coefficient for type 1 surface (m^{-1})
λ^2	filter coefficient for type 2 surface (m^{-1})
μ	fluid viscosity Pa . s
μ_n	a quantity defined by Equation (7.5.7)
ρ_g	filter grain density $(kg\ m^{-3})$
σ	specific deposit (vol/vol)
σ_d	specific deposit due to deposition on particle dendrite
σ_f	specific deposit due to deposition on fibers
σ_i	specific deposit of the i-th segment
σ_{tran}	value of σ corresponding to the transition from 1st stage to 2nd stage of deposition
σ_{ult}	ultimate (maximum) value of σ
σ^1	specific deposit due to deposition on type 1 surface
σ^2	specific deposit due to deposition on type 2 surface
ϕ	grain–particle interaction energy potential
ϕ_{Lo}	grain–particle London–van der Waals energy potential
ϕ_{DL}	grain–particle double-layer energy potential
ϕ_s	grain sphericity factor, –
ϕ_{s_0}	intial value of ϕ_s

The practical motivation of deep bed filtration research is to develop a body of knowledge and information, which can be used as a basis for the design, operation, and control of granular and fibrous filtration systems. In the preceding sections, we have presented a macroscopic description of deep bed filtration and discussed, in some detail, the significance of the parameters used in the description (see Chapter 5). In addition, analyses and experiments aimed at estimating and determining these parameters and the results obtained are described and summarized (see Chapter 6).

In principle, models describing the dynamic behavior and performance of deep bed filters can be formulated by combining the information presented in Chapters 5 and 6. This is, in fact, the common approach used by many earlier investigators. A major problem with these earlier studies is their omission of the effect of deposition on media structures as manifested in the increase in the required pressure drop for constant rate operation or the reduction of the filtration rate for constant pressure operation. Furthermore, the macroscopic description given in Chapter 5 may not apply to all situations (for example, if there is significant re-entrainment of deposited particles) and the correlations presented in Chapter 6 does not cover all cases or with insufficient accuracy for the estimation of the effluent concentration history. As a result, extraneous assumptions are often introduced in model formulations.

The present chapter is intended as an introduction of deep bed filtration modeling and as a complement of the materials presented in 6.4. The basic principles used in formulating models are aimed at predicting filter performance including pressure-drop increase for constant rate operation and incorporating the effects due to changes of filter grain surface geometry and characteristics. Applications of these principles are discussed and illustrated through examples.

7.1   Experimental Results of Filtration Performance

For the purpose of demonstrating the complexities of the dynamics of deep bed filtration, some typical aerosol/hydrosol filtration results are shown in Fig. 7.1. Fig. 7.1(a) gives the turbid water filtration data obtained by Deb (1969), which include the suspension particle concentration as a function of time and position and the total pressure drop across filter bed vs. time. The pressure drop was shown to be a monotonic increasing function of time. However, the suspension particle concentration history does not follow any simple pattern. In fact, it decreases with time and then increases. Furthermore, the extent of change decreases with the increase of filter bed height.

A different kind of behavior on the evolution of suspension particle concentration is shown in Fig. 7.1(b). The data were those of filtration of destabilized aqueous solutions of latex particles in shallow beds (Tobiason et al., 1993). The pressure drop vs. time results is similar to those of Deb's. Unlike Fig. 7.1(a), the suspension particle concentration was found to decrease with time except initially. Whether this rather anomalous initial behavior has certain physical significance or is merely an artifact cannot be ascertained at this time.

Two other sets of hydrosol filtration data shown in Fig. 7.1(c) and (d) display different behavior. The experimental data of Bai and Tien (1999) show that the effluent particle concentration either remains essentially constant for up to 120 min or increases with time, depending upon the electrolyte concentration of the feed stream. Since both the latex particles and filter grains used in these experiments were negatively charged, and the repulsive double-layer force barriers can be suppressed with the presence of electrolytes, the observed phenomenon confirms that unfavorable surface interaction plays an important role in determining the dynamics of deep bed filtration.

The results shown in Fig. 7.1(d) further confirm the importance of the surface interaction effect. The data were those of Liu et al. (1995). The effluent particle concentration history was shown to be a strong function of the electrolyte concentrations of the feed streams used in experiments.

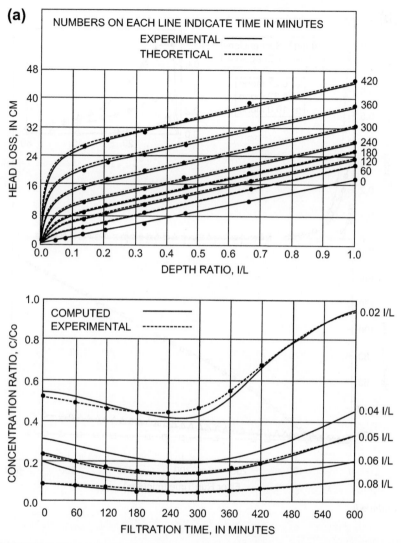

FIGURE 7.1 (a) Water Filtration Data obtained by Deb (1969). *Reprinted from A.K. Deb, "Theory of sand filtration", J. of the sanitary engineering division of the American Society of Civil Engineers, 95 (1969), 399–422, with permission from ASCE.* **(b)** Particle removal and head loss Results. Filtration of destabilized aqueous suspension (Tobiason et al. 1993)($\mu s = 4.8$ m/hr, Bed Height 17 cm, 0.4 mm media). *Reprinted from J.E. Tobiason, G.S. Johnson, P.K. Westerhoft and B. Vigneswaran, "Particle size and chemical effects on contact filtration performance", J. Environmental Engineering 119 (1993), 520-539, with permission from ASCE.* **(c)** Effect of Electrolyte concentration on filtration Performance (Bai & Tien 1999). *Reprinted from J. Colloid Interface Science, 218, R. Bai and C. Tien, "Particle deposition under unfavorable surface interaction, 488–499, 1999, with permission of Elsevier.* **(d)** Filtration data of Liu et al. (1995); effect of electrolyte concentration. *Reprinted with permission from D. Liu, P.R. Johnson and M. Elimelech, "Colloid deposition dynamics in flow through porous media: Role of electrolyte concentration", Environmental Sci. Technol., 29, 2963–2973, 1955, American Chemical Society.* **(e)** Dynamic behavior of Aerosol Filtration in Fibrous Media (1) Regular Media (2) Charged Media. *Reprinted from Filtration and Separation, D.C. Walsh, "Recent understanding of fibrous filtration behavior under solid particle load", 501–506, June 1996, with permission of Elsevier.*

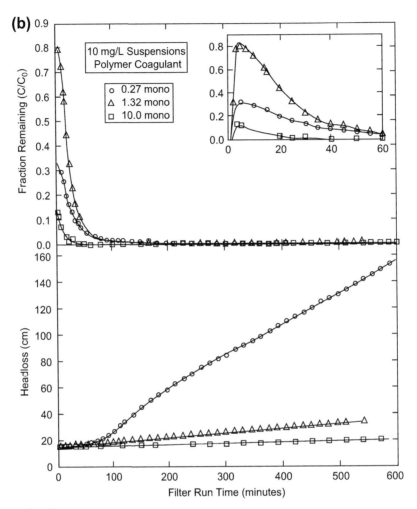

FIGURE 7.1 (continued).

The fibrous aerosol filtration data of Fig. 7.1(e), however, exhibit more complex behavior. The data presented by Walsh (1996) are of two types: those obtained using regular fibrous media and those with electrically charged media. For the former, similar to the results shown in Fig. 7.1(b), penetration (defined as c_{eff}/c_{in}) was found to decrease with time while the pressure drop was found to increase. With charged media, penetration first increases with time and then decreases while pressure drop is a monotonic increasing function of time.

The above examples show rather convincingly the complexed nature of the deep bed filtration process. Practically speaking, it is clear that for model formulations, the knowledge of the expected consequence of deposition (namely, its effect on subsequent

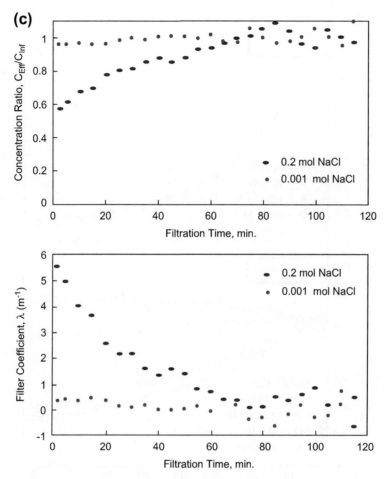

FIGURE 7.1 *(continued)*.

deposition or the dynamic behavior of the filtration process) is required or is often assumed. It is on this premise that we will present a number of models in the following sections.

7.2   Models Based on the Kozeny–Carman Equation

For fluid flow through porous media, such as granular filter beds, the pressure gradient–flow rate relationship according to the Kozeny–Carman equation [i.e., Equation (5.5.1)] is

$$-\frac{\partial p}{\partial z} = k_1 \frac{(1-\varepsilon)^2}{\varepsilon^3} \frac{\mu u_s}{[\phi_s d_g]^2} \tag{7.2.1}$$

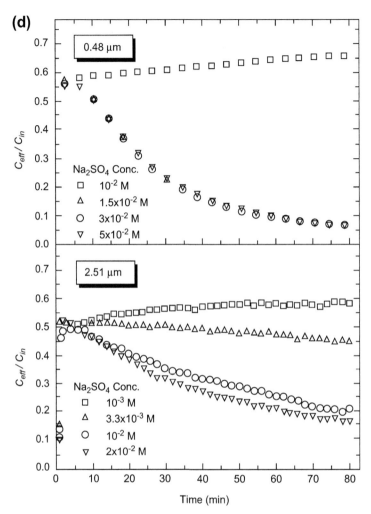

FIGURE 7.1 (continued).

Among the variables present on the right-hand side of the above equation, the bed porosity, ε, the filter grain diameter, d_g, and its sphericity factor, ϕ_s, as well as the effective superficial velocity, u_s, may change as a result of deposition. Assuming that Equation (7.2.1) is valid for media with or without deposited particles, the change of the required pressure gradient for constant rate filtration, namely, $u_s S = $ constant where S is the filter bed cross-sectional area available to suspension flow, may be written as

$$\frac{-\partial p/\partial z}{(-\partial p/\partial z)_0} = \frac{(1-\varepsilon)^2}{(1-\varepsilon_0)^2}\left(\frac{\varepsilon_0}{\varepsilon}\right)^3\left(\frac{d_{g_0}}{d_g}\right)^2\left(\frac{\phi_{s_0}}{\phi_s}\right)^2\left(\frac{S_0}{S}\right)^2 \qquad (7.2.2)$$

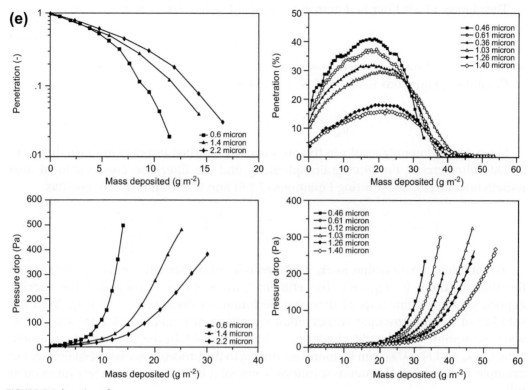

FIGURE 7.1 (*continued*).

where the subscript "0" refers to the initial state (free of deposited particles). The change of the pressure drop, $-\Delta p$, is

$$\frac{-\Delta p}{(-\Delta p)_0} = \frac{1}{L}\int_0^L \left(\frac{1-\varepsilon}{1-\varepsilon_0}\right)^2 \left(\frac{\varepsilon_0}{\varepsilon}\right)^3 \left(\frac{d_{g_0}}{d_g}\right)^2 \left(\frac{\phi_{s_0}}{\phi_s}\right)^2 \left(\frac{S_0}{S}\right)^2 \cdot dz \tag{7.2.3}$$

In other words, if the changes of these variables can be quantified in terms of the extent of deposition (i.e., σ, the specific deposit) and the specific deposit profile across the filter bed, σ vs. z is known, $(-\Delta p)$ can be, in principle, readily estimated according to Equation (7.2.3).

7.2.1 Uniform Deposit Layer Hypothesis

If one assumes that deposition results in the formation and growth of uniform deposit outside filter grains such that the filter grain diameter increases from d_{g_0} to d_g corresponding to a specific value of σ, by definition,

$$\frac{d_g}{d_{g_0}} = \left(\frac{1-\varepsilon}{1-\varepsilon_0}\right)^{1/3} \tag{7.2.4}$$

The change of the filter bed porosity, on the other hand, can be expressed as

$$\varepsilon = \varepsilon_0 - \frac{\sigma}{1 - \varepsilon_d} \tag{7.2.5}$$

where ε_d is the porosity of the deposit layer.

Combining Equations (7.2.4) and (7.2.5) yields

$$\frac{d_g}{d_{g_0}} = \left[1 + \frac{\sigma}{(1 - \varepsilon_0)(1 - \varepsilon_d)}\right]^{1/3} \tag{7.2.6}$$

Since deposition only results in the presence of deposited layers of uniform thickness outside filter grains, the filter grain sphericity and the filter bed cross-sectional area remain unchanged. Substituting Equations (7.2.5) and (7.2.6) into (7.2.3), one has

$$(-\Delta p) = (-\Delta p)_0 \int_0^L \left[1 + \frac{\sigma}{(1 - \varepsilon_0)(1 - \varepsilon_d)}\right]^{4/3} \left[1 - \frac{\sigma}{\varepsilon_0(1 - \varepsilon_0)}\right]^{-3} \cdot dz \tag{7.2.7}$$

To complete the modeling work, the knowledge of the specific deposit profile across the filter medium is required. The effluent particle concentration and the specific deposit profile as functions of time can be found, as discussed previously, from the solution of the macroscopic conservation equations of Equations (5.1.4b) and (5.2.4) with appropriate initial and boundary conditions. To obtain the solution, the required values of λ_0 and $F(= \lambda / \lambda_0)$ can be found according to the procedures discussed before. For example, for filtration of aqueous solutions in granular beds, λ_0 and F can be estimated as follows:

1. Assuming favorable particle–filter grains surface interactions, λ_0 can be estimated according to Equation (6.4.5).
2. As an alternative, if the dominant mechanism of deposition is known, for λ_0 (or η_0), one may use any appropriate equations given in 6.1 such as Equation (6.1.18) (for interception) or (6.1.28) (for the Brownian diffusion).
3. If the suspensions are destabilized, F can be estimated according to Equation (6.4.6) or (6.4.7).

With λ_0 and F known, Equations (5.1.4b) and (5.2.4) may be solved using the method given in 5.7 if the filter bed is free of deposited particles initially. As an alternate, an approximate procedure given below may be applied. A filter bed of height L may be considered as a series of N segments of depth ΔL ($= L/N$) and the deposition within each segment uniform. If c_{i-1} and c_i denote, respectively, the influent and effluent particle concentration of the i-th segment, from Equation (5.2.1) with λ being constant, c_i may be expressed as

$$c_i = c_{i-1} \exp[-\lambda_i(\Delta L)] \tag{7.2.8}$$

where λ_i is the filter coefficient of the i-th segment and may be expressed as

$$\lambda_i = \lambda_0 F(\sigma_i) \tag{7.2.9}$$

with λ_0 being the initial (or clean) filter coefficient. F is the filter coefficient correction factor (see Equation (5.2.3a)). σ_i is the specific deposit of the *i*-th segment at a given time.

The specific deposit increases with time as filtration proceeds, by particle mass balance, one has

$$[\sigma_i(\theta + \Delta\theta) - \sigma_i(\theta)] \cdot \Delta L = u_s(c_{i-1} - c_i)\Delta\theta$$

or

$$\sigma_i(\theta + \Delta\theta) = \sigma_i(\theta) + c_{i-1}[1 - \exp(-\lambda_i\Delta L)](u_s)(\Delta\theta)/\Delta L \qquad (7.2.10)$$

Accordingly, with appropriate initial and boundary conditions and with λ_0 and F known, Equations (7.2.10) and (7.2.7) may be used to estimate the specific deposit profile, which, in turn, can be used to determine the pressure-drop increase according to Equation (7.2.7). Similarly, from Equation (7.2.8), the effluent concentration history ($c_N = c_{\text{eff}}$) can be determined.

■ ■ ■ ━━

Illustrative Example 7.1

Estimate the pressure-drop increase $(-\Delta p)/(-\Delta p)_0$ vs. time according to Equation (7.2.7) for water filtration carried out under the following conditions

Filter Height	$L = 0.5$ m
Filtration (Flow Rate)	$u_s = 2 \times 10^{-3}$ m s^{-1}
Medium Porosity (initial)	$\varepsilon_0 = 0.45$
Influent Particle Concentration	$c_{\text{in}} = 5 \times 10^{-5}$ vol/vol
Initial Filter Coefficient	$\lambda_0 = 20$ m^{-1}

It is assumed that the particle deposit porosity ε_d may be taken as 0.8. The effect of deposition on the filter coefficient is negligible.

Solution

If there is no deposition effect on λ, the specific deposit profile for a filter initially free of deposited particles, subject to a constant influent concentration c_{in}, can be found easily from the solution of Equations (5.1.4) and (5.2.2) as shown in Illustrative Example 5.2. For $\theta > 0$, σ is given as

$$\sigma = u_s\, c_{\text{in}}\lambda_0\theta \exp[-(\lambda_0 L)(z/L)]$$

Under the given conditions, σ is found to be

$$\sigma = (2 \times 10^{-3})(5 \times 10^{-5})(20)\theta \exp[-10z^*] \quad \text{with } z^* = z/L$$
$$= (2 \times 10^{-6})\theta \exp[-10z^*]$$

Substituting the above expressions into Equation (7.2.14), the pressure-drop ratio as a function of time is

$$\frac{(-\Delta p)}{(-\Delta p)_0} = \int_0^1 \left[1 + \frac{2 \times 10^{-6}\theta}{(0.55)(0.2)} \exp(-10z^*)\right]^{4/3} \left[1 - \frac{2 \times 10^{-6}\theta}{(0.45)(0.55)} \exp(-10z^*)\right]^{-3} dz$$

$$= \int_0^1 \frac{[1 + 1.82 \times 10^{-5}\theta \exp(-10z^*)]^{4/3}}{[1 - 8.08 \times 10^{-6}\theta \exp(-10z^*)]^3} dz^*$$

To simplify the computation,

Let $y = \exp(-10\,z^*)$

and $dz^* = -0.1 \exp(10\,z^*)\, dy = -0.1\, dy/y$

and $\dfrac{-\Delta p}{(-\Delta p)_0} = \displaystyle\int_{\exp(-10)}^1 \frac{(1 + 1.82 \times 10^{-5} \cdot \theta \cdot y)^{4/3} dy}{(1 - 8.08 \times 10^6 \theta y)^3 dy}$

The numerical results are shown in the following table and figure:

	$(-\Delta p)/(-\Delta p)_0$	
θ (s)	A	B
100	1.0049	1.00053
500	1.00249	1.00264
1,000	1.00509	1.00532
10,000	1.05593	1.06242
20,000	1.18838	1.14215
A	obtained using $\Delta y = 0.05$	
B	obtained using $\Delta y = 0.1$	

The results are shown in the following figures.

For $\Delta\theta = 0.1$

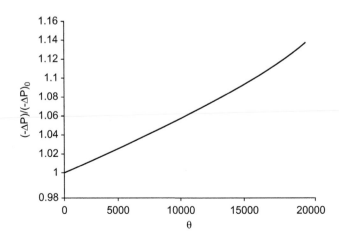

For $\Delta\theta = 0.05$

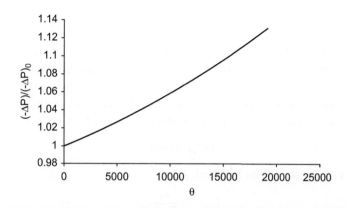

7.2.2   Pore-Blocking Hypothesis

Another hypothesis is to consider the effect of deposition in terms of the extent of the blockade of the pore constrictions. With this hypothesis, a filter medium undergoing filtration may be viewed to be consisting of two parts: a blocked part and an unblocked part, with the former no longer accessible to suspension flow while the latter functioning the same way as a fresh medium. In other words, all variables present in Equation (7.2.2) remain unchanged except S. S may be expressed as

$$S_0/S = \frac{u_{s_0}}{u_s} = \frac{1}{1-f} \tag{7.2.11}$$

where f is the fraction of blocked pore constrictions. The pressure gradient ratio becomes

$$\frac{(\partial p/\partial z)}{(\partial p/\partial z)_0} = \frac{1}{(1-f)} \tag{7.2.12}$$

The fraction of blocked constrictions can be estimated as follows: According to the constricted-tube model for granular media, for a unit bed element (UBE) with a periodicity of ℓ, the number of the constricted tubes (their pore constrictions) present in the UBE, free of deposited particle, N_c, is (Payatakes et al., 1973; Tien and Ramarao, 2007)

$$N_c = \frac{6\varepsilon_0^{1/3}(1 - S_{w_i})^{1/3}(1 - \varepsilon_0)^{2/3}}{\pi d_g^2} \tag{7.2.13}$$

where ε_0 is the initial medium porosity and S_{w_i} is the fraction of irreducible saturation. The length of periodicity, ℓ, is given by Equation (5.3.1a) or

$$\ell = \left[\frac{\pi}{6(1 - \varepsilon_0)}d_g^3\right]^{1/3} \tag{7.2.14}$$

And the pore constriction diameter $\tilde{d}_c$ may be assumed to be

$$\tilde{d}_c = \frac{d_g}{3} \tag{7.2.15}$$

A pore constriction can be blocked by a single particle if the size of the particle is greater than that of the constriction or by a particle aggregate composed of smaller particles. In general, the amount of particles necessary to block a constriction may be written as $\beta(\pi/6)d_p^3$, where β is an arbitrary parameter. The number of constrictions which are blocked corresponding to deposition of σ may be expressed as

$$\ell\sigma/\left[\beta(\pi/6)d_p^3\right]$$

and the fraction of constrictions blocked, f, is

$$f = \frac{\ell\sigma}{(\pi/6)d_p^3\beta N_c} \tag{7.2.16}$$

Substituting Equations (7.2.13) and (7.2.14) into (7.2.16), after rearrangement, one has

$$f = \frac{(\pi/6)^{1/3}\sigma(d_g/d_p)^3}{\beta(1 - S_{w_i})^{1/3}\varepsilon_0^{1/3}(1 - \varepsilon_0)} = A\sigma \tag{7.2.17}$$

with

$$A = \frac{(\pi/6)^{1/3}}{\beta(1 - S_{w_i})^{1/3}\varepsilon_0^{1/3}(1 - \varepsilon_0)(d_p/d_g)^3} \tag{7.2.18}$$

The pressure gradient ratio can then be written as

$$\frac{(\partial p/\partial z)}{(\partial p/\partial z)_0} = (1 - A\sigma)^{-1} \tag{7.2.19}$$

and the pressure-drop history is

$$-(\Delta p) = \frac{(-\Delta p)_0}{L} \int_0^L \frac{dz}{(1 - A\sigma)} \tag{7.2.20}$$

To obtain the pressure-drop history for the above equation, the specific deposit profile at various times is required and can be obtained from the macroscopic conservation and filtration rate equations corrected for the constriction blockade effect. An approximate procedure similar to that of the preceding section is given below for the purpose of illustration.

For a filter medium of depth L approximated as a series of N segments of height ΔL with $N = L/\Delta L$, following the discussion given above, the total number of pore constrictions present in each segment is $N_c(\Delta L/\ell)$, with N_c and ℓ being given by Equations (7.2.13) and (7.2.15). If the specific deposit of the i-th segment is σ_i, the number of blocked constrictions is $(\Delta L)(\sigma_i)/[\beta(\pi/6)d_g^3]$. The function of blocked constriction, f_i, is

$$f_i = \frac{(\Delta L)\sigma_i}{\beta(\pi/6)d_p^3(\Delta L/\ell)N_c} = \frac{\ell\sigma_i}{(\pi/6)d_p^3\beta N_c} = A\sigma_i \tag{7.2.21}$$

which is the same as Equation (7.2.16) except f and σ now referring to the value of the i-th segment. A is a constant given by Equation (7.2.15). The effluent particle concentration of the i-th segment may be estimated according to Equation (7.2.8) or

$$c_i = c_{i-1} \exp[-\lambda_i \Delta L] \qquad (7.2.8)$$

The increase of σ_i over a time increment of $\Delta\theta$ may be approximated as

$$[\sigma_i(\theta + \Delta\theta) - \sigma_i(\theta)](\Delta L)[1 - A\sigma_i(\theta)] = (c_{i-1} - c_i)(\Delta\theta)u_s$$

or

$$\sigma_i(\theta + \Delta\theta) = \sigma_i(\theta) + c_{i-1}\left[1 - e^{-\lambda_i \Delta L}\right] \frac{(\Delta\theta)u_s}{(\Delta L)[1 - A\sigma_i(\theta)]} \qquad (7.2.22)$$

where the values of c_{i-1} and λ_i may be taken as the values at $\theta = \theta$.

Equation (7.2.22) is similar to Equation (7.2.10) except the presence of the term $[1 - A\sigma_i(\theta)]$ which is introduced in order to account for the presence of the blocked part of the medium. The values of λ_i, as before, are given by Equation (7.2.9).

To calculate the pressure-drop increase, let the pressure drop across a segment be δP. δP_i denotes the pressure drop across the i-th segment. The pressure drop across the filter is

$$\Delta p = \sum_{i=1}^{N} (\delta P)_i \qquad (7.2.23)$$

and

$$(\Delta p)_0 = N(\delta P)_0 \qquad (7.2.24)$$

with $N = L/\Delta L$ as defined before.

From Equation (7.2.12), one has

$$\frac{(-\Delta p)}{(-\Delta p)_0} = \frac{1}{N} \sum_{i=1}^{N} \left(\frac{1}{1 - f_i}\right) \qquad (7.2.25)$$

■ ■ ■ ━━━━━━━━━━━

Illustrative Example 7.2

For the same operating conditions given in Illustrative Example 7.1 and with $d_p = 10\ \mu m$, $d_g = 500\ \mu m$, and $S_{w_i} = 0.1$, calculate the pressure-drop increase according to Equation (7.2.25) assuming $\beta = 10^3$ and 10^4.

Solution

The conditions given are

$L = 50$ cm,	$u_s = 2 \times 10^{-3}\ m\,s^{-1}$	$\varepsilon_0 = 0.45$
$c_{in} = 5 \times 10^{-5}$ vol/vol	$\lambda_o = 20\ m^{-1}$	$F = 1.0$
$d_p = 10\ \mu m$	$d_g = 500\ \mu m$	$S_{w_i} = 0.1$
$\beta = 10^3, 10^4$		

Assuming that the filter may be considered as a series of 10 segments with $\Delta L = 5$ cm or $N = 50/5 = 10$.

According to Equation (7.2.25), we have

$$\frac{(-\Delta p)}{(-\Delta p)_0} = \frac{1}{10} \sum_{i=1}^{10} \left(\frac{1}{1-f_i}\right)$$

f_1 can be found from Equations (7.2.21) and (7.2.17)

$$f_i = A\sigma_i$$

and

$$A = \frac{(\pi/6)^{1/3}(d_g/d_p)^3}{\beta \varepsilon_0^{1/3}(1-\varepsilon_0)^{2/3}(1-S_{w_i})^{1/3}} = \frac{(0.806)(50)^3}{\beta(0.45)^{1/3}(0.55)(0.9)^{1/3}} = 24.78 \quad \text{for } \beta = 10^4$$

and $= 247.8 \quad \text{for } \beta = 10^3$

To obtain the specific deposit of the various segments, from Equation (7.2.22), one has

$$\sigma_i(\theta + \Delta\theta) = \sigma_i(\theta) + c_{i-1}\left[1 - e^{-\lambda_i \Delta L}\right] \frac{\Delta\theta(u_s)}{(\Delta L)[1 - A\sigma_i(\theta)]}$$

The initial conditions are

$c_0 = c_{in} = 5 \times 10^{-5}$ (vol/vol)
$\sigma_i(0) = 0$
$\lambda_i = \lambda_0 = 20$ m^{-1}

Using $\Delta\theta = 50s$, values of σ_i vs. θ, f_i vs. θ and $\dfrac{-\Delta p}{(-\Delta p)_0}$ vs θ are obtained and are given in the following table.

The numerical results obtained are as follows:

θ (s)	$(-\Delta p)/(-\Delta p)_0$	
	$\beta = 10^3$	$\beta = 10^4$
0	1.000	1.000
300	1.016	1.001
600	1.035	1.003
900	1.062	1.004
1200	1.103	1.006
1500	1.184	1.008
1800	1.537	–
3000	–	1.016
6000	–	1.037
9000	–	1.067
12000	–	1.117
15000	–	1.260
16500	–	1.910
16800	–	8.194

The results are shown in the following figures:

$\beta = 10^4$; $A = 24.78$

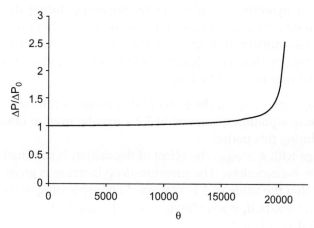

$\beta = 10^3$; $A = 247.8$

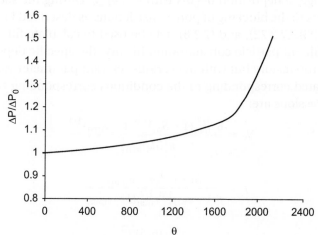

The effect of β may be stated as follows: f, by definition, is the fraction of the blocked pore constrictions. Therefore, f cannot be greater than unity or $1 - f$ should be always greater than zero. The specific deposit corresponding to $1 - A\sigma = 0$ gives the maximum value of σ, or σ_{max}, representing the state of deposition when a bed segment is totally blocked. As σ_{max} is inversely proportional to A, which, in turn, is inversely proportional to β, a larger value of β gives a larger σ_{max} or a longer time for the complete blockade taking place as shown in the results given above.

■ ■ ■

7.2.3   A Two-Stage Deposition Hypothesis

In previous discussions, two simple hypotheses were used to account for the effect of deposition on filter performance. While both the uniform deposit hypotheses and pore constriction blocking hypothesis have their physical significances, it is also possible or even probable that their validity may depend upon the extent of deposition. Specifically, the uniform deposit hypothesis is likely to be operative during the initial period of filtration, corresponding to small values of σ, and pore constriction blockade may take place when σ becomes sufficiently large. With this reasoning, Tien et al. (1979) presented a two-stage model for describing the dynamics of granular filtration. The main features of the two-stage model may be described as follows.

A. For the first stage or $\sigma < \sigma_{trans}$, the effect of deposition is the same as that given in 7.2.1. The relevant equation presented in 7.2.1 may be used to obtain filtration performance during this period.

B. For the 2nd stage with $\sigma > \sigma_{tran}$, the effect of deposition is confined mainly to the blocking of pore constrictions. The pressure-drop increase is given by Equation (7.2.20) except that the value of A is evaluated corresponding to the condition of $\sigma = \sigma_{tran}$ (in other words, d_g is given as $d_g = d_{g_0}[(1 - \varepsilon_{tran})/(1 - \varepsilon_0)]^{1/3}$ with $\varepsilon_{tran} = \varepsilon_0 - \sigma_{tran}/(1 - \varepsilon_d)$ and $\lambda_0 = \lambda_{tran}$.)

C. The second stage of deposition begins with $\sigma = \sigma_{tran}$. During the second stage, deposition leads to the blocking of pore constrictions as described in 7.2.2. Expression similar to (7.2.25), (7.2.22), and (7.2.8) may be used to calculate filter performance, namely, the effluent particle concentration history, the specific deposit profile, and the pressure-drop history but with the various relevant parameter values (namely, d_g, N_c, etc.) evaluated corresponding to the conditions corresponding to $\sigma = \sigma_{tran}$. The corrected expressions are:

$$N_c = \frac{6\,\varepsilon_{tran}^{1/3}(1 - S_{w_i})^{1/3}(1 - \varepsilon_{tran})^{2/3}}{\pi\left[(d_g)_{tran}\right]^2} \tag{7.2.26}$$

$$f = \frac{(6/\pi)^{5/3}(\sigma - \sigma_{tran})}{\beta(1 - s_{w_i})^{1/3}\varepsilon_{tran}^{1/3}(1 - \varepsilon_{tran})} \tag{7.2.27}$$

$$A = \frac{(6/5)^{5/3}}{\beta(1 - S_{w_i})^{1/3}(\varepsilon_{tran})^{1/3}(1 - \varepsilon_{tran})} \tag{7.2.28}$$

with

$$\varepsilon_{tran} = \varepsilon_0 - \frac{\sigma_{tran}}{1 - \varepsilon_d} \tag{7.2.29}$$

$$(d_g)_{tran} = d_{g_0}\left(\frac{1 - \varepsilon_{tran}}{1 - \varepsilon_0}\right)^{1/3} \tag{7.2.30}$$

The pressure drop of Equation (7.2.20) is modified to be

$$-(\Delta p) = (-\Delta p)_{tran} \int_0^L \frac{dz}{1 - A(\sigma - \sigma_{tran})} \tag{7.2.31}$$

and the specific deposit can be found from the following equation

$$\sigma_i(\theta + \Delta\theta) = \sigma_i(\theta) + c_{i-1}\left[1 - e^{-\lambda_i \Delta L}\right] \frac{\Delta\theta \cdot u_s}{\Delta L[1 - A(\sigma_i - \sigma_{\text{tran}})]} \qquad (7.2.32)$$

with the initial condition

$$\theta > \theta_{\text{tran}}, \quad \sigma = \sigma_{\text{tran}} \qquad (7.2.33)$$

and λ_i is based on $d_g = d_{g_o}\left(\dfrac{1 - \varepsilon_{\text{tran}}}{1 - \varepsilon_0}\right)^{1/3}$

A number of parameters, ε_d, β, and σ_{tran}, are present in the models discussed in the preceding sections. While one may attach certain physical significance to these parameters, in reality, they are extraneous quantities and their determinations are difficult if not impossible. Therefore, if the models are to be used, the values of these parameters must be assumed even though the accuracy of the assumed values may be questionable.

As a crude guide in the estimation of ε_d, β, and σ_{tran}, the following suggestions are offered:

1. The values of the deposit porosity, ε_d, may be taken as 0.7 according to Tien et al. (1979). Deb (1969), based on his experimental data of floc porosities, gave $\varepsilon_d = 0.75$. Similarly, Hutchinson and Sutherland (1965) found $\varepsilon_d = 0.8$ from their coagulation simulation results. It is also known that the porosity of granular beds at the incipience of fluidization falls within the range of 0.45–0.60. The value suggested by Tien et al. is a compromise among these values.

2. The parameter β gives an indication of the particle aggregate size necessary to block a pore constriction. If one assumes that the pore constriction diameter is 1/3 of the grain diameter, d_g (see Equation (7.2.14)), a particle aggregate capable to block a constriction should have a size comparable to that of the constriction, or

$$\left\{\left[\beta(\pi/6)d_p^3\right]\big/(1 - \varepsilon_d)\right\}^{1/3} \cong (d_g/3)$$

and

$$\beta = \frac{1 - \varepsilon_d}{\pi/6}\left(\frac{d_g}{3d_p}\right)^3 = (2.12 \times 10^{-2})\left(\frac{d_g}{d_p}\right)^3 \quad \text{if } \varepsilon_d = 0.8 \qquad (7.2.34)$$

In other words, β depends on the relative particle grain size, or $N_R = d_p/d_g$. For $d_p/d_g = 10^{-1}$, β is approximately 2.1. For $d_p/d_g \cong (1/5)$, β is roughly 2.7. On the other hand, for $d_p/d_g \leq 10^{-2}$, β may be of the order of 10^3 or even greater.

3. In contrast, it is more difficult to assign a value to σ_{tran}. The quantity σ_{tran} signifies the transition of deposition mode from one in which deposition results in an increase of effective particle grain size to that of pore blocking. While this transition assumption is plausible, the transition, if it occurs, is likely to take place over a range of values of σ instead of a specific threshold value.

A number of experimental studies (Camp, 1964; Deb, 1969; Ives, 1961) showed that the filter coefficient first increases with σ and then decreases. σ_{tran} therefore could be the value at which λ (or F) reaches a maximum. Based on the data then available, σ_{tran} is found to be (Tien et al., 1979)

$$\sigma_{tran} = 0.05$$

It is clear that this selection is, at best, an approximation since the behavior that λ first increases with σ and then decreases is not universally observed.

Tien et al. carried out a number of calculations of filter performance based on the two-stage model and compared their predictions with experiments. One set of results based on data reported in Camp's work (1964) is shown in Fig. 7.2. The experimental conditions used are listed in Table 7.1. The model parameters used are $\varepsilon_d = 0.7$, $\beta = 2$ or 4, and $\sigma_{tran} = 0.05$. The predicted pressure-drop history agrees reasonably well with experiments. However, there are significant differences between the predicted effluent concentration histories and experiments.

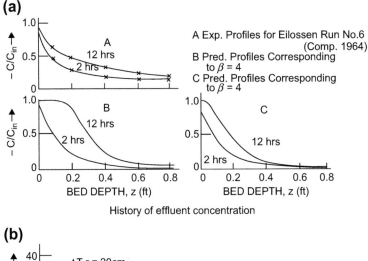

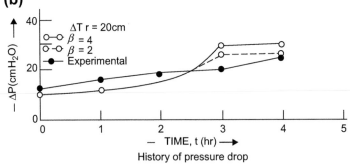

FIGURE 7.2 Comparisons of Experiments with Predictions based on Two-stage Model. *Reprinted from C. Tien, R.M. Turian and H. Pendse, "Simulation of dynamic behavior of deep bed filtration", AIChE J., 25, 385–395, 1979, with permission from John Wiley and Sons, Inc.*

Table 7.1   Experimental Conditions Used by Camp (1964). Data shown in Fig. 7.2

Quantity	Camp (1964)
Filter medium	Sand
Particles in suspension	Hydrous ferric oxide floc
Bed porosity, ε_0	0.41[a]
Superficial velocity, u_s (cm/s)	0.136[a]
Grain diameter, d_{g_0} (cm)	0.0514[a]
Particle diameter, d_p (cm)	0.00062[a]
Particle density, ρ_p (g cm^{-3})	3.6[b]
Feed concentration, c_{in} (vol/vol)	150×10^{-6a}
Transition sp. Deposit σ_{tran}	0.04[c]
Number of constrictions, N_{c_0} (cm^{-2})	352[d]
Length of UBE, ℓ (cm)	0.0494[d]
Constriction diameter, d_c (cm)	0.0184[d]

[a]Reported Values

[b]Assumed Values

[c]These σ_{tran} values are based on empirical data on filter coefficient versus sp. deposit

[d]Calculated Values

■ ■ ■ ▬▬▬▬▬▬▬▬▬▬▬▬▬▬▬▬▬▬▬▬▬▬▬▬▬▬▬▬

Illustrative Example 7.3

For granular filtration of suspensions of submicron particles, assuming that the Brownian diffusion is the dominant deposition mechanism, obtain the filter coefficient correction function, $F(= \lambda/\lambda_0)$ (a) if the deposition effect is based on the uniform deposit hypothesis, (b) if the deposition effect is described by the pore blocking hypothesis, and (c) how the results obtained in (a) and (b) can be incorporated in the two-stage model as discussed in 7.2.3.

Solution

If the dominant deposition mechanism is the Brownian diffusion, the single collector efficiency expression of Equation (6.1.28) is applicable

$$(\eta_s)_{BM} = 4A_s^{1/3}N_{Pe}^{-2/3} \tag{i}$$

with

$$N_{Pe} = d_g u_s / D_{BM}$$
$$A_s = 2(1-p^5)/w$$
$$p = (1-\varepsilon)^{1/3}$$
$$w = 2 - 3p + 3p^5 - 2p^6$$

Generally speaking, deposition may cause changes in the values of u_s, d_g, and ε, which, in turn, leads to changes of A_s, p, and w. Let the subscript o denote the initial (clean filter) state. F may be expressed as

$$F = \lambda/\lambda_0 = \eta_s/(\eta_s)_0 = \left(\frac{A_s}{A_{s_0}}\right)^{1/3}\left(\frac{d_g}{d_{g_0}}\right)^{-2/3}\left(\frac{u_s}{u_{s_0}}\right)^{-2/3} \tag{ii}$$

(a) If the deposition effect is described by the uniform deposit hypothesis, u_s remains unchanged. On the other hand, the porosity decreases from being ε_0 to ε:

$$\varepsilon = \varepsilon_0 - \frac{\sigma}{1 - \varepsilon_d}$$

Therefore,

$$\frac{1 - \varepsilon}{1 - \varepsilon_0} = 1 + \frac{\sigma}{(1 - \varepsilon_0)(1 - \varepsilon_d)}$$

$$(d_g/d_{g_0}) = \left(\frac{1 - \varepsilon}{1 - \varepsilon_0}\right)^{1/3} = \left[1 + \frac{\sigma}{(1 - \varepsilon_0)(1 - \varepsilon_d)}\right]^{1/3}$$

$$p_0 = (1 - \varepsilon_0)^{1/3}$$

$$p = (1 - \varepsilon)^{1/3} = (1 - \varepsilon_0)^{1/3}\left[1 + \frac{\sigma}{(1 - \varepsilon_0)(1 - \varepsilon_d)}\right]^{1/3}$$

Substituting the above expressions into Equation (ii), the ratio λ/λ can be expressed as

$$F = \lambda/\lambda_0 = \left[1 + \frac{\sigma}{(1 - \varepsilon_0)(1 - \varepsilon_d)}\right]^{-2/9}\left(\frac{1 - p^5}{1 - p_0^5}\right)^{1/3}\left(\frac{w_0}{w}\right)^{1/3}$$

$$= \left[1 + \frac{\sigma}{(1 - \varepsilon_0)(1 - \varepsilon_d)}\right]^{-2/9}\left\{\frac{1 - (1 - \varepsilon_0)^{5/3}\left[1 + \frac{\sigma}{(1 - \varepsilon_0)(1 - \varepsilon_d)}\right]^{5/3}}{1 - (1 - \varepsilon_0)^{5/3}}\right\}^{1/3}$$

$$\times\left\{\frac{2 - 3(1 - \varepsilon_0)^{1/3} + 3(1 - \varepsilon_0)^{5/3} - 2(1 - \varepsilon_0)^2}{2 - 3(1 - \varepsilon_0)^{1/3}\left[1 + \frac{\sigma}{(1 - \varepsilon_0)(1 - \varepsilon_d)}\right]^{1/3} + 3(1 - \varepsilon_0)^{5/3}\left[1 + \frac{\sigma}{(1 - \varepsilon_0)(1 - \varepsilon_d)}\right]^{5/3} + 2(1 - \varepsilon_0)^2\left[1 + \frac{\sigma}{(1 - \varepsilon)(1 - \varepsilon_0)}\right]^2}\right\}^{1/3} \tag{iii}$$

(b) If the deposition effect is described by the pore-blocking hypothesis, one part of the filter medium is blocked and not accessible to suspension flow. The state of the unblocked part remains the same as its initial state. Therefore, the porosity (and then p, w, and A_s) remains the same. The superficial velocity, u_s, and its initial value, $(u_s)_0$, are related by Equation (7.2.11). From Equation (ii), F is given as

$$F = \lambda/\lambda_0 = [1/(1 - f)]^{-2/3} = [1 - A\sigma]^{2/3} \tag{iv}$$

with A given by Equation (7.2.18) or

$$A = \frac{(\pi/6)^{1/3}}{\beta(1 - S_{w_i})^{1/3}\varepsilon_0^{1/3}(1 - \varepsilon_0)(d_p/d_g)^3}$$ (v)

(c) To incorporate the results given above, F given by Equation (i) can be readily applied to the first stage calculation. For the second stage, Equation (iv) should be modified by replacing σ with $\sigma - \sigma_{tran}$ and ε_0 by ε_{tran}. Also instead of λ_0, the value of λ_{tran} should be used with $\lambda_{tran} = \lambda_0 F(\sigma_{tran})$, with F given by Equation (iii).

■ ■ ■

7.3 Models Based on Assumption that Deposited Particles Function as Collectors

It is reasonable to assume that during the initial stage of filtration, deposition leads to the presence of individual particles attached to filter grains. As these individual particles are exposed directly to the suspension flow, they may function as additional particle collectors and experience drag forces on them. The contributions of these deposited particles to particle collection and pressure drop reflect the effect of deposition on filter performance. A number of filtration models have been formulated on this premise. In the following, the principles used in such formulations are discussed using two specific examples.

7.3.1 Deposited Particles as Satellite Collectors

Payatakes and Tien (1976) proposed a dendrite growth model for aerosol filtration in which deposition onto previously deposited particles results in the growth of particle dendrites. This approach was later used by O'Melia and Ali (1978) in describing the so-called filter ripening behavior. Since these earlier studies were done, a number of models, based on this basic idea, have appeared in the literature (Payatakes, 1976; Payatakes and Gradon, 1980; Payatakes and Okuyama, 1982; Vigneswaran, 1980; Chang, 1985; Vigneswaran and Tulachan, 1988; Tobiason and Vigneswaran, 1994). For the purpose here, we present the work of Vigneswawran and Tulachan with some minor modifications because of the relative simplicity of Vigneswawran–Tulachan's work and its inclusion of nonretentive filtration feature.

The starting point of Vigneswaran and Tulachan's work is to consider deposition on a filter grain to be the sum of particles deposited directly on the grain and those on deposited particles. If N denotes the number of particles deposited directly on the filter grain, by definition, $\partial N/\partial \theta$, may be expressed as

$$\frac{\partial N}{\partial \theta} = \frac{N_{max} - N}{N_{max}} \eta_{s_0}\left(\frac{\pi}{4}\right)d_g^2 u_s \tilde{c}$$ (7.3.1)

where $\tilde{c}$ is the suspension particle number concentration. $N_{\max}$ is the maximum value of N. $(N_{\max} - N)/N_{\max}$ therefore represents the fraction of grain surface available for direct deposition. η_{s_0} is the initial single collector efficiency.

$N_{\max}$, the maximum value of N, may be estimated as follows. Assuming that the filter grain is spherical with diameter d_g, the surface area of the grain is πd_g^2. The projected surface area of a spherical particle d_p is $(\pi/4)d_p^2$. $N_{\max}$ can be expected to be proportional to $4\pi d_g^2/\pi d_p^2$ or

$$N_{\max} = 4\alpha_1(d_g/d_p)^2 \tag{7.3.2}$$

where α_1 is a fitting parameter to account for the so-called "excluded area" effect.

If N_t denotes the total number of particles attached to a filter grain, namely, the sum of particles deposited directly to the grain and those deposited on the directly deposited particles, the relationship between N_t and the specific deposit, σ is

$$\sigma = N_t \frac{(1 - \varepsilon_0)}{(\pi/6)d_g^3}(\pi/6)d_p^3 = (d_p/d_g)^3 N_t(1 - \varepsilon_0) \tag{7.3.3}$$

where ε_0 is the initial filter bed porosity and $(1 - \varepsilon_0)/[(\pi/6)d_g^3]$ is the number of filter grains per unit medium volume.

The deposition rate, $\partial N_t/\partial \theta$, may be written as

$$\frac{\partial N_t}{\partial \theta} = \left(\frac{\pi}{4}\right)d_g^2 u_s \tilde{c} \cdot \eta_s \tag{7.3.4}$$

The single collector efficiency of a grain plus deposited particle may be assumed to be

$$\eta_s = \eta_{s_0}\frac{N_{\max} - N}{N_{\max}} + (N_t - N)(\eta_s)_p(d_p/d_g)^2 \frac{\varepsilon - \varepsilon^*}{\varepsilon_0 - \varepsilon^*} \tag{7.3.5}$$

where $(\eta_s)_p$ is the single collector efficiency of the deposited particles. It is assumed that $(\eta_s)_p$ is the same for all deposited particles. $(d_p/d_g)^2$ is introduced in order to account for the difference between the projected areas used in defining η_s and $(\eta_s)_p$. The term $(\varepsilon - \varepsilon^*)/(\varepsilon_0 - \varepsilon^*)$ is introduced by Vigneswaran and Tulachan in order to display the experimentally observed nonretentive behavior, namely, the medium has a finite particle retention capacity such that when ε reaches a limiting value, ε^*, deposition ceases. Alternatively, one may assume that there exists an ultimate specific deposit, σ_{ult}, which is related to ε^* by the expression

$$\varepsilon^* = \varepsilon_0 - \frac{\sigma_{\text{ult}}}{1 - \varepsilon_d} \tag{7.3.6}$$

Combining Equations (5.2.1), (5.4.20), and (7.3.5), the macroscopic conservation equation may be written as

$$\frac{\partial \tilde{c}}{\partial z} + \frac{3}{2}\frac{1 - \varepsilon_0}{d_g}\left[\eta_{s_0}\frac{N_{\max} - N}{N_{\max}} + (N_t - N)(\eta_s)_p(d_p/d_g)^2\left(\frac{\varepsilon - \varepsilon^*}{\varepsilon_0 - \varepsilon^*}\right)\right]\tilde{c} = 0 \tag{7.3.7}$$

Equation (7.3.7) together with Equation (7.3.1) and (7.3.4) constitute the governing equation according to Vigneswaran and Tulachan. With appropriate initial and boundary conditions (for a clean filter with constant influent concentration, $c = c_{in}$ at $z = 0$, $N = N_t = 0$, $\theta = 0$) and assuming that all the model parameter values are known, the solution of the system of equations provides a complete description of the filtration dynamics.

The approximate method used by Vigneswaran and Tulachan assumes that N and N_t (or σ) of Equation (7.3.7) can be replaced by their average values throughout the entire filter medium, namely, the uniform deposition assumption. Simple integration of Equation (7.3.7) along the z-axis yields

$$ln\frac{\hat{c}}{\hat{c}_{in}} = ln\frac{c}{c_{in}} = -(3/2)\left(\frac{1-\varepsilon_0}{d_g}\right)\left[\eta_{s_0}\frac{N_{max}-N}{N_{max}} + (N_t - N)(\eta_s)_p(d_p/d_g)^2\frac{\varepsilon-\varepsilon^*}{\varepsilon_0-\varepsilon^*}\right]z \qquad (7.3.8)$$

Similarly, from Equations (7.3.1) and (7.3.4), the following approximate expressions are obtained:

$$N^i = N^{i-1} + \eta_{s_0}(\pi/4)d_g^2 u_s\tilde{c}^{i-1}\frac{N_{max}-N^{i-1}}{N_{max}}\Delta\theta \qquad (7.3.9)$$

$$N_t^i = N_t^{i-1} + \eta_s^{i-1}(\pi/4)d_g^2 u_s\tilde{c}^{i-1}\Delta\theta \qquad (7.3.10)$$

where the superscript i denotes the value at $\theta = i\Delta\theta$.

Thus, from Equations (7.3.8)–(7.3.10), the particle concentration profile ($\tilde{c}$ vs. z) and N_t (or σ) vs. z at various times, $\theta = \Delta\theta, 2\Delta\theta, \ldots$ can be calculated for a given set of initial and boundary conditions and with η_i^{i-1} given as [from Equation (7.3.5)]

$$\eta_s^{i-1} = \eta_{s_0}\frac{N_{max}-N^{i-1}}{N_{max}} + (N_t^{i-1} - N^{i-1})(\eta_s)_p(d_p/d_g)^2\frac{\varepsilon^{i-1}-\varepsilon^*}{\varepsilon_0-\varepsilon^*} \qquad (7.3.11)$$

To obtain the pressure-drop increase, the Kozeny–Carman equation may be applied. Combining Equations (7.2.1) and (7.2.2), one has

$$\frac{(\partial p/\partial z)}{(\partial p/\partial z)_0} = \left(\frac{1-\varepsilon}{1-\varepsilon_0}\right)^2\left(\frac{\varepsilon_0}{\varepsilon}\right)^3\left(\frac{S}{S_0}\right)^2 \qquad (7.3.12)$$

where $S = s_p/v_p$, the specific surface area with s_p and v_p being the surface area and volume of a filter grain.

The specific surface area of a single filter grain is $\pi d_g^2/(\pi/6)\cdot d_g^3 = 6/d_g$ which may be taken as S_0. For a grain with N deposited particles, the specific surface area S can be estimated to be

$$S = \frac{\pi d_g^2 + N\pi d_p^2}{(\pi/6)d_g^3 + N(\pi/6)d_p^3} = \frac{\left[1 + N(d_p/d_g)^2\right]}{1 + N(d_p/d_g)^3}$$

or

$$\frac{S}{S_0} = \frac{1 + N(d_p/d_g)^2}{1 + N(d_p/d_g)^3} \qquad (7.3.13)$$

Substituting Equation (7.3.13) into (7.3.12) yields

$$\frac{(\partial p/\partial z)}{(\partial p/\partial z)_0} = \left(\frac{1-\varepsilon}{1-\varepsilon_0}\right)^2 \left(\frac{\varepsilon_0}{\varepsilon}\right)^3 \frac{1+N(d_p/d_g)^2}{1+N(d_p/d_g)^3} \tag{7.3.14}$$

Based on the above expression, it was argued that as an approximation, the total pressure drop may be expressed as

$$\frac{\Delta p}{(\Delta p)_0} \cong \frac{1+\beta'\overline{N}(d_p/d_g)^2}{1+\overline{N}(d_p/d_g)^3} \tag{7.3.15}$$

where $\overline{N}$ is the spatial average of particles deposited per filter grain and can be calculated as

$$\overline{N} = \frac{u_s \cdot \tilde{c}_{in} \pi d_g^3}{6L(1-\varepsilon_0)} \int_0^\theta \left(1 - \frac{c_{eff}}{c_{in}}\right) \cdot d\theta \tag{7.3.16}$$

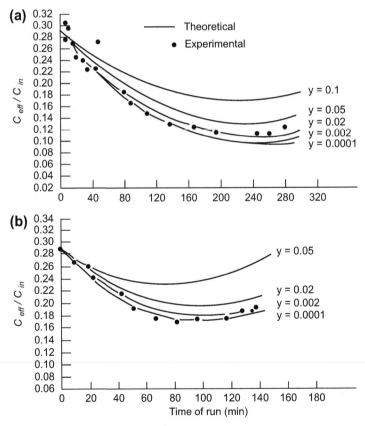

FIGURE 7.3 Fitting of Experimental data to the model of Vigneswaran and Tulachan Model (1988) with different η values (a) $u_s = 5$ m/hr, $d_p = 0.27$ μm, $d_g = 200$ μm, $n_{sp} = 0.003$, $\varepsilon^* = 0.25$, (b) $u_s = 10$ m/hr, $d_p = 0.27$ μm, $d_g = 200$ μm, $n_{sp} = 0.0015$, $\varepsilon^* = 0.29$. *Reprinted from Water Research, S. Vigneswaran and R.K. Tulachan, "Mathematical modeling of transient behavior of deep bed filtration", 72, 1093–1100, 1988, with permission of Elsevier.*

In other words, Equation (7.3.15) is obtained by ignoring the first two quantities at the right hand side of Equation (7.3.14), replacing N with $\overline{N}$, and introducing an arbitrary empirical constant β'. With the results of c_{eff} vs. θ obtained from the solution of Equation (7.3.8), the pressure-drop increase, (Δp), as a function of time can be estimated according to Equation (7.3.15) provided the empirical constant β' is known.

There are altogether six model parameters, α_1, ε_d, σ_{ult}, η_{s_0}, η_{s_p}, β', some of which are not known a priori. Thus, the model cannot be used for predictions. Vigneswaran and Tulachun, however, argued that by carrying out filtration experiments using laboratory-scale apparatus and fitting the data obtained to the model, the values of the model parameters can be obtained, which, in turn, can be used to simulate plant size operations as a basis of ration design.

The ability of the model to describe full cycle filtration behavior is shown in Fig. 7.3 in which predicted effluent particle concentration history is compared with experiments. The experiment was carried out using a small filter of $L = 2$ cm at two different flow rates. The value of η_{s_0} was determined from the initial filtration data while the other parameters were considered as fitting parameters.

■ ■ ■ ▬▬▬▬▬▬▬▬▬▬▬▬▬▬▬▬▬▬▬▬▬▬▬▬▬▬▬▬▬▬▬▬▬▬▬

Illustrative Example 7.4

The results given in 7.3.1 are based on the assumption of uniform deposition throughout filter medium. Modify the results so that they may be applicable to filters with sufficiently large height and uniform deposition assumption may not be valid.

Solution

The approach used in 7.2.1, namely, a filter bed of height L may be viewed as a series of N segments of height $\Delta L = L/N$ can be used to modify the results of Vigneswaran and Tulachan. Consider the i-th segment with the influent and effluent concentrations being $\tilde{c}_{i-1}$ and $\tilde{c}_i$. From Equation (7.3.8), $\tilde{c}_i$ is related to $\tilde{c}_{i-1}$ by the expression

$$\ell n \frac{\tilde{c}_i}{\tilde{c}_{i-1}} = -(3/2)\left(\frac{1-\varepsilon_0}{d_g}\right)\left[\eta_{s_0}\frac{N_{\max}-N_i}{N_{\max}} + (N_{t_i}-N_i)\,(\eta_s)_p(d_p/d_g)^2\frac{\varepsilon_i-\varepsilon^*}{\varepsilon_0-\varepsilon^*}\right]\Delta L \qquad \text{(i)}$$

where the subscript i denotes the quantity of the i-th segment. From Equation (7.3.9), the values of N_i at $\theta = j\Delta\theta$, or N_i^j is given as

$$N_i^j = N_i^{j-1} + \eta_{s_0}(\pi/4)d_g^2 u_s \tilde{c}_i^{j-1}\frac{N_{\max}-N_i^{j-1}}{N_{\max}}\Delta\theta \qquad \text{(ii)}$$

From Equation (7.3.10), the values of $(N_j)_i^j$ is given as

$$(N_t)_i^j = (N_t)_i^{j-1} + (\pi/4)d_g^2 u_s \tilde{c}_i^{j-1}(\eta_s)_i^{j-1} \qquad \text{(iii)}$$

The single collector efficiency of the i-th segment at time $\theta = j\Delta\theta$ is [from Equation (7.3.11)]

$$(\eta_s)_i^j = \eta_{s_0}\frac{N_{max} - N_i^j}{N_{max}} + \left[(N_t)_i^j - N_i^j\right](\eta_s)_p(d_p/d_g)^2\frac{\varepsilon_i^j - \varepsilon^*}{\varepsilon_0 - \varepsilon^*} \tag{iv}$$

The effluent concentration can be found from the following expression

$$\frac{\tilde{c}_{eff}}{\tilde{c}_{in}} = \prod_{i=1}^{N}\frac{\tilde{c}_i^j}{\tilde{c}_{i-1}^j} \tag{v}$$

The pressure drop across the i-th segment is [from Equation (7.3.15)]

$$\frac{(\Delta p)_i^j}{(\Delta p_0)/N} = \frac{1 + \beta' N_i^j}{1 + N_i^j(d_p/d_g)^2}$$

and

$$\frac{\Delta p}{(\Delta p_0)} = \frac{1}{N}\sum_{i=1}^{N}\frac{1 + \beta N_i^j}{1 + N_i^j(d_p/d_g)^2} \tag{vi}$$

■ ■ ■

7.3.2 Deposited Particles as Additional Collectors

In fibrous filtration of aerosols, deposition leads to the formation and growth of particle dendrite from fibers surface into the media void space. As an approximation, these dendrites may be viewed as additional fiber collectors and contribute to the removal of particles from the gas streams. Models based on this premise have been suggested by a few investigators (see Bergman et al., 1978; Thomas et al., 2001). A simplified version of Thomas et al.'s work is given below to demonstrate the principles used.[1]

If one considers that particle deposition leads to the formation and growth of particle dendrites, for a given degree of deposition, a fibrous medium, in fact, is composed of fibers as well as particle dendrites in the form of strings of particles. For a medium with packing density of α with specific deposit of σ, the effective porosity, ε, becomes

$$\varepsilon = 1 - (\alpha + \sigma) \tag{7.3.17}$$

Particle retention now takes place on both fibers and dendrites. Consistent with Equation (5.4.34b), increase in σ due to deposition onto fibers may be written as

$$\frac{\partial\sigma_f}{\partial\theta} = \frac{4\alpha}{\pi d_f}\frac{u_s}{1 - (\alpha + \sigma)}\eta_{f_s}c \tag{7.3.18a}$$

where σ_f denotes the amount (volume) of particles deposited on fibers and η_{f_s} the single fiber efficiency.

---

[1]Thomas et al. assumed that the size of aerosols of the feed stream follows the log-normal distribution. The treatment given below assumes that the aerosols are uniform in size.

For the dendrites formed by collected particles, assume that the dendrites may be approximated as fibers of diameter d_p, the rate of particles deposited to form dendrite, $\partial \sigma_d / \partial \theta$, may be written as[2]

$$\frac{\partial \sigma_d}{\partial \theta} = \frac{4\alpha}{\pi d_p} \frac{u_s}{1 - (\alpha + \sigma)} \eta_{p_s} c \tag{7.3.18b}$$

The total rate of particle retention, $\partial \sigma / \partial \theta$, is

$$\frac{\partial \sigma}{\partial \theta} = \frac{\partial \sigma_f}{\partial \theta} + \frac{\partial \sigma_d}{\partial \theta} \tag{7.3.19}$$

where η_{f_s} and η_{p_s} are the single fiber efficiencies of fiber of diameter d_f and the equivalent single fiber efficiency of diameter d_p for particle dendrite. Both η_{f_s} and η_{p_s} can be estimated from existing correlations (see Chapter 6). α is the fiber packing density and equal to $1 - \varepsilon_0$. σ, as before, is the specific deposit. The macroscopic conservation equation is

$$u_s \frac{\partial c}{\partial z} + \frac{4\alpha}{\pi d_f} \frac{u_s}{1 - (\alpha + \sigma)} \eta_{f_s} \cdot c + \frac{4\alpha}{\pi d_p} \frac{u_s}{1 - (\alpha + \sigma)} \eta_{p_s} c = 0 \tag{7.3.20}$$

For a fibrous medium free of deposited particles initially, the initial and boundary conditions are

$$c = c_{in} \quad \text{at} \quad z = 0 \quad \theta > 0$$
$$\sigma = \sigma_f = \sigma_d = 0, \quad z > 0, \quad \theta \le 0 \tag{7.3.21}$$

A simple procedure similar to that given in Illustrative Example 7.4 was used by Thomas et al. to solve the above set of equations. Consider a fibrous bed to be a series of segments of height Δz and let the subscripts i and j denote the value at the i-th segment at $\theta = j\Delta\theta$. From Equation (7.3.20), one has

$$\ell n \frac{c_{i+1,j}}{c_{i,j}} = \exp \left[-\frac{4\alpha \eta_{f_s}}{\pi d_f} - \frac{4\sigma_{ij} \eta_{d_s}}{\pi d_p} \right] \frac{\Delta z}{1 - \alpha - \sigma_{i,j}} \tag{7.3.22a}$$

From Equations (7.3.18a) and (7.3.18b), one has

$$\sigma_{i,j} = \left[\frac{4\alpha \eta_{f_s}}{\pi d_f} + \frac{4\sigma_{i,j-1} \eta_{p_s}}{\pi d_p} \right] \frac{u_s c_{i,j-1}}{1 - (\alpha + \sigma_{ij-1})} \Delta\theta \tag{7.3.22b}$$

and

$$\sigma_{i0} = 0$$

Equations (7.3.22a) and (7.3.22b) together with Equations (7.3.21) give the effluent concentration history as well as the specific deposit profile as functions of time. To estimate the pressure drop, the empirical expression of Davis (1973) [also see Equation (5.5.6)] was used as the starting point or

---

[2]With this assumption, the dendrite diameter remains constant throughout filtration.

$$-\Delta p = 16(1-\varepsilon)^{1.5}\left[1 + 56(1-\varepsilon)^3\right]Lu_s\mu/r_f^2$$

$$\simeq 1.6(1-\varepsilon)^{1.5}L\,u_s\mu/r_f^2 \quad \text{for small values of } 1-\varepsilon$$

(7.3.23)

If one assumes that deposited particles are present as dendrites which may be considered as cylindrical collectors of diameter d_p and furthermore, there is no interaction between fibers and dendrites, the pressure drop across a fibrous medium with an initial porosity ε_0 and with a specific deposit, σ, may be expressed as

$$-\Delta p = (-\Delta p)_f + (-\Delta p)_d \tag{7.3.24}$$

Namely, the pressure drop is the sum of the pressure drop across a fibrous media free of deposited particles and that across a medium composed of particle dendrites with a packing density of σ. $(-\Delta p)_f$ and $(-\Delta p)_d$ according to Equation (7.3.23) are

$$(-\Delta p)_f = 16(1-\varepsilon_0)^{1.5}L\mu u_s/r_f^2 \tag{7.3.25a}$$

$$(-\Delta p)_d = 16(\sigma)^{1.5}L\mu u_s/r_p^2 \tag{7.3.25b}$$

and

$$(-\Delta p) = 16L\mu\ u_s\left[\frac{(1-\varepsilon_0)^{1.5}}{r_f^2} + \frac{\sigma^{1.5}}{r_p^2}\right] \tag{7.3.26}$$

The total length of fibers present in the bed, ℓ_f, is

$$\ell_f(\pi r_f^2) = L(1-\varepsilon_0)$$

or

$$\ell_f = \frac{L(1-\varepsilon_0)}{\pi r_f^2} \tag{7.3.27a}$$

Similarly, the total dendrite length, ℓ_d, is

$$\ell_d = \frac{L\sigma}{\pi r_p^2} \tag{7.3.27b}$$

Combining Equations (7.3.26), (7.3.27a), and (7.3.27b) yields

$$(-\Delta p) = 16\pi\mu\ u_s\left[(1-\varepsilon_0)^{0.5}\ \ell_f + \sigma^{0.5}\ \ell_d\right]$$

$$= 16\ \mu\ u_sL\left[\frac{(1-\varepsilon_0)^{1.5}}{r_f^2} + \frac{\sigma^{1.5}}{r_p^2}\right]$$

$$= 64\ \mu\ u_sL\left[\frac{(1-\varepsilon_0)^{1.5}}{d_f^2} + \frac{\sigma^{1.5}}{d_p^2}\right] \tag{7.3.28}$$

The above expression, as stated before, does not consider any possible interactions between fibers and dendrites. In order to obtain better agreement with experiments, Bergman et al. (1978) introduced corrections to $(-\Delta p)_f$ and $(-\Delta p)_d$ by dividing the value of $(-\Delta p)_f$ by $\left(\dfrac{\ell_f}{\ell_f + \ell_d}\right)^{1/2}$ and $(-\Delta p)_d$ by $\left(\dfrac{\ell_d}{\ell_f + \ell_d}\right)^{1/2}$. $(-\Delta p)$ then becomes

$$(-\Delta p) = 64\mu\ u_s L \left(\frac{1-\varepsilon_0}{d_f} + \frac{\sigma}{d_p}\right)\left(\frac{1-\varepsilon_0}{d_f^2} + \frac{\sigma}{d_p^2}\right)^{1/2} \tag{7.3.29}$$

The $(-\Delta p)$ expression of Equation (7.3.29) is obtained on the basis of uniform specific deposit throughout the entire medium. For a filter medium of sufficiently large height, since the extent of deposition is far from being uniform, Equation (7.3.29) should be rewritten on a local basis as

$$-\frac{\partial p}{\partial z} = 64\mu\ u_s \left(\frac{1-\varepsilon_0}{d_f} + \frac{\sigma}{d_p}\right)\left(\frac{1-\varepsilon_0}{d_f^2} + \frac{\sigma}{d_p^2}\right)^{1/2} \tag{7.3.30}$$

where σ is the local value of the specific deposit.

If a fibrous bed of height L is considered as a series of N segments of height $\Delta L = L/N$ with ΔL being a relatively small value, the pressure drop across the bed is

$$\frac{-\Delta p}{64\mu\ u_s L} = \frac{1}{N}\sum_{i=1}^{N}\left(\frac{1-\varepsilon_0}{d_f} + \frac{\sigma_i}{d_p}\right)\left(\frac{1-\varepsilon_0}{d_f^2} + \frac{\sigma_i}{d_p^2}\right)^{1/2} \tag{7.3.31}$$

where σ_i is the specific deposit value of the i-th segment.

Comparisons of predicted particle penetration profile and pressure drop vs. deposited particle mass with experiments are shown in Fig. 7.4 Pressure drop vs. particles retained and profile of particle retention. On the whole, good agreement is observed.

There are two major features of this model; its relative simplicity and, perhaps more important, the absence of any extraneous model parameter. Both η_{f_s} and η_{p_s}, as stated before, can be readily estimated using the existing correlations such as those listed in Chapter 6. In this sense, it is superior to any of those discussed earlier.

One may well ask, what is the reason that the approach used by Thomas et al. cannot be extended for modeling deep bed filtration of hydrosols? The answer is rather simple. For aerosol filtration in fibrous media, the ratio of d_p/d_g is often of the order of 1/10. Furthermore, the fibrous medium packing density, in most cases, is not more than 0.1. The large void space and the electric charges present at particle surfaces favor dendrite growth to the degree that the dendrites may be treated as fibers. On the other hand, for hydrosol deposition in granular media, the porosity of granular medium is relatively low and the size difference between filter grains and particles is likely to be large (perhaps more than two orders of magnitude). Consequently, the approach described here is less applicable to hydrosol filtration.

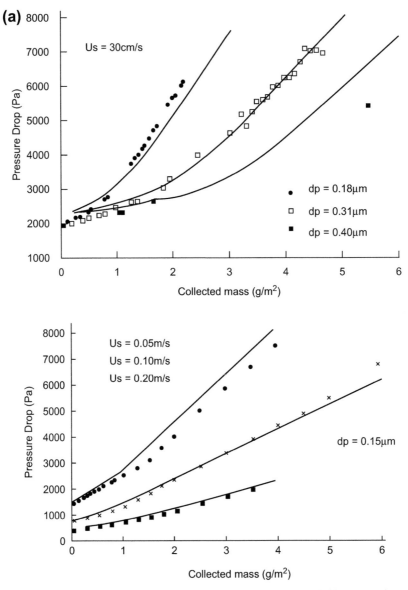

FIGURE 7.4 Comparison of experimental data with the model of Thomas et al. (2001). (a) Pressure drop vs. deposited particle mass, (b) Rate of particle retained in each layer to the amount of particle retained. *Reprinted from Chemical Eng. Sci., D. Thomas, P. Penicot, P. Costal and V. Vendel, "Clogging of fibrous filters by solid aerosol particles: Experimental and modeling study", 56, 3549–3561, 2001, with permission of Elsevier.*

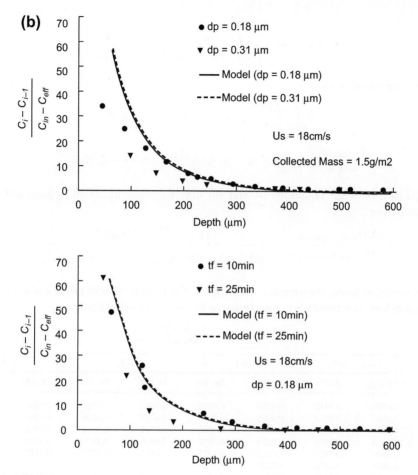

FIGURE 7.4 *(continued)*.

Illustrative Example 7.5

Compare the pressure-drop predictions according to Equations (7.3.28) and (7.3.29) for fibrous media with packing density $(1 - \varepsilon_0)$ being 0.05, 0.10, 0.15, and 0.20, $d_f = 25$ μm, and $d_p = 2$ μm.

Solution

From Equations (7.3.28) and (7.3.29), one has

$$(-\Delta p) = 64\mu \; u_s L \left[\left(\frac{(1 - \varepsilon_0)^{1.5}}{d_f^2} + \frac{\sigma^{1.5}}{d_p^2} \right) \right] \qquad \text{(i)}$$

and

$$(-\Delta p) = 64\mu \ u_s L \left(\frac{1-\varepsilon_0}{d_f} + \frac{\sigma}{d_p}\right)\left(\frac{1-\varepsilon_0}{d_f^2} + \frac{\sigma}{d_p^2}\right)^{1/2} \tag{ii}$$

They may be rewritten as

$$\frac{(-\Delta p)(d_f)^2}{64\mu \ u_s L} = (1-\varepsilon_0)^{1.5} + \sigma^{1.5}(d_f/d_p)^2 \tag{iii}$$

$$\frac{(-\Delta p)(d_f)^2}{64\mu \ u_s L} = \left[(1-\varepsilon_0) + \sigma(d_f/d_p)\right]\left[(1-\varepsilon_0) + \sigma(d_f/d_p)\right]^{1/2} \tag{iv}$$

A comparison between Equations (i) and (ii) may be made by dividing Equation (iii) by Equation (iv), or

$$\frac{(-\Delta p)_{\text{from Equation (i)}}}{(-\Delta p)_{\text{from Equation (ii)}}} = \frac{(1-\varepsilon_0)^{1.5} + \sigma^{1.5}(d_f/d_p)^2}{\left[(1-\varepsilon_0) + \sigma(d_f/d_p)\right]\left[(1-\varepsilon_0) + \sigma(d_f/d_p)^2\right]^{1/2}} \tag{v}$$

For the condition given, the pressure-drop ratio is found to be a function of σ and $(1-\varepsilon_0)$ as tabulated below. The results are also shown in the accompanying figure.

$$\frac{(-\Delta p)(d_f^2)}{64\mu \ u_s L}$$

σ	$(1-\varepsilon) = 0.05$		0.1		0.15		0.2	
	Eq (iii)	Eq (iv)	Eq (iii)	Eq (iv)	Eq (iii)	Eq (iv)	Eq (iii)	Eq (iv)
0	0.0112	0.0112	0.0316	0.0316	0.0581	0.0581	0.0894	0.0894
0.001	0.0161	0.0284	0.0366	0.0569	0.0630	0.0899	0.0944	0.1268
0.002	0.0252	0.0452	0.0456	0.0803	0.0721	0.1190	0.1034	0.1611
0.003	0.0369	0.0630	0.0573	0.1037	0.0838	0.1475	0.1151	0.1942
0.004	0.0507	0.0822	0.0712	0.1277	0.0976	0.1761	0.1290	0.2271
0.005	0.0664	0.1026	0.0869	0.1525	0.1133	0.2051	0.1147	0.2600
0.006	0.0838	0.1242	0.1042	0.1783	0.1307	0.2346	0.1621	0.2933
0.007	0.1027	0.1471	0.1231	0.2049	0.1496	0.2649	0.1810	0.3270
0.008	0.1230	0.1710	0.1434	0.2324	0.1699	0.2558	0.2012	0.3612
0.009	0.1446	0.1961	0.1650	0.2608	0.1915	0.3275	0.2229	0.3961
0.010	0.1674	0.2222	0.1872	0.2901	0.2143	0.3599	0.2457	0.4315

7.4   Models Based on Changing Particle–Collector Surface Interactions

The models discussed so far are based on considering media structural changes due to deposition. However, particle deposition may also result in other types of changes as well. Specifically, with deposition, the nature and magnitude of the surface interactions between collectors and particles present in the suspension can be altered and may exert significant effect on filter performance.

In the following sections, we present three different models of hydrosol filtration in granular media in which the particle–collector interactions undergo changes during the course of filtration. These models are presented in order to demonstrate the different approaches one may use in analyzing problems of this type.

7.4.1 A Model Based on Filter Grain Surface Charge Changes

First, we will discuss the work of Wnek et al. (1975). In formulating their model, Wnek et al. considered the effect of electric charge transfer associated with particle deposition. If J denotes the particle flux (in numbers) over a single filter grain and q_p being the electric charge carried per particle, the rate of the accumulation of the surface charge of a filter grain, q_g, may be written as

$$\frac{\partial q_g}{\partial \theta} = q_p J = q_p \pi a_g^2 u_s \cdot \eta_s \tilde{c} \tag{7.4.1}$$

where $\tilde{c}$ is the particle concentration in numbers and J is given as

$$J = \pi a_g^2 u_s \tilde{c} \eta_s$$

where $\hat{q}_p$ is the effective particle surface charge η_s in the single collector efficiency and the other symbols have the same meanings as defined before.

The relationship between q_p and $\hat{q}_p$ is given as (Wnek, 1973)

$$\hat{q}_p = q_p - 4a_p^2 q_g'[1 - \exp(-\kappa a_p)] \tag{7.4.2}$$

where q_g' is $q_g/4\pi a_g^2$ and κ, the reciprocal double layer thickness [given by Equation (6.5.4)].

With the inclusion of the surface charge effect, the dynamics of deep bed filtration is described by the following set of equations:

$$u_s \frac{\partial c}{\partial z} + \frac{\partial \sigma}{\partial \theta} = 0 \tag{7.4.3}$$

$$\frac{\partial \sigma}{\partial \theta} = u_s \lambda c = (3/4)\big[(1 - \varepsilon)/a_g\big]\eta_s u_s c \tag{7.4.4}$$

$$\frac{\partial q_g}{\partial \theta} = \pi a_g^2 u_s \tilde{c}\ \eta_s \Big[q_p - (q_g/\pi)(a_p/a_g)^2(1 - e^{-\kappa a_p})\Big] \tag{7.4.5}$$

with

$$
\begin{aligned}
c &= c_{\text{in}}, & z &= 0, \\
c &= 0, & & \\
\sigma &= 0 & z &> 0, \quad \theta \le 0 \\
q_g &= q_g(z)
\end{aligned}
\tag{7.4.6}
$$

Equations (7.4.3) and (7.4.4) are the macroscopic mass conservative equations and rate equations derived before, and c and σ are expressed in terms of volume. Equation

(7.4.5) is obtained by combining Equations (7.4.1) and (7.4.2) and $\tilde{c}$ is the number concentration and equal to $6c/\pi d_p^3$. In the above expressions, η_s is a function of σ, q_g, and q_p.

For the solution of the above set of equations, the relationship among η_s, q_g, and σ is required. First, consider that media structural changes lead to a decrease in media porosity and an increase in the effective grain size as mentioned before [see Equations (7.2.4) and (7.2.9)], or

$$\varepsilon = \varepsilon_0 - \frac{\sigma}{1 - \varepsilon_d} \tag{7.4.7a}$$

$$d_g = d_{g_0} \left[\frac{1 - \varepsilon}{1 - \varepsilon_0} \right]^{1/3} \tag{7.4.7b}$$

The base value of η_s [namely, the initial single collector efficiency] can be estimated using some of the correlations given in Chapter 6. Wnek et al. suggested that $(\eta_s)_0$ can be taken as the sum of the collector efficiencies due to interception and diffusion, assuming that there was no repulsive surface interaction effect. Alternatively, η_s may be obtained from the correlation of Equation (6.4.5) and possibly corrected by Equation (6.5.7) in case the initial surface interaction is unfavorable.

To account for the effect due to grain surface charges, the condition at collector/particle surface can be characterized by their surface potentials (through electrophoretic measurements). The surface potential–surface charge density relationships are:

For suspended particles

$$\hat{q}_p = \hat{\varepsilon}(1 + \kappa a_p)\zeta_p/a_p \tag{7.4.8a}$$

For filter grains

$$\hat{q}_g = \hat{\varepsilon}\kappa\,\zeta_g \tag{7.4.8b}$$

where $\hat{\varepsilon}, \zeta_p, \zeta_g$ are the permittivity, particle, and grain surface potentials, respectively (see 6.5).

Equations (7.4.3)–(7.4.5) together with expressions of the collector efficiency (including relationships relating the change of η_s with the change of filter grain surface potential) and appropriate initial and boundary conditions provide a complete description of the dynamics of deep bed filtration involving charge transfer. The solutions provide not only filtration performance (namely, effluent concentration history and specific deposit profile) but also the charge distribution through the bed as a function of time.

The specific procedure used by Wnek et al. in estimating η_s is as follows: For submicron particles, the dominant mechanism of deposition is diffusion. The effect of surface interactions may be approximately expressed as

$$\frac{\eta_s}{\eta_s^{(0)}} = \frac{1}{1 + I^{(0)}/I_s^{(0)}} \tag{7.4.9}$$

where η_s^0 is the single collector efficiency without surface interaction effect. $I^{(0)}$ denotes the particle flux over a filter grain due to diffusion only and $I_s^{(0)}$ is the flux due to surface reaction which can be used to represent the surface charge effect. The two fluxes are given as

$$I^{(0)} = \pi \, a_g^2 u_s \tilde{c} \, \eta_s^{(0)} \tag{7.4.10a}$$

$$I_s^{(0)} = 4\pi \, a_g^2 K \cdot \tilde{c} \tag{7.4.10b}$$

where K is the virtual rate constant which accounts for the surface interaction effect (Ruckenstein and Prieve, 1973; Spielman and Friedlander, 1974). K can be expressed approximately as

$$K = \frac{D_{BM}}{f_r^t(\delta_m) \exp(E/kT)} \sqrt{\frac{\gamma}{2\pi kT}} \tag{7.4.11}$$

where D_{BM} is the Brownian diffusivity [given by Equation (6.1.21)]. E is the maximum of the grain–particle interaction energy potential curve $\phi = \phi_{Lo} + \phi_{DL}$ vs. $\delta^+ = \delta/a_p$ at $\delta^+ = \delta_m^+$) and γ is $\phi''(\delta_m^+) \cdot f_r^t$ is the hydrodynamic retardation factor. An approximate expression of f_r^t given by Dahneke (1974) is

$$f_r^t(\delta^+) \cong \frac{\delta^+}{1 + \delta^+} \tag{7.4.12}$$

The results of one sample calculation (filtration of ferric oxide through packed beds) of Wnek et al.'s are shown in Fig. 7.5. Also shown in the figure are experimental data of Heertjes and Lerk (1962). The condition used to obtain the results is given in Table 7.2. Good agreement between experiments and prediction was observed although the same degree of agreement was not found in all cases considered by Wnek et al.

7.4.2   Expressing Surface Interaction Effect in terms of Particle Re-Entrainment

In Chapter 5 (see 5.2), the inclusion of the re-entrainment of deposited particles in expressing filtration rate is mentioned. If one considers that the extent of re-entrainment is proportional to the amount of deposited particles, the net rate of filtration can be expected to decrease with time during a filtration cycle. Rajagopalan and Chu (1982) proposed an analogy between particle re-entrainment and the change of the filter grain–particle surface interactions from being favorable to unfavorable due to deposited particles. Based on this analogy, the classical Thomas solution for fixed-bed adsorption may be applied to describe the breakthrough behavior of effluent particle concentration history.

Particle re-entrainment may be viewed as the opposite of deposition. Therefore, if re-entrainment is included in expressing filtration rate, particle deposition/re-entrainment may be viewed as a reversible chemical reaction. For this purpose, the extent of deposition may be described by the number of deposited particles per unit grain surface area, c_s. c_s is related to specific deposit, σ, by the expression

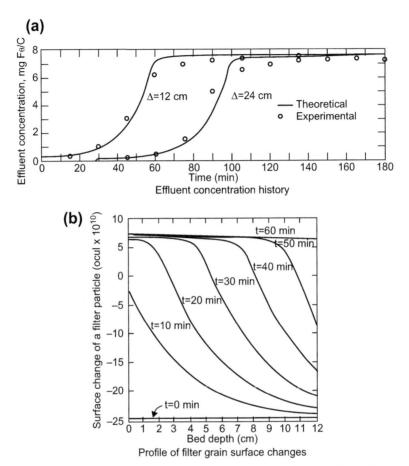

FIGURE 7.5 Comparison of Prediction based on Electric Charge Transfer (Wnek et al., 1995) with experiments (a) Effluent Concentration History, (b) Surface Charge Profiles. *Reprinted from Chemical Eng. Sci., W.J. Wnek, D. Gidaspow and D.T. Wasen, "The role of colloid chemistry in modeling deep bed liquid filtration", 30, 1035–1047, 1975, with permission of Elsevier.*

$$\sigma = (1 - \varepsilon)\rho_g a_s c_s \left(\frac{4}{3} \pi \, a_p^3\right) \tag{7.4.13}$$

where ρ_g is the filter grain density and a_s the specific surface area per unit filter grain mass with filter grain being spherical and uniform in size (with radius a_p).

The rate of filtration including the re-entrainment effect may be written as

$$\frac{\partial c_s}{\partial t} = k_f \tilde{c} - k_r c_s \tag{7.4.14}$$

where $\tilde{c}$ is the particle concentration (in numbers) of the suspension to be treated, k_f and k_r are the deposition (forward reaction) and re-entrainment (reverse reaction) rate constants, respectively.

Table 7.2   Experimental Conditions of Heertjes and Lerk (1967) *

Mass concentration of Suspension	7.7 mg Fe/1
Radius of Suspension Particles	200 Å
Density of Suspension Material [Fe(OH)$_3$]	3.61 g cm^{-3}
Particle Concentration of Suspension	1.2×10^{11} particles cm^{-3}
Radius of Filter Particles	250 μm
Superficial Velocity	0.07 cm s^{-1}
Height of Filter	12 and 24 cm
Porosity	0.36
Temperature	25 °C
Conductivity	5.6×10^{-5} ohm^{-1} cm^{-1}
Ionic Strength	4.43×10^{-4} mol/1
Hamaker Constant	0.5 to 1.5×10^{-12} erg
Initial Zeta Potential of Filter Particles	−55 mv
Initial Surface Charge of a Filter Particle	-2.5×10^{-9} coul
Zeta Potential of Suspension Particles	+34.5 mv

*These conditions are also used for the predictions shown in Fig. 7.5.

The macroscopic conservation equation and filtration rate equation expressed in $\tilde{c}$ and c_s are

$$u_s \frac{\partial \tilde{c}}{\partial z} + \rho_g a_s (1 - \varepsilon) \frac{\partial c_s}{\partial t} = 0 \tag{7.4.15}$$

$$\frac{\partial c_s}{\partial t} = k_f \tilde{c} - k_r c_s \tag{7.4.16}$$

and

$$\begin{aligned} \tilde{c} &= \tilde{c}_{in} & z &= 0, \quad \theta > 0 \\ \tilde{c} &= c_s = 0, & z &> 0, \quad \theta \le 0 \end{aligned} \tag{7.4.17}$$

Introducing the following dimensionless quantities

$$c^+ = c/c_{in} \tag{7.4.18a}$$

$$c_s^+ = \frac{c_s}{c_0 (k_f/k_r)} \tag{7.4.18b}$$

$$\theta^+ = k_r \left(t - \frac{z}{u_s/\varepsilon} \right) \tag{7.4.18c}$$

$$z^+ = \frac{\rho_g a_s (1 - \varepsilon) k_f z}{u_s} \tag{7.4.18d}$$

the above set of equations become

$$\frac{\partial c^+}{\partial z^+} + \frac{\partial c_s^+}{\partial \theta^+} = 0 \tag{7.4.19}$$

$$\frac{\partial c_s^+}{\partial \theta^+} = c^+ - c_s^+ \tag{7.4.20}$$

and

$$c^+ = 1, \qquad z^+ = 0, \quad \theta^+ > 0$$
$$c^+ = c_s^+ = 0, \quad z^+ > 0, \quad \theta \leq 0 \tag{7.4.21}$$

The solution of Equations (7.4.19)–(7.4.21), known as the Thomas solution of the linear case, is given as

$$c^+(z^+, \theta^+) = J(z^+, \theta^+) \tag{7.4.22}$$

$$c_s^+(z^+, \theta) = 1 - J(\theta^+, z^+) \tag{7.4.23}$$

and

$$J(\alpha, \beta) = 1 - e^{-\beta} \int_0^\alpha e^{-\zeta} I_0(2\sqrt{\beta\zeta}) d\zeta \tag{7.4.24}$$

Tabulated values of $J(\alpha, \beta)$ are given in the text of Sherwood et al. (1975). Asymptotic expressions of $J(\alpha, \beta)$ have been given by several investigators, a complete listing of which $J(\alpha, \beta)$ can be found in Beveridge (1962).

According to Thomas (1944), for large values of α and β, $J(\alpha, \beta)$ may be approximated as

$$J(\alpha, \beta) \cong \begin{array}{ll} (1/2)\text{erf}(\sqrt{\alpha} - \sqrt{\beta} & \text{if} \quad \beta < \alpha \\ (1/2)[1 + \text{erf}(\sqrt{\alpha} - \sqrt{\beta})] & \text{if} \quad \beta > \alpha \end{array} \tag{7.4.25}$$

for $\alpha\beta > 3{,}600$

As the breakthrough behavior of deep bed filtration is similar to that of fixed-bed adsorption; namely, the effluent concentration increases with time and approaches to the influent concentration, it is not surprising that Equation (7.4.25) can be applied to describe deep bed filtration performance. An example of using Equation (7.4.25) describing one set of data obtained by Chu is shown in Fig. 7.6. The filtration experiment was conducted using a filter bed of relatively shallow height, 7.8 cm, packed with glass beads ($a_g = 200$ µm) with $\varepsilon = 0.35$, test suspension of chromium (111) ion particle ($a_p = 0.15$ µm) and $\tilde{c}_{in} = 1.55 \times 10^8$ particles/cm^{3}s at $u_s = 0.14$ cm s^{-1}. The particle and glass bead surface potentials were found to be $\zeta_g = -30$ mV and $\zeta_p = 38$ mV

The prediction shown in the figure was obtained by treating the two rate constants as fitting parameters with $k_f = 1.80 \times 10^{-4}$ cm s^{-1} and $k_r = 7.69 \times 10^{-5}$ s^{-1}. If the Thomas solution is to be used as a tool of prediction or for scale-up, rational methods which may be used to estimate the rate constants (either from experimental data or estimated based

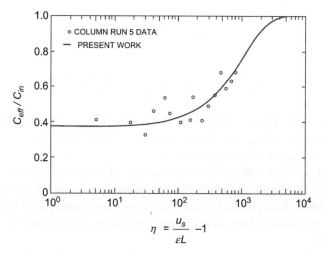

FIGURE 7.6 Adoption of Thomas solution for predicting the effect of changing surface interactions in deep bed filtration C_{eff} Vs η. Reprinted from *J. Colliod Interface Sci.*, R. Rajagopalan and R.Q. Chu, "Dynamics of adsorption of colloid particles in packed beds", 86, 299–317, 1982, with permission of Elsevier.

on theories or correlations as discussed before) are required. An outline of these methods is given below.

A. Relationship between Rate Constant and Saturation Surface Concentration. An approximate relationship between the rate constants and certain filtration parameters discussed before will be established first. It is simple to show that the forward rate constant, k_f, is directly related to the single collection efficiency, η_s. By definition, in the absence of re-entrainment, one has

$$4\pi a_g^2 \frac{dc_s}{dt} = \left(4\pi a_g^2\right) k_f \tilde{c}_s = \left(\pi a_g^2\right) u_s \tilde{c} \eta_s \tag{7.4.26a}$$

or

$$k_f = \left(\frac{1}{4}\right) u_s \cdot \eta_s \tag{7.4.26b}$$

Next referring to Equation (7.4.14), deposition ceases when $\tilde{c} \to \tilde{c}_{in}$ and $c_s \to c_{s,sat}$, where $c_{s,sat}$ is the saturation surface concentration. $c_{s,sat}$ is given as

$$c_{s,sat} = (k_f/k_r)\tilde{c}_{in} \tag{7.4.27}$$

The above expression provides a relationship among k_f, k_r, and $c_{s,sat}$. In the context of the present discussion, if one of the rate constants and $c_{s,sat}$ are known, the other rate constant can be readily determined.

B. Determination of $c_{s,sat}$ for Deep Bed Filtration. The saturation surface concentration, by definition, can be expressed as

$$c_{s,sat} = \frac{3L(1-\varepsilon)}{a_g}(u_s) \cdot \int_0^{t_\infty} (\tilde{c}_{in} - \tilde{c}_{eff})dt$$

$$= \frac{3L(1-\varepsilon)}{a_g}(u_s)(\tilde{c}_{in}) \int_0^{t_\infty} \left[1 - \left(\frac{c_{eff}}{c_{in}}\right)\right]dt$$

(7.4.28)

where a_g is the filter grain radius, L the bed height, t_∞ is the time when the effluent concentration is indistinguishable from the influent concentration, and the other symbols have the same meaning as before.

To account for the effect of the change of surface interactions on particle deposition due to charge transfer, the relationships between the surface charge density and the surface potential for both filter grains and particles must be known. The simplest expression between $\hat{q}_s$ and ζ given by Gouy and Chapman (Shaw, 1970, p. 138) is

$$\hat{q}_s = (8 n_0 \hat{\varepsilon}_0 \varepsilon_r k_B T)^{1/2} \sinh \frac{ze\zeta}{2k_B T}$$

(7.4.29)

where n_0 is the bulk concentration of the ionic species, z the charge number, and e the electron charge. $\hat{\varepsilon}_0$ is the permittivity of a vacuum and ε_r is the dielectric constant of the medium. For low surface potential, the above expression reduces to

$$\hat{q}_s = \hat{\varepsilon}_0 \varepsilon_r \kappa \zeta$$

(7.4.30)[3]

where κ is the reciprocal double-layer thickness.

Assuming that deposition results in a change of filter grain surface potential due to charge transfer, let $(\zeta_g)_{initial}$ and $(\zeta_g)_{final}$ be the initial grain surface potential (i.e., grains free of deposited particles) and the final grain surface potential (corresponding to the completion of breakthrough or $c_{eff} \simeq c_{in}$). The surface charge density change of a filter grain, $(\Delta\hat{q}_s)_g$, (Coulombs per unit grain surface area) according to Equation (7.4.30) is

$$\Delta(\hat{q}_s)_g = (\Delta\zeta_g)(\varepsilon_r\hat{\varepsilon}_0)\kappa$$

(7.4.31a)

with

$$\Delta\zeta_g = (\zeta_g)_{final} - (\zeta_g)_{initial}$$

(7.4.31b)

The surface charge of a single particle is

$$\zeta_p(\varepsilon_r\hat{\varepsilon}_0) \cdot \kappa \cdot 4\pi a_p^2$$

(7.4.32)

---

[3]Wnek suggested a modification of Equation (7.4.30) or $q_s = \varepsilon_r\hat{\varepsilon}_0\kappa\,\zeta\left(1 + \dfrac{1}{\kappa a_p}\right)$.

Therefore, the saturation surface concentration is

$$c_{s,sat} = \frac{(\Delta \zeta_g)}{(\zeta_p)(4\pi a_p^2)} \tag{7.4.33}$$

C. Estimation/Determination of Rate Constants. The reverse rate constant can be determined from Equation (7.4.27) assuming that the values of $c_{s,sat}$ and k_f are known. The forward rate constant is related to the single collection efficiency, η_s, by Equation (7.4.26b), which can be estimated by applying the various filter coefficient correlations [for example, Equations (6.1.28), (6.4.5), or (6.5.7)] depending upon the operating conditions and Equation (5.4.20) which relates η_s to the filter coefficient. Alternatively, one may determine the initial filter coefficient from effluent concentration history using the procedure outlined in 6.2.

■ ■ ■ ───

Illustrative Example 7.6

Obtain an expression which gives the change of the surface potential as a result of charge transfer from deposition of charged particles assuming that the particle surface potential and the initial grain surface potential are given and both are negative as in most cases. State the assumption used clearly.

Solution

The following assumptions are used:

1. The relationships between surface potential and surface charge density are given by Equations (7.4.30) and its modified version (see footnote of p. 378).
2. Simple additive law is used to describe the charge transfer. The surface charge of a filter grain with deposited particles is equal to the initial grain surface charge plus the surface charges of the individual deposited particles.
3. Both particles and grains are negatively charged initially.

The relationship between the surface charge density and surface potential are:

$$q_g = \hat{\varepsilon}_0 \, \varepsilon_r \, \kappa \, \zeta_g \tag{i}$$

$$q_p = \hat{\varepsilon}_0 \, \varepsilon_r \, \kappa \, \zeta_p \left(1 + \frac{1}{\kappa a_p}\right) \tag{ii}$$

The surface charge of a filter grain with N deposited particles, as a result of charge transfer, may be written as

$$(q_g)(\pi d_g^2) = (q_g)_0(\pi d_g^2) + N(q_p)(\pi d_p^2) \tag{iii}$$

where $(q_g)_0$ is the initial value of q_g and N is related to c_s by the expression

$$N = (\pi d_g^2) \cdot c_s \tag{iv}$$

Combining Equations (i)–(iv) yields

$$\hat{\varepsilon}_0 \varepsilon_r \, \kappa \, \zeta_g \;=\; \hat{\varepsilon}\varepsilon_0 \kappa (\zeta_g)_0 + c_s(\pi d_g^2)\left(\frac{d_p}{d_g}\right)^2 \hat{\varepsilon}_0 \varepsilon_r \; \kappa \; \zeta_p\left(1 + \frac{1}{\kappa a_p}\right)$$

or

$$\zeta_g \;=\; (\zeta_g)_0 + c_s(\pi d_g^2)(d_p/d_g)^2\left(1 + \frac{1}{\kappa a_p}\right)\zeta_p \qquad\qquad \text{(v)}$$

One may write c_s in terms of σ as

$$(c_s)\left(\pi d_g^2\right)\left(\frac{\pi}{6}\,d_p^3\right) \;=\; (\sigma)(\pi/6)\left(d_g^3\right)/(1 - \varepsilon)$$

or

$$c_s \pi \;=\; [\sigma/(1 - \varepsilon)](d_g/d_p)^3 \qquad\qquad \text{(vi)}$$

Substituting Equation (vi) into (v), we have the required expression

$$(\zeta_g) \;=\; (\zeta_g)_0 + [\sigma/(1 - \varepsilon)](d_g/d_p)\left(1 + \frac{1}{\kappa a_p}\right)\zeta_p$$

or

$$(\zeta_g)/(\zeta_g)_0 \;=\; 1 + [\sigma/(1 - \varepsilon)](d_g/d_p)(1 + (\kappa a_p)^{-1}(\zeta_p/\zeta_{q_0}) \qquad\qquad \text{(vii)}$$

∎ ∎ ∎

7.4.3   A Model Based on Collector Surface Heterogeneity

In both 7.4.1 and 7.4.2, the transient state of deep bed filtration was analyzed by considering electric charge transport associated with particle deposition on an overall (for entire filter grain) basis. However, as the coverage of deposited particles over filter grains may not be complete, it may be necessary to examine the deposition effect locally; in other words, the surface of a grain should be considered to be consisting of parts with deposited particles and parts free of deposited particles. To account for the surface heterogeneity in this manner and specifically, their differences in surface interaction with suspended particles, Bai and Tien (2000) proposed a set of macroscopic conservation equations different from those given in Chapter 5. A generalization of the Bai–Tien formulation is given below.

Consider the filter grain surface at a given instant to be of two types: Type "1" or the parts free of deposited particles and Type "2" with deposited particles. The specific deposit σ may be written as

$$\sigma = \sigma^1 + \sigma^2 \qquad\qquad (7.4.34)[4]$$

where σ^1 and σ^2 are the amount of deposited particles on filter grains and that on deposited particles.

[4] σ^1 refers to those deposited particles in mono-layer coverage and σ^2 those in multi-layer coverage.

Analogous to the filtration rate equation used before, one may write

$$\frac{\partial \sigma^1}{\partial \theta} = u_s \lambda^1 c (1 - f) \tag{7.4.35a}$$

$$\frac{\partial \sigma^2}{\partial \theta} = u_s \lambda^2 c \, f \tag{7.4.35b}$$

where $\lambda^1 \, \lambda^2$ are the respective filter coefficients of the two parts. f is the fraction of grain surface covered with deposited particles. By geometrical consideration, f can be written as

$$f = \frac{\left[\sigma^1 / \left\{(4\pi/3)(a_p^3)\right\}\right] \pi a_p^2}{\left[(1 - \varepsilon_0)/\left\{(4\pi/3)(a_c^3)\right\}\right](4\pi a_c^2)} \cdot k_1$$

$$= \frac{k_1 a_c \sigma^1}{4 a_p (1 - \varepsilon_0)} \tag{7.4.36}$$

where k_1 is a factor to account for that part of the surface immediately adjacent to deposited particles which may not be available for deposition (i.e., the exclusion effect).

The filter coefficient λ^1 may be considered to be the same as the initial filter coefficient or

$$\lambda^1 = (\lambda^1)_0 \tag{7.4.37}$$

As the extent of deposition increases, λ^2 may be expressed as

$$\lambda^2 = (\lambda^2)_0 F(\underline{\alpha}, \sigma^2) \tag{7.4.38}$$

Table 7.3   Conditions Used to Obtain the Results Shown in Fig. 7.7a–c

$u_s = 3.7$ m/hr		$c_{in} = 50$ mg/ℓ	$\rho_p = 1050$ kg m^{-3}	
$\varepsilon_0 = 0.41$		$d_p = 3.0663$ μm	$d_g = 230$ μm	$K_1 = 11$
Fig. 7.7a	Case (1)	$(\lambda^1)_0 = 15$ m^{-1}	$\lambda^2 = 0$	
	Case (2)	$(\lambda^1)_0 = 3$ m^{-1}	$\lambda^2 = 0$	
Fig. 7.7b	Case (1)	$(\lambda^1)_0 = 5$ m^{-1}	$\lambda^2 = 10$ m^{-1}	
	Case (2)	$(\lambda^1)_0 = 15$ m^{-1}	$\lambda^2 = 15$ m^{-1}	
	Case (3)	$(\lambda^1)_0 = 10$ m^{-1}	$\lambda^2 = 5$ m^{-1}	
Fig. 7.7c	$F = 1 - k_3\sigma^2$			
	Case (1)	$(\lambda^1)_0 = 15$ m^{-1}	$\lambda_0^2 = 15$ m^{-1}	
	Case (2)	$(\lambda^1)_0 = 10$ m^{-1}	$\lambda_0^2 = 5$ m^{-1}	
	Case (3)	$(\lambda^1)_0 = 5$ m^{-1}	$\lambda_0^2 = 10$ m^{-1}	

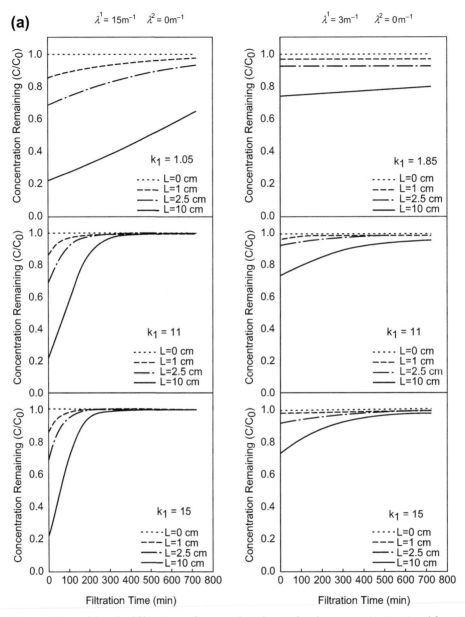

FIGURE 7.7 Predictions of deep bed filtration performance based on surface heterogeneity. *Reprinted from Colliods and Surfaces A: Physicochemical and Engineering Aspects, R. Bai and C. Tien, "Transient behavior of particle deposition in granular media under various surface interactions", 165, 95–114, 2000, with permission of Elsevier.*

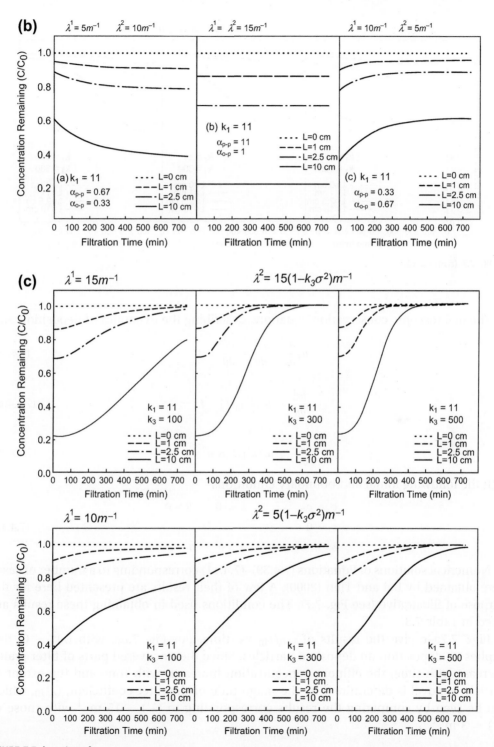

FIGURE 7.7 (*continued*).

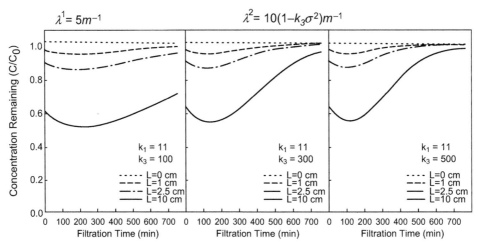

FIGURE 7.7 *(continued).*

The macroscopic conservation equations describing the dynamics of deposition are

$$u_s \frac{\partial c}{\partial z} + \frac{\partial \sigma^1}{\partial \theta} + \frac{\partial \sigma^2}{\partial \theta} = 0 \tag{7.4.39}$$

$$\frac{\partial \sigma^1}{\partial \theta} = u_s(\lambda^1)_0 c(1 - f) \tag{7.4.40a}$$

$$\frac{\partial \sigma^2}{\partial \theta} = u_s(\lambda^2)_0 F(\alpha, \sigma^2) c \tag{7.4.40b}$$

with the following initial and boundary conditions

$$
\begin{aligned}
c &= c_{\text{in}}, & z &= 0 & \theta &> 0 \\
c &= 0 \\
\sigma^1 &= \sigma^2 = 0 & z &> 0 & \theta &\le 0
\end{aligned}
\tag{7.4.41}
$$

Numerical solutions of Equations (7.4.39)–(7.4.11) corresponding to a number of cases were obtained by Bai and Tien (2000). A few of their results are presented here for the purpose of illustrative (see Fig. 7.7). The conditions used in obtaining these results are listed in Table 7.3.

Figs. 7.7a–c give the results of $c_{\text{eff}}/c_{\text{in}}$ vs. time. For Fig. 7.8a, with $(\lambda)^2 = 0$, this implies no deposition on deposited particles. Since the uncovered parts of filter grains decrease with time, the effluent concentration increases with time and the extent of particle removal is determined by the magnitude of the filter coefficient, $(\lambda^1)_0$, which can be seen by comparing the results corresponding to $(\lambda^1) = 15 \text{ m}^{-1}$ with those of $(\lambda^1) = 3 \text{ m}^{-1}$.

In Fig. 7.7b, the effect of the relative magnitude of $(\lambda)^1$ to $(\lambda)^2$ is shown. With $F(\alpha, \sigma^2) = 1$ and $\lambda^1 = \lambda^2$, the rate of filtration $\partial(\sigma^1 + \sigma^2)/\partial\theta$ is constant and the effluent concentration remains unchanged with time. For $(\lambda)^1 > (\lambda)^2$ the decrease of $(\partial\sigma^1/\partial\theta)$ with time due to the increase of f is not sufficiently compensated by $(\partial\sigma^2/\partial\theta)$. As a result, the effluent concentration increases with time. The contrary is found in the case of $(\lambda)^2 > (\lambda)^1$ as $\partial(\sigma^1 + \sigma^2)/\partial\theta$ now increases with time.

Fig. 7.7c gives the results corresponding to more complexed relationships between $\partial\sigma^1/\partial\theta$ and $\partial\sigma^2/\partial\theta$. With $\lambda^1 = 15\ \text{m}^{-1}$ and $\lambda^2 = 15\,(1 - k_3\sigma^2)\text{m}^{-1}$, for small values of σ^2, $\partial\sigma/\partial\theta$ is nearly constant but then decreases with the increase of σ^2 (or time). The S-shaped effluent concentration history curve is, therefore, expected. On the other hand, with $\lambda^1 = 10\ \text{m}^{-1}$ and $\lambda^2 = 5\,(1 - k_3\sigma^2)\text{m}^{-1}$, $\partial\sigma/\partial\theta$ decreases rapidly with time (or σ^2) and the effluent history curve is quite similar to what is shown in Fig. 7.7a. For the case of $\lambda^1 = 5\ \text{m}^{-1}$ and $\lambda^2 = 10(1 - k_3\sigma^2)$, $\partial\sigma/\partial\theta$ first increases with time (or σ^2) and then decreases. The effluent concentration first decreases with time until a minimum is reached and then increases.

The few examples discussed above demonstrate that one may simulate a variety of filtration behavior through the use of different rate expressions. It also shows the limitation of the lumped parameter macroscopic approach discussed here, namely, the same kind of behavior may be obtained through the use of different rate expressions. The validity of an assumed rate expression cannot be assured even if there is good agreement between the predicted performance based on it with experiments.

■ ■ ■ ▬▬▬▬▬▬▬▬▬▬▬▬▬▬▬▬▬▬▬▬▬▬▬▬▬▬▬▬▬▬▬

Illustrative Example 7.7

Compare the model of Bai and Tien based on surface heterogeneity and that of Vigneswaran and Tulachan.

Solution

The filtration rate according to the Bai–Tien model is expressed as [see Equations (7.4.34), (7.4.40a), and (7.4.40b)]

$$\frac{\partial\sigma}{\partial\theta} = \frac{\partial\sigma^1}{\partial\theta} + \frac{\partial\sigma^2}{\alpha\partial\theta}$$

$$= u_s(\lambda^1)_0 c(1 - f) + u_s(\lambda^2)_0 F(\alpha, \sigma^2) cf \tag{i}$$

Vigneswaran and Tulachan express filtration rate as [see Equations (7.3.1), (7.3.4), and (7.3.6)]

$$\frac{\partial\sigma}{\partial\theta} = (d_p/d_g)^3(1 - \varepsilon_0)\frac{\partial N_t}{\partial\theta}$$

$$= (d_p/d_g)^3(1 - \varepsilon_0)(\pi/4)d_g^2 u_s \tilde{c}\left[\eta_{s_0}\frac{N_{\max} - N}{N_{\max}} + N_t(\eta_s)_p(d_p/d_g)^2\frac{\varepsilon - \varepsilon^*}{\varepsilon_0 - \varepsilon}\right] \tag{ii}$$

The first term of the above expression

$$(d_p/d_g)^3(1 - \varepsilon_0)(\pi/4)d_g^2 u_s \tilde{c} \; \eta_{s_0} \frac{N_{max} - N}{N_{max}} = \left(\frac{3}{2}\right)(1 - \varepsilon_0)\frac{u_s}{d_g}(\pi/6)d_p^3 \tilde{c}\eta_{s_0}\frac{N_{max} - N}{N_{max}}$$

is equivalent to the first terms of Equation (i) since $c = (\pi/6)d_p^3\tilde{c}(3/2)(1 - \varepsilon_0)(\eta_{s_0}/d_g)$, is equivalent to $(\lambda')_0$, namely, the filter coefficient of filter grain free of deposited particles and $(N_{max} - N)/N_{max}$ may be considered as the fraction of the surface area uncovered with deposited particles.

The second term of Equation (ii) may be rewritten as

$$(d_p/d_g)^3(1 - \varepsilon_0)(\pi/4)d_g^2 u_s \tilde{c}(\eta_s)_p(d_p/d_g)^2[(\varepsilon - \varepsilon^*)/(\varepsilon_0 - \varepsilon^*)] \left(\frac{N_t}{N_{max}}\right) N_{max}$$

$$= \left(\frac{3}{2}\right)(1 - \varepsilon_0)\frac{u_s}{d_g}c(\eta_s)_p(d_p/d_g)^2[(\varepsilon - \varepsilon^*)/(\varepsilon_0 - \varepsilon^*)]\frac{N}{N_{max}}\frac{N_{max}}{(N/N_t)} \qquad \text{(iii)}$$

In the work of Vigneswaran and Tulachun, N is assumed, as an approximation, to be a fixed fraction of N_t or $N/N_t = \gamma$.

Since $(3/2)(1 - \varepsilon_0)(\eta_{s_p})/d_g$ may be taken as $(\lambda^2)_0$ and $N/N_{max} = 1 - f$.

Equation (iii) is the same as the second term of Equation (i) if F is taken to be

$$F = (d_p/d_g)^2[(\varepsilon - \varepsilon^*)/(\varepsilon_0 - \varepsilon^*)](N_{max}\gamma)^{-1} \qquad \text{(iv)}$$

The only difference is that F of Equation (i) is assumed to be a function of σ^2 or $N_t - N$. On the other hand, in Equation (iv) F depends upon $\sigma = \sigma^1 + \sigma^2$ or N_t. We may summarize the corresponding quantities given above as follows.

Vigneswaran and Tulachun	Bai and Tien
η_{s_0}	$\lambda_0^1 = (3/2)(1 - \varepsilon_0)(\eta_{s_0})/d_g$
η_{s_p}	$\lambda_0^2 = (3/2)(1 - \varepsilon_0)\eta_p/d_g$
N_t	$\sigma = \sigma^1 + \sigma^2 = (1 - \varepsilon_0)(d_p/d_g)^3 N_t$
N	$\sigma^1 = (1 - \varepsilon_0)(d_p/d_g)^3 N$
$(N_{max} - N)/N_{max}$	$f = 1 - (N/N_{max})$
$(d_p/d_g)^2[(\varepsilon - \varepsilon^*)](N_{max}\gamma)^{-1}$	F

7.5   Modeling Filtration as a Stochastic Process

In the above discussions, deep bed filtration is treated as a deterministic process. A specific filtration rate expression is assumed and filtration performance is definitive and predictable. This deterministic treatment has, to a large degree, been used by most workers in the field although studies, which examine deep bed filtration using nondeterministic approach, have also appeared in the literature in the past.

The use of nondeterministic approach to describe granular filtration was first considered by Litwiniszyn (1963) and was the basis used by Hsiung in his correlation and prediction of the performance of sand filters (1967). The problem was further examined by Fan and coworkers (Hsu and Fan, 1984; Fan et al., 1985). Their treatment of filter pressure-drop history as a birth–death process is discussed below.

If a granular medium is viewed as a large number of interconnected pores, then filtration through the medium causes deposits to accumulate within the pores, ultimately blocking some of them. At the same time, as the pressure drop increases, some of the deposits or parts of them may be reentrained and the blocked pores reopened (or scoured). Thus, the pressure-drop increase may be viewed as resulting from two processes, blockage and scouring, which occur simultaneously and are stochastic in nature.

A stochastic process consists of, by definition, a family of random variables describing an empirical process whose development is governed by probabilistic laws. In treating filtration (or the consequence of filtration) as a stochastic process, Litwiniszyn (1963) considered the number of blocked pores in a unit filter volume at time t, $N(t)$, as the random variable. A specific value of $N(t)$ will be denoted by n. Given $N(t) = n$, it is assumed that for the birth–death process,

(1) The conditional probability that a pore will be blocked (a birth event) during the interval $(t, t + \Delta t)$ is $\lambda_n \Delta t + 0(\Delta t)$, where λ_n is a function of n;

(2) The conditional probability that a blocked pore will be scoured (a death event) during the interval $(t, t + \Delta t)$ is $\mu_n \Delta t + 0(\Delta t)$, where μ_n is a function of n;

(3) The conditional probability that more than one event will occur in the interval $(t, t + \Delta t)$ is $0 (\Delta t)$, where $0 (\Delta t)$ signifies

$$\lim_{\Delta t \to 0} \frac{0(\Delta t)}{\Delta t} = 0 \tag{7.5.1}$$

It is obvious that the probability that there is no change in the interval $(t, t + \Delta t)$ is

$$1 - \lambda_n \Delta t - \mu_n \Delta t - 0(\Delta t)$$

The probability that exactly n pores are blocked is denoted as

$$P(t) = Pr[N(t) = n] \quad n = 0, 1, 2, ...n \tag{7.5.2}$$

Consider two adjacent time intervals, $(0, t)$ and $(t, t + \Delta t)$. The blocking of exactly n pores during the interval $(0, t + \Delta t)$ can be realized in the following mutually exclusive ways:

(1) All n pores are blocked during $(0, t)$ and none during $(t, t + \Delta t)$ with probability $P_n(t)[1 - \lambda_n \Delta t - \mu_n \Delta t - 0(\Delta t)]$.

(2) Exactly $(n - 1)$ pores are blocked during $(0, t)$, and one pore is blocked during $(t, t + \Delta t)$ with probability $P_{n-1}(t)[\lambda_{n-1} \Delta t + 0(\Delta t)]$.

(3) Exactly $(n+1)$ pores are blocked during $(0, t)$ and one blocked pore is scoured during $(t, t + \Delta t)$ with probability $P_{n-1}(t)[\mu_{n-1} \Delta t + 0(\Delta t)]$.

(4) Exactly $(n - j)$ pores, where $2 \leq j \leq n$, are blocked during $(0, t)$, and j pore are blocked during $(t, t + \Delta t)$ with probability $0(\Delta t)$.

(5) Exactly $(n + j)$, where $2 \leq j \leq (n_0 - n)$, are blocked during $(0, t)$, and j blocked pores are scoured during $(t, t + \Delta t)$ with probability $0(\Delta t)$.

Thus, the probability of having n pores blocked at $(t, t + \Delta t)$ is

$$P_n(t + \Delta t) = P_n(t)[1 - \lambda_n \Delta t - \mu_n \Delta t] + P_{n-1}(t)\lambda_{n-1}\Delta t + P_{n+1}(t)\mu_{n+1}\Delta t + 0(\Delta t), \quad n \geq 1$$

and

$$P_0(t + \Delta t) = P_0(t)[1 - \lambda_0 \Delta t] + P_1(t)\mu_1 \Delta t + 0(\Delta t) \tag{7.5.3}$$

Rearranging the above expression and taking the limit as $\Delta t \to 0$ yields the so-called master equations:

$$\frac{dP_n(t)}{dt} = \lambda_{n-1}P_{n-1}(t) - (\lambda_n + \mu_n)P_n(t) + \mu_{n+1}P_{n+1}(t), \quad n \geq 1 \tag{7.5.4}$$

$$\frac{dP_n(t)}{dt} = -\lambda_0 P_0(t) + \mu_1 P_1(t) \tag{7.5.5}$$

Litwiniszyn (1963) assumed that

$$\lambda_n = \alpha(n_0 - n) \quad n = 0, 1, 2, ..., n_0 \tag{7.5.6}$$

$$\mu_n = \beta n \tag{7.5.7}$$

where n_0 is the total number of open pores in a clean filter.

Equations (7.5.4) and (7.5.5) become

$$\frac{dP_n(t)}{dt} = \alpha[n_0 - (n - 1)]P_{n-1}(t) - \alpha(n_0 - n) + \beta n]P_n(t) + \beta(n + 1)P_{n+1}(t), \quad n \geq 1 \tag{7.5.8}$$

$$\frac{dP_0(t)}{dt} = -\alpha n_0 P_0(t) + \beta P_1(t) \tag{7.5.9}$$

For a clean filter, all the pores are open. Accordingly, the initial conditions of the above two equations are

$$P_n(0) = 0, \quad n = 1, 2, ..., n \tag{7.5.10}$$

$$P_0(0) = 1 \tag{7.5.11}$$

The solution of Equations (7.5.8) and (7.5.9) subject to the initial condition of Equations (7.5.10) and (7.5.11) yields the distribution of the pore-blockage probabilities. For the problem at hand, we are not interested in such detailed information, but rather in the most likely event, namely, the expected number of blocked pores at time t, $E[N(t)]$, which is given as

$$E[N(t)] = \sum_{n=0}^{n_0} n P_n(t) \tag{7.5.12}$$

To evaluate this quantity, solving Equations (7.5.8) And (7.5.9) is unnecessary. Instead, one can use the method of probability-generating function, defined as

$$g[s, t] = \sum_{n=0}^{n_0} s^n P_n(t) \tag{7.5.13}$$

Examining the definitions of $g(s,t)$ and $E[N(t)]$ reveals that

$$E[N(t)] = \left. \frac{\partial g(s, t)}{\partial s} \right|_{s=1} \tag{7.5.14}$$

By combining Equations (7.5.13), (7.5.8), and (7.5.9) and applying the definition of $g(s,t)$, it can be shown (Fan et al., 1985) that $g(s,t)$ is the solution to the following equation:

$$\frac{\partial g(s, t)}{\partial t} = \frac{\partial g(s, t)}{\partial s} \left[\beta + (\alpha - \beta)s - \alpha s^2 \right] + g(s, t)[\alpha n_0(s - 1)] \tag{7.5.15}$$

with the following initial and boundary conditions:

$$g(s, 0) = 1 \tag{7.5.16a}$$

$$g(l, t) = 1 \tag{7.5.16b}$$

$$g(s, t) = \left[\frac{(\alpha s + \beta) - \alpha(s - 1)e^{-(\alpha+\beta)t}}{\alpha + \beta} \right]^{n_0} \tag{7.5.17}$$

Substituting Equations (7.5.17) into Equation (7.5.14), one has

$$E[N(t)] = \alpha n_0 \left[\frac{1 - e^{-(\alpha+\beta)t}}{\alpha + \beta} \right] \tag{7.5.18}$$

The above results can now be incorporated with the pressure drop–flow rate relationship discussed previously [namely, the Kozeny–Carman equation, or Equation (7.2.1) with $\phi_s = 1$]. Since the consequence of particle deposition is to block some of the pores, the effective cross-sectional area of the filter flow decreases as the number of blocked pores increases. Therefore, under the constant flow condition, the effective superficial velocity is inversely proportional to the number of pores remaining unblocked. Applying Equation (7.5.18), we see that the pressure-drop across a filter bed is proportional to the number of unblocked pores. Accordingly, the ratio of the pressure of a clogged filter to that of a clean filter is

$$\frac{(-\Delta p/L)}{(-\Delta p/L)_0} = \frac{n_0}{n_0 - E[N(t)]} = \frac{\alpha + \beta}{\beta + \alpha e^{-(\alpha+\beta)t}} \tag{7.5.19}$$

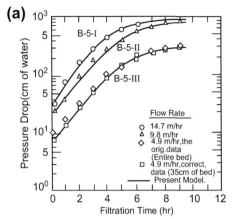

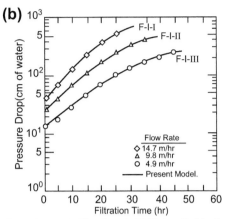

Fitting Equation (2.82) model to Huang's (1972) data. Sand size, 0.092 cm; bed depth, 50.8 cm; waste water suspension, 12.5 mg/L
Runs: B-5-I($\alpha = 0.787$ hr^{-1}, $\beta = 0.0033$ hr^{-1}, $R^2 = 0.996$)
B-5-II($\alpha = 0.690$ hr^{-1}, $\beta = 0.019$ hr^{-1}, $R^2 = 0.997$)
B-5-III($\alpha = 0.698$ hr^{-1}, $\beta = 0.017$ hr^{-1}, $R^2 = 0.999$)

Fitting Equation (2.82) to Huang's (1972) data. Bed depth, 60.96 cm: anthracite, 30.48 cm, $d_a = 0.184$cm; sand, 30.48 cm, $d_s = 0.055$ cm; waste water suspension 12.5 mg/L
Runs: F-1-I ($\alpha = 0.122$ hr^{-1}, $\beta = 0.0054$ hr^{-1}, $R^2 = 0.999$)
F-1-II ($\alpha = 0.100$ hr^{-1}, $\beta = 0.0045$ hr^{-1}, $R^2 = 0.992$)
F-1-III ($\alpha = 0.0089$ hr^{-1}, $\beta = 0.0040$ hr^{-1}, $R^2 = 0.998$)

FIGURE 7.8 Pressure drop change with time as a birth–death process. *Reprinted from L. Fan, R. Nassar, S.H. Hwang, and S.T. Chou, "Analysis of deep bed filtration data: Modeling as a birth-death process", AIChE J., 31, 1781–1790, 1985, with permission from John Wiley and Sons, Inc.*

The ability of Equation (7.5.19) to represent the pressure-drop history of granular filters is shown in Fig. 7.8. These data are those reported by Huang (1972) on the filtration of a wastewater suspension through both single medium and multimedia filters. The effect of scanning becomes significant only at later times.

It should be mentioned that constants α and β present in Equation (7.5.19) are adjustable parameters. The stochastic model discussed above provides no information about the magnitudes of α (or β) nor the relationships between α (or β) with relevant operating variables. Similar to the parameter vector α of the filtration rate correction function, F of Equation (5.2.3) [or β of the pressure-drop correction function of Equation (5.2.5)]. α and β can only be evaluated from appropriate experimental data.

Problems

7.1 Models discussed in 7.2.1 and 7.2.2 were formulated under the constant rate condition; namely, u_s is constant. Modify the model equations which give filtration rate as a function of time if filtration is carried out under constant pressure.

7.2 Obtain equations for the effluent concentration history and pressure-drop history of fibrous filtration under constant rate condition and the uniform deposit hypothesis.

7.3 Rework Illustrative Example 7.1 if $\lambda = \lambda_0(1 + k_1\sigma)$ with $\lambda_0 = 20$ m^{-1} and $k_1 = 50$.

7.4 Rework Illustrative Example 7.2, if the filter coefficient is not independent of σ. The correction factor, $F(\sigma)$ is given by Equation (iii) of Illustrative Example 7.3.

7.5 Estimate filtration performance using the two-stage model corresponding to the condition given in Illustrative Examples 7.1 and 7.2 and assuming $\sigma_{tran} = 0.04$.

7.6 Based on the results of 7.2.2, discuss the sensitivity of the estimated pressure-drop history with the parameter β assuming uniform deposition within the filter medium.

7.7 If one considers that deposited particles in granular filtration form particle dendrites, which may be approximated as fibers, a granular filter with significant deposition may be considered as a filter medium composed of spherical and cylindrical collectors. Estimate the pressure-drop increase if the assumptions used in obtaining Equation (7.3.24) are applicable.

References

Bai, R., Tien, C., 1999. J. Colloid Interface Sci. 179, 631.

Bai, R., Tien, C., 2000. Colloids Surf. A 165, 95.

Bergman, W., Taylor, R.D., Miller, H.H., 1978. 15th DOE Nuclear Air Cleaning Conference, CONF-780819, Boston.

Beveridge, G.S.G., 1962. A Survey of Interphase Reaction and Exchange in Fixed and Moving Beds. Dep't. of Chem. Eng., University of Minneosota.

Camp, T.R., 1964. Proc. ASCE J. Sanitary Eng. Div. 90, 3.

Chang, J.W., 1985. Mathematical Modeling of Deep Bed Filtration: Microscopic Approach, M.S. Thesis, Asian Institute of Technology, Bangkok.

Davis, C.N., 1973. Air Filtration. Academic Press.

Deb, A.K., 1969. Proc. ASCE J. Sanitary Eng. Div. 95, 399.

Fan, L.T., Nassar, R., Hwang, S.H., Chou, S.T., 1985. AIChE J. 31, 1781.

Heertjes, R.M., Lerk, C.F., 1962. Some Aspects of the Removal of Irons from Ground Water, Interaction between Fluids and Particles, Inst. Chem. Engrs (London), p. 269.

Heertjes, R.M., Lerk, C.F., 1967. Trans. Inst. Chem. Eng. 45, T138.

Hsiung, K., 1967. Prediction of Performance of Granular Filters for Water Treatment, Ph.D. Dissertation, Iowa State University.

Hsu, E.H., Fan, L.T., 1984. AIChE J. 30, 267.

Huang, J.Y., 1972. Granular Filters for Tertiary Waste Water Treatment, PhD Thesis, Iowa State University, Ames, Iowa.

Hutchinson, H.P., Sutherland, D.N., 1965. Nature 206, 1063.

Ives, K.J., 1961. Proc. ASCE J. Sanitary Eng. Div. 87, 23.

Litwiniszyn, J., 1963. Bulletin, De L'Academic Polonaise Des Science. Serie des Technique 11, 117.

Liu, D., Johnson, P.R., Elimelech, M., 1995. Environ. Sci. Technol. 29, 2963.

O'Melia, C.R., Ali, W., 1978. Prog. Water Technol. 10, 167.

Payatakes, A.C., Turian, R.M., Tien, C., 1973. AIChE J. 19, 58.

Payatakes, A.C., Tien, C., 1976. J. Aerosol Sci. 67, 85.

Payatakes, A.C., 1976. Powder Technol. 14, 267.

Payatakes, A.C., Gradon, L., 1980. Chem. Eng. Sci. 35, 1083.

Payatakes, A.C., Okuyama, K., 1982. J. Colloid Interface Sci. 88, 55.

Rajagopalan, R., Chu, R.Q., 1982. J. Colloid Interface Sci. 86, 299.

Ruckenstein, E., Prieve, D., 1973. J. Chem. Soc. Faraday Trans. 69, 1522.

Shaw, D.J., 1970. Introduction to Colloid and Surface Chemistry, second ed. Butterworths.

Sherwood, T.K., Pigford, R.L., Wilke, C.R., 1975. Mass Transfer. McGraw-Hill.

Spielman, L.A., Friedlander, S.K., 1974. J. Colloid Interface Sci. 46, 22.

Thomas, D., Penicot, P., Contal, P., Leclerc, D., Vendel, J., 2001. Chem. Eng. Sci. 56, 3549.

Thomas, H.C., 1944. J. Am. Chem. Soc. 66, 1664.

Tien, C., Turian, R.M., Pendse, H., 1979. AIChE J. 25, 385.

Tien, C., Ramarao, B.V., 2007. Granular Filtration of Aerosols and Hydrosols, second ed. Elsevier.

Tobiason, J.E., Johnson, G.S., Westerhoff, P.K., Vigneswaran, B., 1993. J. Environ. Eng. ASCE 119, 520.

Tobiason, J.E., Vigneswaran, B., 1994. Water Res. 28, 335.

Vigneswaran, S., 1980. Contribution a la Modelisation dans la Masse D. Eng. Thesis, Montpelier.

Vigneswaran, S., Tulachan, R.K., 1988. Water Res. 22, 1093.

Walsh, D.C., June 1996. Filtration and Separation, p. 501.

Wnek, W.J., 1973. The Role of Surface Phenomenon and Colloid Chemistry in Deep Bed Filtration, PhD Dissertation, Illinois Institute of Technology.

Wnek, W.J., Gidaspow, D., Wasan, D.T., 1975. Chem. Eng. Sci. 30, 1035.

Index

Note: Page numbers followed by *f* indicate figures, *t* indicate tables and *b* indicate boxes.

Printed and bound by CPI Group (UK) Ltd, Croydon, CR0 4YY

08/05/2025

01864822-0003